Tracking Environmental Change Using Lake Sediments.
Volume 3: Terrestrial, Algal, and Siliceous Indicators

Developments in Paleoenvironmental Research

VOLUME 3

Tracking Environmental Change Using Lake Sediments Volume 3: Terrestrial, Algal, and Siliceous Indicators

Edited by

John P. Smol

Department of Biology,
Queen's University

H. John B. Birks

Botanical Institute,
University of Bergen

and

William M. Last

Department of Geological Sciences,
University of Manitoba

KLUWER ACADEMIC PUBLISHERS

DORDRECHT / BOSTON / LONDON

A C.I.P. Catalogue record for this book is available from the Library of Congress.

ISBN 1-4020-0681-0

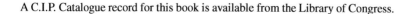

Published by Kluwer Academic Publishers,
P.O. Box 17, 3300 AA Dordrecht, The Netherlands.

Sold and distributed in North, Central and South America
by Kluwer Academic Publishers,
101 Philip Drive, Norwell, MA 02061, U.S.A.

In all other countries, sold and distributed
by Kluwer Academic Publishers,
P.O. Box 322, 3300 AH Dordrecht, The Netherlands.

Cover Photos: East Basin Lake, Victoria, Australia.
Inset: SEM photograph of the diatom *Luticola cohnii*
(from: David G. Mann).

Printed on acid-free paper

DEDICATION

Dedicated to the memory of Dr. Julian M. Szeicz.

CONTENTS

Introduction
Biology
Distribution and ecology
Taphonomy and preservation
Field sampling and coring
Laboratory procedures
Data analysis and interpretation
Summary
Acknowledgements
References

Introduction
Taxonomy and nomenclature
Methods
Paleolimnological applications
Future research directions
Summary
Acknowledgements
References

Introduction
Morphology, taxonomy and preservation in the sediments
Methodological aspects
Brief history of use of ebridians in palaeoecological research
Indicator value and future research priorities
Summary
Acknowledgements
References

Introduction and history
Phytolith production and taxonomy
Laboratory methods
Applications of phytolith analysis in lake sediments
Summary of the major results
Other potential applications of phytoliths in lake sediments
Summary
Acknowledgments
References

PREFACE

Paleolimnology is a rapidly developing science that is now being used to study a suite of environmental and ecological problems. This volume is the third handbook in the *Developments in Paleoenvironmental Research* book series. The first volume (Last & Smol, 2001a) examined the acquisition and archiving of sediment cores, chronological techniques, and large-scale basin analysis methods. Volume 2 (Last & Smol, 2001b) focused on physical and chemical methods. This current volume, along with Volume 4 (Smol et al., 2001), summarize the many biological methods and techniques that are available to study long-term environmental change using information preserved in sedimentary profiles. A subsequent volume (Birks et al., in preparation) will deal with statistical and data handling procedures. It is our intent that these books will provide sufficient detail and breadth to be useful handbooks for both seasoned practitioners as well as newcomers to the area of paleolimnology. These books will also hopefully be useful to non-paleolimnologists (e.g., limnologists, archeologists, palynologists, geographers, geologists, etc.) who continue to hear and read about paleolimnology, but have little chance to explore the vast and sometimes difficult to access journal-based reference material for this rapidly expanding field. Although the chapters in these volumes target mainly lacustrine settings, many of the techniques described can also be readily applied to fluvial, glacial, marine, estuarine, and peatland environments.

This current volume focuses on terrestrial, algal, and siliceous indicators, whilst Volume 4 deals primarily with zoological indicators. The taxonomic divisions between these two books, however, are not exact, as some zoological indicators are discussed in this book (e.g., protozoa, rotifers, sponges) because these groups are typically studied using techniques often associated with palynologists or diatomists. Hence, it was more practical to cover these topics here.

Many people have helped with the planning, development, and final production of this series. In addition to the hard work provided by the authors, this publication benefitted from the technical reviews furnished by our scientific colleagues, many of whom remain anonymous. Each chapter was typically examined by two external referees as well as the editors. In order to assure readability for the major target audience, we asked many of our graduate students to also examine selected chapters; their insight and questioning during the reviewing and editorial process are most gratefully acknowledged. The staff of the Environmental, Earth and Aquatic Sciences Division of Kluwer Academic Publishers are commended for their diligence in production of the final presentation. In particular, we would also like to thank Ad Plaizier, Anna Besse-Lototskaya (Publishing Editor, Aquatic Science Division), and Rene Mijs (former Publishing Editor, Biosciences Division) for their long-term support of this new series of monographs and their interest in paleoenvironmental research. John Glew (Queen's University, PEARL) designed our logo. Finally, we would like to thank our respective universities and colleagues for support and encouragement during this project.

THE EDITORS

John P. Smol is a professor in the Department of Biology at Queen's University (Kingston, Ontario, Canada), with a cross-appointment at the School of Environmental Studies. He co-directs the Paleoecological Environmental Assessment and Research Lab (PEARL). Prof. Smol is co-editor of the *Journal of Paleolimnology* and holds the *Canada Research Chair in Environmental Change*.

H. John B. Birks is Professor of Quantitative Ecology and Palaeoecology in the Botanical Institute, University of Bergen, Norway, and ENSIS Professor of Quantitative Palaeoecology, Environmental Change Research Centre, University College London. He is also an Adjunct Professor at PEARL, Queen's University, Kingston, Canada. He is a member of the Board of Advisors for the *Journal of Paleolimnology*.

William M. Last is a professor in the Department of Geological Sciences at University of Manitoba (Winnipeg, Manitoba, Canada) and is co-editor of the *Journal of Paleolimnology*.

AIMS AND SCOPE OF *DEVELOPMENTS IN PALEOENVIRONMENTAL RESEARCH* SERIES

Paleoenvironmental research continues to enjoy tremendous interest and progress in the scientific community. The overall aims and scope of the *Developments in Paleoenvironmental Research* book series is to capture this excitement and document these developments. Volumes related to any aspect of paleoenvironmental research, encompassing any time period, are within the scope of the series. For example, relevant topics include studies focused on terrestrial, peatland, lacustrine, riverine, estuarine, and marine systems, ice cores, cave deposits, palynology, isotopes, geochemistry, sedimentology, paleontology, etc. Methodological and taxonomic volumes relevant to paleoenvironmental research are also encouraged. The series will include edited volumes on a particular subject, geographic region, or time period, conference and workshop proceedings, as well as monographs. Prospective authors and/or editors should consult the series editors for more details. The series editors also welcome any comments or suggestions for future volumes.

CONTENTS OF VOLUMES 1 TO 4 OF THE SERIES

Contents of Volume 3: *Tracking Environmental Change Using Lake Sediments: Terrestrial, Algal, and Siliceous Indicators.*

Contents of Volume 4: *Tracking Environmental Change Using Lake Sediments: Zoological Indicators.*

SAFETY CONSIDERATIONS AND CAUTION

Paleolimnology has grown into a vast scientific pursuit with many branches and subdivisions. It should not be surprising, therefore, that the tools used by paleolimnologists are equally diverse. Virtually every one of the techniques described in this book requires some familiarity with standard laboratory or field safety procedures. In some of the chapters, the authors have made specific reference to appropriate safety precautions; others have not. The responsibility for safe and careful application of these methods is yours. Never underestimate the personal risk factor when undertaking either field or laboratory investigations. Researchers are strongly advised to obtain all safety information available for the techniques they will be using and to explicitly follow appropriate safety procedures. This is particularly important when using strong acids, alkalies, or oxidizing reagents in the laboratory or many of the analytical and sample collection/preparation instruments described in this volume. Most manufacturers of laboratory equipment and chemical supply companies provide this safety information, and many Internet and other library resources contain additional safety protocols. Researchers are also advised to discuss their procedures with colleagues who are familiar with these approaches, and so obtain further advice on safety and other considerations.

The editors and publisher do not necessarily endorse or recommend any specific product, procedure, or commercial service that may be cited in this publication.

LIST OF CONTRIBUTORS

Richard W. Battarbee
Environmental Change Research Centre
Department of Geography
University College London
26 Bedford Way
London WC1H 0AP
United Kingdom
e-mail: rbattarb@geography.ucl.ac.uk

Keith D. Bennett
Department of Quaternary Geology
Uppsala University
Geocentrum, Villavägen 16
S-752 36 Uppsala, Sweden
e-mail: keith.bennett@geo.uu.se

Helen Bennion
Environmental Change Research Centre
Department of Geography
University College London
26 Bedford Way
London WC1H 0AP
United Kingdom
e-mail: hbennion@geography.ucl.ac.uk

Louis Beyens
Department of Biologie
RUCA (UA)
Groenenborgerlaan 171
B-2020 Antwerpen
Belgium
e-mail: lobe@ruca.ua.ac.be

H. John B. Birks
Botanical Institute
University of Bergen
Allégaten 41, N-5007
Bergen, Norway
e-mail: John.Birks@bot.uib.no

Hilary H. Birks
Botanical Institute
University of Bergen
Allégaten 41, N-5007
Bergen, Norway
e-mail: Hilary.Birks@bot.uib.no

xxii

Nigel G. Cameron
Environmental Change Research Centre
Department of Geography
University College London
26 Bedford Way
London WC1H 0AP
United Kingdom
e-mail: ncameron@geog.ucl.ac.uk

Laurence Carvalho
Centre for Ecology and Hydrology Edinburgh
Bush Estate, Penicuik
Midlothian
EH26 0QB
United Kingdom
e-mail: laca@ceh.ac.uk

Daniel J. Conley
Department of Marine Ecology & Microbiology
National Environmental Research Institute
P.O. Box 358
DK-4000 Roskilde
Denmark
e-mail: conley@hami1.dmu.dk

Marianne S. V. Douglas
Department of Geology
University of Toronto
22 Russell St.
Toronto, Ontario
M5S 3B1, Canada
e-mail: msvd@geology.utoronto.ca

Roger J. Flower
Environmental Change Research Centre
Department of Geography
University College London
26 Bedford Way
London WC1H 0AP
United Kingdom
e-mail: rflower@geography.ucl.ac.uk

Tom M. Frost
deceased

Dominic A. Hodgson
British Antarctic Survey
Cambridge, CB3 0ET
United Kingdom
e-mail: daho@pcmail.nerc-bas.ac.uk

Vivienne J. Jones
Environmental Change Research Centre
Department of Geography
University College London
26 Bedford Way, London WC1H 0AP
United Kingdom
e-mail: vjones@geography.ucl.ac.uk

Stephen Juggins
Department of Geography
University of Newcastle
Newcastle upon Tyne NE1 7RU
United Kingdom
e-mail: stephen.juggins@newcastle.ac.uk

Atte Korhola
Division of Hydrobiology
Department of Ecology and Systematics
P.O. Box 17 (Arkadiankatu 7)
FIN-00014 University of Helsinki
Helsinki, Finland
e-mail: atte.korhola@helsinki.fi

Christopher P. S. Larsen
Department of Geography
State University of New York at Buffalo
Buffalo, NY 14261-0023, USA
e-mail: larsen@seneca.geog.buffalo.edu

William M. Last
Department of Geological Sciences
University of Manitoba, Winnipeg
Manitoba, R3T 2N2, Canada
e-mail: WM_Last@UManitoba.ca

Peter R. Leavitt
Department of Biology
University of Regina
Regina, Saskatchewan
S4S 0A2, Canada
e-mail: Peter.Leavitt@uregina.ca

Glen M. MacDonald
Department of Geography
University of California at Los Angeles (UCLA)
405 Hilgard Ave.
Los Angeles, CA 90095-1524, USA
e-mail: macdonal@geog.sscnet.ucla.edu

Ralf Meisterfeld
Institute for Biology II (Zoology)
RWTH Aachen
Kopernikusstrasse 16
D- 52056 Aachen, Germany
e-mail: meisterfeld@rwth-aachen.de

Dolores R. Piperno
Smithsonian Tropical Research Institute (Balboa, Republic of Panama)
Unit 948
APO AA (Miami) 34002-0948
e-mail: pipernod@stri.org

Claire L. Schelske
Department of Geological Sciences
Land Use and Environmental Change Institute
University of Florida
Gainesville, FL 32611
USA
e-mail: schelsk@ufl.edu

Bas van Geel
Hugo de Vries Laboratory
Department of Palynology and Paleoecology
University of Amsterdam
1098 SM Amsterdam
The Netherlands
e-mail: vangeel@bio.uva.nl

Cathy Whitlock
Department of Geography
University of Oregon
Eugene, OR 97403 U.S.A.
e-mail: whitlock@oregon.uoregon.edu

Katherine J. Willis
School of Geography, University of Oxford
Mansfield Road, Oxford OX1 3TB
United Kingdom
e-mail: kathy.willis@geog.ox.ac.uk

Barbara A. Zeeb
The Royal Military College of Canada
Environmental Sciences Group
Kingston, Ontario
K7K 5L0, Canada
e-mail: zeeb-b@rmc.ca

1. USING BIOLOGY TO STUDY LONG-TERM ENVIRONMENTAL CHANGE

JOHN P. SMOL (SmolJ@Biology.QueensU.Ca)

Paleoecological Environmental Assessment
and Research Lab (PEARL)
Department of Biology
Queen's University
Kingston, Ontario
K7L 3N6, Canada

H. JOHN B. BIRKS (John.Birks@bot.uib.no)

Botanical Institute
University of Bergen
Allégaten 41, N-5007
Bergen, Norway, and
Environmental Change Research Centre
University College London
London WC1H 0AP
United Kingdom, and
Paleoecological Environmental Assessment
and Research Lab (PEARL)
Department of Biology
Queen's University
Kingston, Ontario
K7L 3N6, Canada

WILLIAM M. LAST (mlast@Ms.UManitoba.CA)

Department of Geological Sciences
University of Manitoba
Winnipeg, Manitoba
R3T 2N2, Canada

A central theme in ecology has always been the attempt to classify ecosystems by the biota they support. For example, presenting even a school-age child with a picture of polar bears and arctic hare will immediately suggest a cold, arctic environment, whereas a picture of camels and cacti will indicate an arid, hot, desert environment. Much of what biological paleolimnology does is simply an extension of these types of relationships. However, the

1

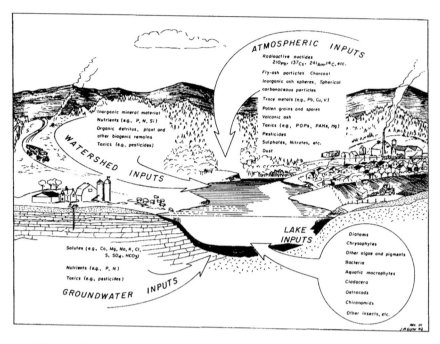

Figure 1. Schematic diagram showing the accumulation of allochthonous and autochthonous indicators used by paleolimnologists to track long-term environmental change (modified from Charles et al., 1994).

degree of quantification, accuracy, and precision that these relationships and inferences have reached has brought paleolimnology to the forefront of the study of many environmental change problems.

Broadly defined, paleolimnology is the study of the physical, chemical, and biological information stored in lake deposits. In most cases, lake sediments are considered the primary archives, however other deposits (e.g., sediments from rivers, bogs, wetlands, marshes, estuaries, and so forth) also contain important proxy data of past environmental change, and many of the approaches discussed in this book can be equally applied to these environments. The amount of information stored in sediments is staggering, with both allochthonous and autochthonous sources of proxy data (Fig. 1), and so it soon became clear to the series editors that several books would be required to summarize some of the commonly used methods. These books build on the foundation set by previous compilations of paleoenvironmental techniques, such as Gray (1988), Warner (1990), and especially Berglund (1986). The latter has been the standard reference for about 15 years.

As we hope to capture in these current volumes, many new approaches and techniques are now available to paleolimnolgists. The first two books in this series deal with physical and chemical techniques. Volume 1 (Last & Smol, 2001a) includes chapters on field work, such as core collection and sectioning, basin analysis techniques, and dating sedimentary profiles. Volume 2 (Last & Smol, 2001b) summarizes the myriad of physical and chemical techniques that paleolimnologists use to describe and interpret sedimentary profiles.

Volumes 3 (this volume) and 4 (Smol et al., 2001) focus on the biological information we can glean from sediments, with some example applications. This book (Volume 3) contains

15 chapters that deal primarily with terrestrial, algal, and siliceous indicators. Following this short introduction, we begin with a cluster of six chapters that deal with primarily terrestrial and allochthonous indicators. This first section includes chapters on topics such as pollen, stomates, macrofossils, charcoal, fungi and other non-pollen palynomorphs, as well as semi-terrestrial indicators, such as testate amoebae. Knowledge of changes occurring in a lake's catchment are very often needed to interpret the changes in lake systems, and lake-catchment interactions often form a central theme in paleolimnology. This is followed by six chapters dealing with siliceous microfossils, such as diatoms, chrysophytes, ebrideans, phytoliths, sponges, and siliceous protozoan plates, all of which can often be studied using standard diatom microscope slides. The volume concludes with two chapters that use chemical techniques to track biological populations (i.e., biogenic silica, fossil pigments), followed by a glossary and index.

Modern data sets (e.g., surface-sediment calibration sets or training sets) provide the basis for much interpretation of fossil data in terms of calibration, transfer functions, and analogues. The field and laboratory methods for surface-sediment samples are essentially the same as for down-core studies.

Although paleolimnological applications are discussed throughout this book, only cursory comments are made on the statistical and quantitative approaches that have been developed to interpret the proxy information gleaned from sediments. A separate volume on statistical and data handling approaches is currently in preparation (Birks et al., in preparation).

Paleolimnology has progressed at a tremendous rate over the last few decades, with biological approaches often leading many new initiatives. These are exciting times for paleoenvironmental research. We hope these current volumes capture this progress, and will aid the next generation of paleolimnologists in their investigations.

References

Berglund, B. E. (ed.), 1986. Handbook of Holocene Palaeoecology and Palaeohydrology. John Wiley & Sons, Chichester, 869 pp.

Birks, H. J. B., S. Juggins, A. F. Lotter & J. P. Smol (eds.), In preparation, Tracking Environmental Change Using Lake Sediments. Volume 5: Data Handling and Statistical Techniques. Kluwer Academic Publishers, Dordrecht.

Charles, D. F., J. P. Smol & D. E. Engstrom, 1994. Paleolimnological approaches to biomonitoring. In Loeb S. & A. Spacie (eds.) Biological Monitoring of Aquatic Systems. Lewis Press, Ann Arbor: 233–293.

Gray, J. (ed.), 1988. Paleolimnology: Aspects of Freshwater Paleoecology and Biogeography. Elsevier, Amsterdam, 678 pp.

Last, W. M. & J. P. Smol (eds.), 2001a. Tracking Environmental Change Using Lake Sediments. Volume 1: Basin Analysis, Coring, and Chronological Techniques. Kluwer Academic Publishers, Dordrecht, 548 pp

Last, W. M. & J. P. Smol (eds.), 2001b. Tracking Environmental Change Using Lake Sediment. Volume 2: Physical and Geochemical Methods. Kluwer Academic Publishers, Dordrecht, 504 pp

Smol, J. P., H. J. B. Birks & W. M. Last (eds.), 2001. Tracking Environmental Change Using Lake Sediments. Volume 4: Zoological Indicators. Kluwer Academic Publishers, Dordrecht, 217 pp

Warner, B. G. (ed.), 1990. Methods in Quaternary Ecology. Geological Association of Canada, St. John's, 170 pp.

2. POLLEN

K. D. BENNETT (Keith.Bennett@geo.uu.se)
Department of Earth Sciences
Uppsala University
Villavägen 16
SE-752 36 Uppsala, Sweden

K. J. WILLIS (kathy.willis@geography.oxford.ac.uk)
School of Geography
University of Oxford
Mansfield Rd
Oxford OX1 3TB, UK

Keywords: pollen, principles, laboratory technique, identification, interpretation

Introduction

Analysis of the pollen content of sediment samples is the principal technique available for determining vegetation response to past terrestrial environmental change. The technique has been in use for nearly a century, initially as a method for investigating past climatic changes (von Post, 1946). More recently, the importance for vegetation change of processes such as human impact, successional change and other biotic and abiotic factors have been recognised. Pollen analysis can be used to examine these factors, although separating them remains difficult.

Pollen grains are plant parts, found in angiosperms and gymnosperms, that contain a male nucleus for fertilization with the female nucleus in an ovule. Spores are equivalent parts of ferns and mosses, although the reproductive process is somewhat different. Pollen and spores are usually considered together within the broad term 'palynology'. Pollen is produced in anthers, often in different flowers (sometimes on different plants) from ovules. Different plant taxa disperse pollen in different ways, but most pollen is either wind-dispersed (anemophilous) or insect-dispersed (entomophilous) (Faegri & van der Pijl, 1979; Proctor et al., 1996). Plants with wind-dispersed pollen tend to produce more than those with insect-dispersed pollen. Pollen grains are usually spherical or elliptical, and vary in diameter from about $10\,\mu$m (0.01 mm) to about $100\,\mu$m (0.1 mm), depending on species: most are 20–$30\,\mu$m. The spores of ferns and mosses are somewhat different from pollen in structure (see below), but fall within the same size range. They are wind-dispersed.

J. P. Smol, H. J. B. Birks & W. M. Last (eds.), 2001. *Tracking Environmental Change Using Lake Sediments.*
Volume 3: Terrestrial, Algal, and Siliceous Indicators. Kluwer Academic Publishers, Dordrecht, The Netherlands.

Pollen is liberated into the environment from flowers, but most of it is 'waste', in the sense that it does not fertilize any ovules. This 'surplus' pollen becomes mixed in the atmosphere and is eventually washed, or falls, to the ground where some of it will accumulate in the sediments of lakes, bogs, rivers, or the sea.

The walls of pollen grains are made of a mixture of cellulose and a complex substance called sporopollenin. The precise molecular structure of sporopollenin is unclear, but it is a polymer with saturated and unsaturated hydrocarbons and phenolics (Southworth, 1990). Sporopollenin is unusually resistant to most forms of chemical (including enzymatic) and physical degradation, except oxidation. This has the important consequences that, first, pollen may be preserved indefinitely (as far as we know) in anaerobic environments; and, second, strong chemicals can be used to remove other components of sediment, thus concentrating pollen and facilitating analysis.

The outer walls of pollen grains have different shapes and patterns, depending on the taxon. This makes it possible to identify grains to the taxon from which they originated, although the degree of precision for identification varies somewhat. Identification is achieved by comparing unknown fossil grains with reference material of pollen collected from flowers of identified plants.

The principles of pollen analysis (based on Birks & Birks, 1980) may be summarized as follows:

1. Pollen and spores are produced by plants in abundance.

2. Most of these fall to the ground.

3. Pollen and spores are preserved in anaerobic environments, such as
   bogs, lakes, fens, and the ocean floor.

4. Pollen and spores in the atmosphere are mixed by atmospheric turbulence, resulting in a uniform pollen rain over a given area.

5. The proportion of each pollen type in the pollen rain depends upon the abundance of its parent plants, so the composition of the pollen rain is a function of the composition of the vegetation. A sample of the pollen rain is thus a snap-shot of vegetation at a particular point in space and time.

6. Pollen is identifiable to various taxonomic levels.

7. When a sample of pollen from sediment of known age is examined, the result is an index of the vegetation surrounding the site of deposition at that point in space and time.

8. When pollen spectra are obtained from several samples through a sequence of sediment, they provide a picture of vegetation changes through the period of time represented by the sediments.

9. When two or more series of pollen spectra are obtained from several separate sediment sequences, it is possible to compare changes in vegetation through time at different places.

There is, therefore, a clear progression from the release of pollen from the parent plant to changes in vegetation through space and time. Pollen analysis is highly versatile tool that can provide a long-term perspective to key ecological questions.

Where is pollen found?

If pollen happens to be incorporated in an anaerobic environment before it decays, then it will be preserved and persist. The living contents of the grain probably die within hours or days of release, but the outer wall can survive for considerable periods of time. The oldest known spores date from about 470 Myr (Ordovician) and the oldest known gymnosperm pollen from 360 Myr (Carboniferous) (Traverse, 1988). The first angiosperm pollen is 140 Myr old (Cretaceous) (Hughes, 1976; Crane et al., 1995). Such early finds are important parts of the evidence for the antiquity of these plant groups. Anaerobic sediments are found wherever there is permanent waterlogging: the bottoms of lakes, bogs, and some river or stream sediments (away from the influence of moving, aerated water). In these sediments, pollen accumulates in the same way as the other materials, organic and inorganic, that make up the sediment.

The environments that may contain pollen are diverse. The pollen-analytical technique has multiple uses, and can work well in many sediment types, including ice (McAndrews, 1984; Bourgeois, 1986), middens of hyraxes (Carrión et al., 1999) and packrats (Anderson & van Devender, 1991), marine sediment (Sánchez Goñi et al., 1999), faeces (van der Knaap, 1989), cave sediment (Coles et al., 1989; Carrión et al., 1999), and loess (Sun et al., 1997). However, for the purpose of reconstructing past terrestrial environments, the most suitable sediments are those accumulating in small lakes. The most suitable lakes are those with the least mixing (and hence least oxygenation) at the bottom, typically lakes that are deep for their area. Such sites offer simple continuous sedimentation, dominated by the 'rain' of dead planktonic material but incorporating a proportion of particles, such as pollen and spores, that are blown or washed into the lake. With such a uniform sedimentary environment, the investigator is able to concentrate on the main problem (terrestrial environments) and not have to separate changes in that part of the landscape from changes brought about by change in the depositional environment.

Extraction from sediments

The process of extracting pollen from sediments was developed during the first half of the 20th century, but the processes in use today have changed little during the last 40 years. There is a need for a thorough review of these procedures in the light of more recent chemical and biochemical knowledge. This processing technique also concentrates stomata (MacDonald, this volume) and microscopic charcoal (e.g., Clark, 1982, Whitlock & Larsen, this volume), and these components can also be quantified from pollen residues.

The account below is structured around the goal of each process, rather than the means, and is based especially on the accounts of Gray (1965), Berglund & Ralska-Jasiewiczowa (1986), Faegri & Iversen (1989) and Moore et al. (1991). Methods are included here for obtaining measures of the absolute abundance of pollen (concentrations), which have become a useful interpretative tool, complementing the classical pollen percentage presentation technique.

Subsampling

Small (usually $0.5\,cm^3$ or $1.0\,cm^3$) sediment samples are taken at intervals appropriate for the task in hand along the sequence. Whatever sampling resolution is chosen, it is worth using an interval divisible by 2 (e.g., 64, 32, 16, or 8 cm) to facilitate improving the resolution by halving the interval. A typical strategy might be to aim for one sample per hundred years for a Holocene sequence, which would mean about 100 samples at 8 cm intervals for an 8-m core. Additional samples can then taken to improve the resolution as the pollen stratigraphy becomes clear, or as the time control becomes available (often after the initial choices about sampling interval).

Samples can be taken either using a volumetric sub-sampler or by volumetric displacement. It is important that the volume extracted is known accurately, and is consistent between samples. A simple sub-sampler can be made from a plastic syringe with the nozzle removed. Remove the plunger, push the syringe into the sediment, replace the plunger, and withdraw the syringe. Then use the plunger to extrude all sediment *except* the volume desired. Finally, extrude the remaining, measured, portion. Estimates of the error of the sampled volume can be obtained by carrying out the procedure repeatedly on some homogenous sediment or other material (e.g., plasticene), and weighing the resulting samples.

The volumetric displacement method may be useful if sediments are particularly unconsolidated or if they are stiff. Add sediment by spoon to a measuring cylinder containing a known volume of distilled water. It may be necessary to crush dry sediment coarsely with a pestle and mortar.

Pollen concentrations

Two methods have been described, indirect and direct, for the determination of pollen concentrations. The direct method, probably first carried out by Armstrong et al. (1930), was developed by Davis (1965, 1966). It involves processing a known volume (or weight) of sediment and then counting **all** pollen in an aliquot of the residue. The method in general use today is an indirect method determined by adding a known quantity of exotic pollen (e.g., *Eucalyptus* in the northern hemisphere, *Alnus* in the southern hemisphere) or spores (e.g., *Lycopodium*) to the sample before processing by suspension or tablet (pill). The suspension or tablet needs to calibrated to give the number of exotic grains per unit volume, weight, or tablet, as appropriate.

Safety

The method of treating sediments for pollen analysis uses strong acids, of which hydrofluoric acid (HF) is especially dangerous, both in solution and as vapour. Accidents have resulted in fatalities (Chitty, 1995). The following guidelines are strongly recommended:

1. Anyone using HF should wear *always* personal protective equipment including a laboratory coat, neoprene rubber apron, full face mask, rubber gloves, closed shoes (i.e. not sandals) and long trousers.

2. All laboratory activity involving HF must take place within a HF safe fume cupboard. Always check that the fume extraction is working to full capacity and that the sash is

at the correct height. Fume cupboard fans should stay on for at least 30 minutes after using HF. Rubber caps must always be used on the sample tubes when they are outside the fume cupboard.

3. Calcium gluconate gel, an HF antidote, should be available in the laboratory. It needs to be kept refrigerated, and renewed regularly.

4. Before starting to use HF, make up a beaker of sodium carbonate (Na_2CO_3) solution. This should be placed inside the fume cupboard, together with a clean cloth. The cloth, saturated with Na_2CO_3 solution, should be used for wiping up drips and cleaning the outside of the sample tubes before placing them in the water bath. Gloves and contaminated equipment should be rinsed in the Na_2CO_3 solution before washing up.

Laboratory treatment

Preparation of samples for pollen analysis requires little in the way of specialist equipment. Essential items are:

1. Fume cupboard resistant to HF fumes and capable of extracting the fumes safely.

2. A system for disposal of acidic (including HF) and alkaline waste. The best system is a trap in the sink drainage, because it is fast and convenient to use. If this is not available, waste may have to be collected for disposal elsewhere. Take local advice and follow regulations.

3. Centrifuge that can spin 50 ml tubes to 3000 rpm. More time is spent during processing with the samples in the centrifuge than doing anything else, so it is worth obtaining one that is large enough to carry all the samples that will be processed at one time (at least 8, preferably 16, or even more);

4. Hot water bath capable of holding 50 ml tubes at boiling point.

5. A 'whirlimix' for stirring samples. The base of a test-tube is placed in a rotating rubber cup, and gently agitated.

6. Sieves with 180 μm mesh, preferably metal, for coarse sieving, and with 10 μm mesh, usually nylon, for fine sieving. Metal sieves can be cleaned by holding them (with tongs) in a bunsen-burner flame, which is fast and effective. Nylon sieves can be cleaned in an ultrasonic bath.

A basic procedure for the preparation of sediment samples for the analysis of pollen is described below. More details and some additional techniques are covered well in Faegri & Iversen (1989). If, at any stage, samples in an aqueous solution will not spin down well, add a few drops of methanol to reduce specific gravity. There is no harm in doing this routinely. The procedure below assumes that the work is being done in 50 ml polypropylene boiling tubes: adjust volumes appropriately for smaller tubes. HF treatment cannot be carried out in glass tubes.

This basic procedure can be treated as modular, consisting of key stages that can be included or not depending on what is needed for particular sediments. The stages are separated by washing the sample with distilled water and centrifuging. Samples can be safely left in distilled water if it is necessary to break up the processing sequence. Without breaks, the scheme can be completed comfortably in one day with 8–16 samples. It is set out diagrammatically in Figure 1, and it is recommended that the series of stages indicated by the vertical sequence is tried first, as a basic method. The other stages can be added as needed for particular sediments.

1. (a) If using an exotic pollen suspension (e.g., *Eucalyptus*, *Alnus*), stir overnight.

   (b) To each 0.5–2.0 cm^3 sediment sample, add 0.5–2.0 cm^3 exotic pollen suspension, aiming to add about as much exotic pollen as there is pollen in the sample. If using *Lycopodium* tablets, add 1–2 tablets, on the same principle.

2. (a) Add 25 cm^3 of 7% HCl (cautiously if samples are calcareous), and leave either cold overnight, or in the water bath at 90 °C for 30 mins, or until effervescence stops. Stir samples.

   (b) Balance samples, centrifuge and decant.

   (c) Whirlimix, wash in distilled water, balance, centrifuge and decant.

3. (a) Add 10 cm^3 of 10% NaOH and leave for 2–5 mins in water bath at 90 °C.

   (b) Immediately sieve (180 μm mesh) and wash samples with distilled water into second set of tubes, retaining macrofossils.

   (c) Balance samples, centrifuge and decant.

   (d) Whirlimix, wash in distilled water, balance, centrifuge and decant. Repeat until supernatant is clear.

4. (a) Add 10 cm^3 of 10% sodium pyrophosphate (Na$_4$P$_2$O$_7$), stir vigorously and place in a hot water bath for 10–20 mins.

   (b) Balance, centrifuge, decant.

5. (a) Sieve through 10 μm nylon monofilament mesh with either distilled water or Na$_4$P$_2$O$_7$ until filtrate runs clear.

   (b) Wash residue from sieve back into tubes with distilled water.

   (c) Balance, centrifuge, decant.

6. (a) Add 25 cm^3 of ZnCl$_2$ (specific gravity 1.96). Stir and centrifuge.

   (b) Decant supernatant into distilled water with a few drops of HCl.

   (c) Centrifuge and decant.

7. (a) Whirlimix, add 25 cm^3 of 7% HCl, balance, centrifuge and decant.

   (b) Make up a large (e.g., 1 l) beaker of Na$_2$CO$_3$ solution (about 100 g dry powder to 1 l water).

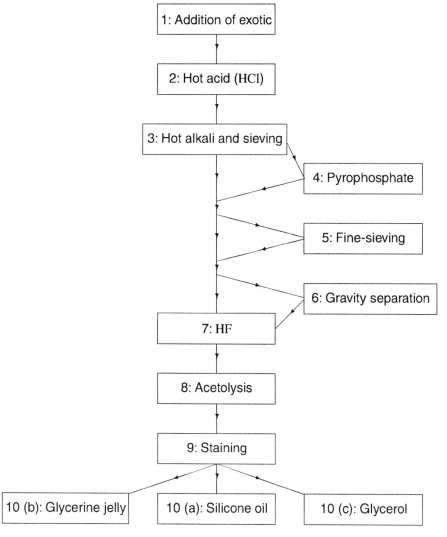

Figure 1. Flowchart of sediment processing for fossil pollen. Recommended default method follows the vertical scheme. Additional methods or alternatives are indicated to the right. See text for explanation and rationale of the numbered stages.

(c) Ensure fume cupboard extraction fans are on; wear HF resistant apron, gloves, and visor (see section on Safety, above).

(d) Add $25\,cm^3$ HF to each tube, pouring above beaker of Na_2CO_3. Stir with a polypropylene rod. Leave tubes in water bath at 90 °C for at least 30 mins.

(e) Wearing protective gear, add methanol to each tube, cap and centrifuge, using methanol to balance buckets.

(f) Carefully decant HF into sink or other waste facility. Allow plenty of water to rinse sink (avoid splashes).

(g) Rinse caps, gloves and any other items with Na_2CO_3, then water.

(h) Immediately after HF, add $25\,cm^3$ of 7% HCl and place in water bath at 90 °C for at least 30 mins.

(i) Balance samples, centrifuge and decant.

(j) Wash in distilled water, centrifuge and decant.

8. (a) In the fume cupboard, wearing rubber gloves, wash with $25\,cm^3$ of concentrated (glacial) acetic acid; balance, centrifuge and decant.

(b) Make up acetolysis mixture in measuring cylinder (9 parts acetic anhydride : 1 part concentrated sulphuric acid). The mixture should have a weak yellow colour, and there should be enough of it for all the samples being processed (i.e., $15\,cm^3 \times$ the number of samples).

(c) Add $15\,cm^3$ of the acetolysis mixture and place in water bath at 90 °C for 2–4 mins only. If adding the mixture causes violent spitting, the sample has not been completely dehydrated, so wash in glacial acetic acid again.

(d) In the fume cupboard, add $25\,cm^3$ glacial acetic acid to the acetolysis mixture. Balance, centrifuge and decant.

(e) Wash with glacial acetic acid, balance, centrifuge and decant.

(f) Wash with distilled water, balance, centrifuge and decant. Repeat once or twice until neutral.

9. (a) Add 2–3 drops of 0.2% aqueous safranin to each sample, whirlimix, wash with distilled water, balance, centrifuge and decant.

10. One of:

(a)   i. Add $25\,cm^3$ of tertiary butyl alcohol (TBA), balance, centrifuge and decant.

    ii. Whirlimix and rinse with TBA into small labelled tubes.

    iii. If tubes are glass, wrap in paper tissue. Balance tubes, centrifuge and decant.

    iv. Add as much silicone oil (2000 cs) as there is residue in the tubes and stir with a cocktail stick.

    v. Cover tubes and leave for at least 24 hours to allow TBA to evaporate (or accelerate the process by placing them in a warm oven).

    vi. Make slides as necessary by placing a drop of the suspension on the slide, distributing evenly over an area just less than the size of a coverslip. Add the coverslip, and press gently to extrude air bubbles. Hold coverslip in position with a drop of nail varnish at each corner.

(b)   i. Melt glycerine jelly in a beaker suspended in a water bath.

    ii. Add to the residue about twice as much jelly as there is residue.

    iii. Keep tube warm while stirring thoroughly to produce a uniform suspension.

iv. Heat a clean microscope slide on a warming plate and add a small amount of the suspension to the slide. Spread out and add coverslip. Seal the edge of the slide with cellulose in acetone (nail varnish is often used for this).

(c)   i. Add a drop of glycerol to the residue.

    ii. Allow remaining water to evaporate.

    iii. Add a small amount of the suspension to the slide. Spread out and add coverslip. Seal the edge of the slide with cellulose in acetone (nail varnish is often used for this).

All equipment used should be thoroughly washed and preferably cleaned by soaking in a commercial cleaning agent, such as 'Decon', to prevent contamination being carried over between samples.

Rationale

Stage 1:   Addition of exotic pollen
This is needed in order to obtain estimates of the concentration of pollen and spores in sediments. It must be the first stage, so that any losses of pollen and spores during the processing affect fossil and exotic pollen equally.

Stage 2:   Hot acid
This process removes carbonates. The reaction products are calcium hydroxide, which is soluble, and carbon dioxide, which is released as a gas (often vigorously). Magnesium-containing carbonate (which may occur in areas with dolomitic limestones) reacts in the same way, but much slower. If its presence is suspected, longer periods in hot HCl are needed. Tablets of exotic pollen are made with calcium carbonate (chalk), so this step is needed to break them down, even if the sediment is free of carbonate.

Stage 3:   Hot alkali and sieving
This process removes 'humic acids' (unsaturated organic soil colloids) by bringing them into solution, and also disaggregates the sediment. The quantity of humic acid may be considerable in organic material that is highly decomposed (such as some peats). Sieving the sample removes particles larger than any known pollen or spores. The residue retained on the sieve can often usefully be examined for smaller identifiable macrofossils (such as fruits of *Potamogeton* or *Betula*, conifer needles).

Stage 4:   Sodium pyrophosphate ($Na_4P_2O_7$)
This removes clay by using a deflocculant. Clay particles remain in the supernatant after centrifugation and are thus removed with the supernatant. The step is somewhat time-consuming, and may not work well with all clays, probably depending on the particular particle size classes present. The use of pyrophosphate in this way was introduced by Bates et al. (1978).

Stage 5:   Fine-sieving
This removes particles smaller than pollen by sieving. It may be useful where there is

abundant clay or fine organic particles. Sieves used are usually made of nylon, and can be obtained in a range of sizes, though the most commonly used is 10 μm. The filtration process may be slow, but can be aided by gently rubbing the underside of the mesh to prevent the pores becoming clogged. The residue may be washed with either distilled water (Glaister, 2000) or sodium pyrophosphate (Cwynar et al., 1979). Ultrasound can also accelerate the process, but may encourage pollen grains to pass between mesh fibres (Jemmett & Owen, 1990). The fine-sieving technique was introduced by Cwynar et al. (1979).

Stage 6:   Gravity separation

This method removes mineral fragments by separating them from organic particles according to their relative density. It is difficult to use and not always successful, for reasons that are not entirely clear. Nakagawa et al. (1998) have obtained excellent results with peat and gyttja. The sample is placed in a liquid at density of 1.88, which is more dense than organic material (specific gravity < 1.7) but less dense than mineral matter (specific gravity > 2). Many different liquids have been proposed, including zinc chloride, sodium polytungstate, and mixtures of potassium iodide and cadmium iodide, but the density is the critical feature (Nakagawa et al., 1998) It is important that, after centrifugation, the density of the liquid used is reduced sufficiently that the pollen grains can then be centrifuged down. In the case of zinc chloride, a few drops of HCl are needed at this stage to prevent precipitation of zinc hydroxide. Gravity separation techniques should be used with care, and only after other methods have proved unsatisfactory (Faegri & Iversen, 1989), perhaps because the quantity of mineral matter is exceptionally high. Dinoflagellate cysts may be lost. This use of zinc chloride was first described by Funkhouser & Evitt (1959).

Stage 7:   HF

This stage removes silica and silicates. The first step (washing with HCl) is intended to ensure that the sample is acidified, and to remove any residual calcium carbonate. The reaction of HF with calcium carbonate is extremely vigorous, and thus dangerous. If magnesium-containing carbonate might be present, special care is needed to ensure its removal before adding HF. The HF step brings silica and some silicates into solution (Table I), but cannot dissolve all minerals. As a general rule, 30 minutes in hot HF will remove everything that is soluble. Adding a large quantity of methanol to the hot HF lowers the density of the suspension, and thus reduces the tendency for finer particles to float during centrifugation. The final HCl step removes colloidal silica and silicofluorides. Depending on the composition of the sediment, this step may be at least as important as the HF step. The reaction between HF and some minerals may produce an insoluble white precipitate, consisting of fluorides such as CaF_2 (Langmyhr & Kringstad, 1966). The use of HF treatment was introduced to pollen analysis by Assarsson & Granlund (1924).

Stage 8:   Acetolysis

This stage removes polysaccharides by hydrolysing the polymer chain into soluble monosaccharide units. Polysaccharides are present on the surface of the grain and in the cytoplasm, so removing them greatly facilitates viewing the grain. Polysaccharides such as cellulose may also be significant components of the sediment, so removing these helps concentrate the pollen in the residue. The technique described here is a 'brute force' technique, used because it is fast. Experiments with cotton (cellulose) showed 100% loss of insoluble

Table I. Decomposition of selected minerals in a 20 ml 1:1 mixture of hydrofluoric and perchloric acids. Of the seven minerals undecomposed after 60 mins, topaz shows some decomposition in a bomb at 250 °C, but the other six remain undecomposed. After Langmyhr & Sveen (1965). Zircon may also be insoluble in HF (KDB, personal observations).

Mineral	Amounts (mg) of mineral remaining at 95 °C after:			
	0 min	20 min	40 min	60 min
Sodium feldspar	200	0		
Potassium feldspar	200	0		
Plagioclase	200	0		
Nepheline	200	0		
Muscovite	200	0		
Biotite	200	0		
Talc	200	0		
Hornblende	200	0		
Cordierite	200	0		
Wollastonite	200	0		
Olivine	200	0		
Aragonite	200	0		
Magnesite	200	0		
Apatite	200	0		
Fluorite	200	0		
Enstatite	200	17	0	
Epidote	200	22	0	
Antophyllite	200	28	0	
Andradite	200	42	0	
Quartz	200	55	0	
Magnetite	200	59	0	
Staurolite	200	80		76
Pyrrhotite	200	134		96
Kyanite	200	140		136
Beryl	200	155		142
Pyrite	200	170		156
Chalcopyrite	200	171		168
Topaz	200	178		176

material after 3 minutes in acetolysis solution (pers. obs.). After more than 30 minutes in the solution, insoluble material begins to reform, possibly due to some repolymerisation. Acetolysis treatment was introduced by Erdtman (1934).

Stage 9: Staining
Staining increases the contrast of sculptural elements on pollen grains and spores.

Stage 10: Mounting medium
The choice of mounting medium depends upon the relative importance given to permanence, ease of preparation, and the ability to move grains within the mount. A good mounting medium should have a refractive index close to sporopollenin (1.48), but not too close. If the refractive indices are too different, the contrast is high, and may obscure features. If

they are too similar, there may be insufficient contrast to see features of the grains. The use of three different media are described here, but there are others (see Traverse, 1988).

Stage 10 (a):   Silicone oil mounts
Tertiary butyl alcohol dehydrates the sample in a single step, replacing the older method of dehydration with a series of increasingly concentrated alcohols. Silicone oil is miscible with tertiary butyl alcohol, but not with water. Its use was introduced by Andersen (1960). The silicone oil used should have a viscosity of at least 2000 cs, but can be up to 40 000–60 000 cs (Faegri & Iversen, 1989). Silicone oil does not set, making it possible to move and rotate pollen grains by exerting gentle pressure on the coverslip. It does not distort pollen grains, and appears not to degrade them over long periods of time. Silicone oil is soluble in diethyl ether, which can be useful if it is necessary to re-process samples, perhaps because a stage was omitted, deliberately or otherwise. Simply wash the residue into a tube with diethyl ether and continue processing from the desired point (KJW, personal obs.).

Stage 10 (b):   Glycerine jelly mounts
Glycerine jelly is easily handled and has excellent optical properties, but has the disadvantage of absorbing water from the atmosphere, causing pollen grains to swell and degrade (Moore et al., 1991). The mount is solid, so grains cannot move, which has the advantage of making it easier to relocate grains, and the disadvantage that grains cannot easily be rotated. Use of glycerine jelly is not recommended in tropical climates, as it may not set reliably (Moore et al., 1991).

Stage 10 (c):   Glycerol liquid mounts
Glycerol is easily handled, but has a refractive index (1.47) very close to sporopollenin, so may not be so good optically. Grains can be turned easily, but the low viscosity also means that the liquid may 'creep' away from the coverslip with time if they are not completely sealed (Faegri & Iversen, 1989).

Identification

Microscope technique

The slide with mounted residue is scanned along traverses spaced regularly across the slide (side to side, or up and down), and all pollen and spores identified and counted. A magnification of about ×400 is used for routine identification and ×1000 for critical identifications, which means that high quality optics are necessary.

Identification depends upon comparison of fossil material with reference material, prepared and mounted in a similar way. The investigator quickly becomes familiar with the most common types, and after some experience, most grains can be identified from memory. Direct comparison between fossil and reference material can usually be supplemented to some extent by the use of keys and photographs. Published examples include *The Northwest European Pollen Flora* (Punt, 1976; Punt & Blackmore, 1991; Punt & Clarke 1980, 1981, 1984; Punt et al., 1988, 1995) and the atlases of Reille (1992, 1995, 1998) for Europe, McAndrews et al. (1973) for central North America, Colinvaux et al.

(1999) and Markgraf & D'Antoni (1980) for South America, Jagudilla-Bulalacao (1997) for the Philippines, Moar (1993) for New Zealand, and Bonnefille & Riollet (1980) for central Africa. A global list of pollen and spore atlases has been produced by Hooghiemstra & van Geel (1998). An annotated list of internet sites with images of pollen grains is available at http://www.geo.arizona.edu/palynology/polonweb.html. Photographs serve well as an *aide memoire* of the available types for someone who is already familiar with most of them, but access to a reference collection is essential at some stage, however, since the three-dimensional aspect of real material can never be completely replaced by photographs. Anyone attempting a pollen investigation should make, or arrange access to, a comprehensive reference collection for their geographic area of study.

Identification of pollen and spores is carried out by means of features such as number and location of apertures and nature of sculpturing on the surface of the grain. The possible combinations, and nomenclature, are complex. The subject is covered thoroughly in books such as Faegri & Iversen (1989), Moore et al. (1991), and Punt et al. (1994). Spores typically carry a scar which may be linear ('monolete') or have three arms ('trilete'). Pollen is more complex, with, in the generalized case, a wall composed of two layers, separated by supporting structures ('columellae'). Both layers can be sculptured and, or, perforated, and either or both may have apertures of pores ('pori') or furrows ('colpi'), or both. Apertures are normally arranged radially around the equator of the spherical grain, often with three per grain, although there is considerable variation from 0 to many tens of apertures. Additionally, there are some other forms, notably the pollen grain of conifers, which has two bladders attached (bisaccate), probably increasing the buoyancy of the grain. Examples of a spore and three pollen grains are shown in Figure 2.

The precision of identification varies from family level (e.g., grasses [Poaceae] and sedges [Cyperaceae]) to specific level (e.g., species of plantain [*Plantago*] in northern Europe). Typically, identification is possible to generic level (most European trees) or to groups of genera within large families. Table II displays terms that may be used to indicate different levels of uncertainty.

Counting

Pollen counts can be tallied by hand, on paper count sheets, or on arrays of mechanical counters, but is greatly facilitated if the counts can be entered directly into a computer (Bennett, 1990). Pollen counting is sufficiently time-consuming that efforts are being made to automate it. Stillman & Flenley (1996), Duller et al. (1999), and France et al. (2000) describe current approaches towards achieving this.

Data-handling

Pollen grains are easily extracted and identified, but the resulting data can present considerable data-handling problems. Firstly, there may be subjective decisions to be made, on the inclusion of taxa in pollen sums, for example. Secondly, there are decisions to make on the use of numerical techniques to handle the material. In this section, we outline some of the basics. More details of simple data-handling techniques, with specific reference to pollen, can be found on the internet at http://www.kv.geo.uu.se/datah.

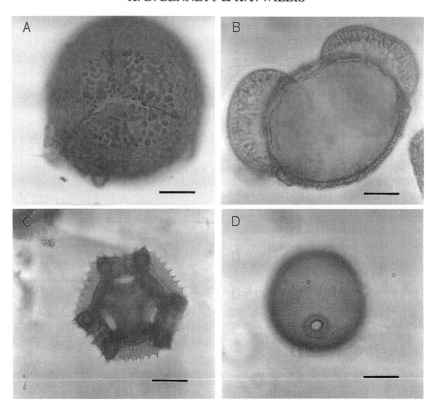

Figure 2. Four pollen and spores, chosen to illustrate some of the features used in identification. Scale bar is 10 μm for each. A: spore of *Osmunda regalis*, showing scar, trilete in this case, typical of spores. B: pollen of *Pinus sylvestris*, showing the bisaccate morphology characteristic of many conifers. C: pollen of *Taraxacum officinale*, with spiny (echinate), fenestrate morphology. D: pollen of *Molinia caerulea*, showing the characteristic monoporate (single pore) morphology of grasses.

Pollen sums

The basic idea is that each pollen and spore type should be presented as a proportion of a sum that includes itself. Results are normally expressed as percentages of a *main sum*, which consists of the total pollen count, excluding pollen of obligate aquatic plants, moss spores, and any grains that are not identifiable.

Taxa within sum: Normally all pollen and spores of vascular plants except obligate aquatics are included in the main sum. Thus, in northern Europe, Cyperaceae would normally be included, because at least some sedges are terrestrial plants. *Potamogeton* would be excluded because all species are aquatic. What it is appropriate to include or exclude will vary regionally, and should be determined with reference to an appropriate flora. Pollen and spores of all included taxa are summed to produce the main sum, and the frequency of each pollen or spore type is presented as a percentage of the main sum.

Table II. Conventions used to indicate the degree of certainty in the identification of pollen. After Birks (1973).

Example	Description
Poaceae	Family identification certain but lower categories either undetermined[1] or indeterminable[2]
Thalictrum	Genus identification certain but lower categories either undetermined or indeterminable
Plantago lanceolata	Species determination certain
Sedum cf. *S. rosea*	Genus identification certain, species identification tentative through poor preservation or close resemblance to other species
Plantago major / *Plantago media*	One pollen type present, two plant taxa considered equally likely alternatives
Angelica-type	One pollen type present, three or more plant taxa considered equally likely alternatives
Rosaceae undiff.	Family identification certain, one or more other morphological types within the family distinguishable and, if present, treated separately
Stellaria undiff.	Genus identification certain, one or more other morphological types within the genus distinguishable and, if present, treated separately

[1] Grains in a good state of preservation, but the analyst did not carry the level of identification lower.
[2] Grains in a poor state of preservation which limited the degree of identification.

Taxa outside sum: These may be grouped (e.g. all aquatics) or left as individual taxa (e.g. *Sphagnum*), and then combined with the main sum. The frequency of each type is presented as a percentage of (main sum + group sum) or (main sum + individual sum).

Pollen concentrations

The amounts of exotic and fossil pollen counted can be used to determine the fossil pollen concentration as follows:

$$\text{fossil pollen concentration} = \frac{\text{exotic pollen added} \times \text{fossil pollen counted}}{\text{exotic pollen counted}} \qquad (1)$$

The aim is to add about as much exotic pollen or spore as there is fossil pollen, to minimize counting effort and maximize precision of results (Maher, 1981). Clearly this requires prior knowledge, but with experience of particular sediments a reasonable estimate can be made. This method was first described by Benninghoff (1962), and developed more fully by Matthews (1969) and Bonny (1972).

Pollen accumulation rates

Pollen concentrations are a combination of the rate at which pollen falls on a sediment surface and the rate at which that sediment accumulates. High concentrations may result from high pollen input or low sediment accumulation rates. If the sediment accumulation rate is known, then pollen accumulation rates can be calculated as:

$$\text{pollen concentration (cm}^3) \, / \, \text{sediment accumulation rate (cm yr}^{-1}) \qquad (2)$$

Sediment accumulation rate is obtained from radiocarbon or other age determinations. Radiocarbon age determinations should be calibrated first (Stuiver et al., 1998). The first estimates of pollen accumulation rates were made by Welten (1944), using annually-laminated sediments as the basis for the age part of the calculation, but the technique did not come to the fore until the advent of radiocarbon dating made estimates of sediment accumulation rates routinely possible (Davis & Deevey, 1964). The quantity 'pollen accumulation rate' is also known, incorrectly, as 'pollen influx' (Thompson, 1980).

Error estimation

A pollen count is only a sample of the total amount of pollen under consideration. Consequently, there are statistical uncertainties attached to it. The greater the pollen count, the fewer these uncertainties become. Since larger counts are more time consuming to carry out, there is clearly a trade-off between time spent and the level of uncertainty.

Pollen percentages: Methods for calculating errors associated with pollen percentage calculations are described by Maher (1972), developing a method originally outlined by Mosimann (1965). The errors depend solely upon the size of the sum and the proportion within the sum of the taxon of interest and confidence intervals can be calculated. The degree of uncertainty decreases with the size of the count, and is about ±2% for a pollen type that represents about 50% of the sum in a count of 1000 grains (Maher, 1972; Traverse, 1988).

Pollen concentrations: Methods for calculating errors associated with pollen concentration calculations are described by Maher (1981). Confidence intervals can be calculated, but the estimation of all the error depends upon being able to assess and combine errors associated with measurement of sample volume, the quantity of exotic added, and the amount of exotic counted, as well as the size of the count for the taxon of interest.

Pollen accumulation rates: A method for calculating errors on pollen accumulation rates is described by Bennett (1994). The errors are as for pollen concentrations, with the additional element of adding the uncertainty attached to the rate of sediment accumulation. The errors may be 30% or more of the calculated value, and it is therefore important to understand their magnitude before attempting to interpret changing pollen accumulation rates.

Count sizes

Consideration of errors leads to the conclusion that count sizes (for the main sum) should be a minimum of 300 grains, but that 500 grains is a more appropriate target, and in some circumstances, the decreased errors associated with a count of 1000 grains is worth the extra trouble. For pollen concentrations, the amount of exotic added should be similar to the amount of fossil pollen and spores in the residue, to minimize time counting exotic grains while maximizing the reduction in error of the calculated concentrations (see Maher, 1981).

Presentation of results

Once an extremely time-consuming drawing exercise, pollen data can now be easily displayed using one of a number of computer programs, including Tilia (Grimm, 1993), psimpoll (Bennett, 2000), and Polpal (Walanus & Nalepka, 1997). An example pollen diagram is presented and described in Figure 3. Details of how to obtain copies of the programs and information on how to run them can be found at the internet site http://www.kv.geo.uu.se/datah. A number of programs have been developed to view pollen data in various ways, and major international databases exist to facilitate the process of synthesising pollen data (thus helping to achieve the ninth in the list of principles of pollen analysis, above). Access to these databases is simplest through an internet site at NOAA (http://www.ngdc.noaa.gov/paleo/pollen.html).

Data analysis

Pollen counting generates a large amount of numerical data that can be most easily understood with the help of numerical analyses. The most useful of these are outlined below. More details can be found in the literature cited, the textbooks of Birks & Birks (1980) and Birks & Gordon (1985), papers of Birks (1986, 1995, 1998), and on the internet at http://www.kv.geo.uu.se/datah. A companion volume to the present book (Birks et al., in preparation) will discuss all these methods fully.

Zonation

In order to facilitate description and correlation of microfossil data, it is usual to divide pollen sequences stratigraphically into 'zones'. There are many different types of zones (see Hedberg, 1976), of which the most useful for Quaternary palaeoecology is the 'assemblage zone'. These may be defined as bodies of sediment that are characterized by distinctive natural assemblages of pollen and, or, spores. Traditionally, sequences have been split into zones 'by eye', but the process can now be done easily by numerical procedures, which are not only fast and repeatable, but reduce considerably the element of subjectivity. The basic principles were established by Gordon & Birks (1972) and Birks & Gordon (1985), and remain unchanged. Bennett (1996) discusses methods for establishing how many zones should be defined.

Ordination

Ordination seeks to represent the trends in multivariate datasets in a smaller number of dimensions, preferably two or three, so that the main directions of variance are displayed.

Dallican Water

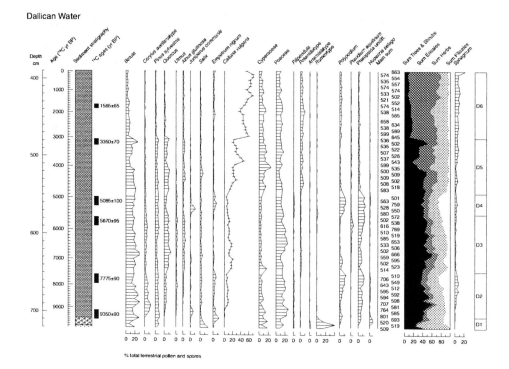

Figure 3. Pollen diagram from Dallican Water, Shetland, UK (1° 6' W, 60° 23.5' N), to illustrate some of the common features of pollen data presentation. The data plotted are selected pollen percentages (values on *x*-axis). The *y*-axis is depth, and a secondary *y*-axis shows a timescale, based on a curve fitted through the radiocarbon age determinations. The shaded column shows sediment stratigraphy, using symbols devised by Troels-Smith (1955). Next, the radiocarbon age determinations (plotted against depth) are displayed. The pollen data are presented as a series of curves, with horizontal lines at the depths of the sample locations. One curve, *Calluna vulgaris*, is drawn with 95% confidence intervals for the percentage values. The main sum, on which calculations are based, is placed after the pollen taxa that are included within it, and is followed by a summary diagram comparing the main ecological groupings of the pollen and spores included within the sum. This is followed by the record for *Sphagnum*, a moss whose spore counts are not part of the main sum. The digram ends with a column showing the zonation scheme for the site. See Bennett et al. (1992) for more information about the site and its interpretation.

An excellent example of palaeoecological use of ordination can be seen in MacDonald (1987). Birks & Gordon (1985) and Maddy & Brew (1995) provide good introductions to methods available and the underlying principles.

Rarefaction analysis

The diversity of pollen types in a sample, and how that diversity differs between samples, is potentially a matter of interest, but cannot be investigated unless the count size of the samples is constant or standardized. The number of pollen types found increases as the pollen count size increases. Rarefaction analysis (see Birks & Line, 1992, for details) provides a way of obtaining this diversity, also known as palynological richness, by estimating the number of pollen taxa that would have been found if the pollen count had been n, where n is less than the actual minimum count among the samples of interest. Conventionally, n is usually taken as 250. If the size of standardized count is kept constant between sequences, and other conditions hold (similarity of pollen taxonomy between analysts, for example), then it may be possible to compare richness between sequences.

Rates of change

Pollen sequences record changes with time: how rapidly do these changes take place? We have chronologies to give us a measure of the passage of time: we also need a measure of change, or difference. Mathematically, this is usually termed 'dissimilarity', and there are many ways of measuring it (see Prentice, 1980, for one list). One approach is to measure the chord distance dissimilarity between any pair of samples (Bennett & Humphry, 1995), and dividing by the age interval between the pair. Another approach is to smooth the sequence first, interpolate to constant time intervals, and then to calculate the dissimilarity measures and divide by the time interval (Jacobson et al., 1987; Jacobson & Grimm, 1988).

Time-series analysis

Time series analysis is the process of taking data from a temporally-based, historical, domain, to a frequency domain. It examines data to determine the proportion of the variance that can be attributed to each of a continuous range of frequencies. This can become an extremely sophisticated and complex analysis, and the full potential cannot be covered here. Specialist advice may be needed to interpret results, and to take analyses further. The technique has been under-used with pollen data, but the potential is shown clearly by Green (e.g., 1981, 1995). Willis et al. (1999a, 1999b) use the technique to identify periodicities in pre-Quaternary pollen records.

Comparison of sequences

A classic problem in palaeoecology has been how to compare two (or more) pollen sequences directly. One approach is known as 'sequence-slotting', which involves amalgamating two or more sequences to produce one combined sequence. The approach was developed by Gordon (1980), and has developed in several directions from there. A number of articles in the INQUA Data-handling Newsletter discuss these developments (http://www.kv. geo.uu.se/inqua). It is also possible to combine datasets for simultaneous analysis by techniques such as ordination (see above).

A major advance in the use of pollen data came with the development of techniques for mapping the results on continental scales. This requires simultaneous handling of many sequences, and combining them on a common time-scale. Davis (1976, 1983) mapped arrival times ('isochrones') of the main forest trees during the Holocene of eastern North America, showing time-transgressive spread. For the same area, Jacobson et al. (1987) generated maps contoured with the abundance of major pollen types ('isopolls'). The two methods use the same data, but emphasize different aspects.

Interpretation

Pollen counts are a measure of the vegetation composition around the site of accumulation at the time of accumulation. Interpretation of the counts may be descriptive or may be directed at 'causes' of either change or stability. The two aspects are quite distinct, and may be treated separately.

Description

The principles of pollen analysis, set out above, make it clear that there is a long progression between release of pollen and its incorporation into sediment. Ideally, the proportions of the different pollen types in the sediment would have an exact relationship to the proportions of the plants in the landscape. This is not the case, for many reasons, including:

1. Differential production. Plants vary greatly in the amount of pollen that they produce.

2. Differential dispersal. Similarly, plants vary greatly in the distance to which pollen grains are dispersed.

3. Differential preservation. Pollen grains are differentially susceptible to degradation by both physical and chemical attack, and this susceptibility appears to be related to sporopollen content (Havinga, 1964, 1984). However, in practice, it is found that anaerobic sediments probably preserve most pollen grains of most types. A few pollen types are badly under-represented despite being produced and dispersed in abundance (e.g., *Populus*: Traverse, 1988), and these seem to be lost almost immediately. Given continuous lacustrine sedimentation of a type that incorporates pollen in an anaerobic environment, there is probably little loss from this factor after the initial losses.

Efforts to improve the interpretation of pollen data in terms of the vegetation from which it originated have involved observational studies and numerical modelling of the processes involved. Observations in modern environments are used to build up relationships between pollen and vegetation that can then be applied to fossil pollen assemblages. This approach is most fully developed in the analysis of Bradshaw & Webb (1985), which showed that different pollen types had different source areas. This means that a single pollen sample is a representation of the abundance of plants with poorly dispersed pollen close to the site, and a representation of the abundance of plants with well-dispersed pollen to much greater distances from the site. Simple integer 'correction factors' have been produced, and used with some success (e.g., Andersen, 1974; Baker et al., 1978). However, comparison of such correction factors between regions emphasizes their dependence on local conditions (Bradshaw, 1981). Most of this work has considered woodlands, but recently Broström et al. (1998) and Sugita et al. (1999) have begun looking at the simple-sounding question of determining the proportion of a landscape that is wooded.

Numerical approaches involve modelling the dispersal of pollen of different types, using their density as a variable. The approach was stimulated by Prentice (1988), and has since been developed further by Sugita (1993, 1994). The observational approaches and numerical approaches complement each other but, so far, have been little used in practice. Most pollen analysts still operate at the level of generally understanding which pollen types are over-represented, and which are not.

Cause

When pollen counts are made from a series of samples along a sequence, changes are found. Generally, we assume that these represent changes in vegetation (though they might also be due to changes in sedimentation). But what brought about the change in vegetation? The earliest pollen analysts assumed that the vegetation was changing with climate. This was the default explanation, and even the entire rationale for carrying out the pollen analyses in the first place (von Post, 1946). But the advent of radiocarbon dating and comparison of the dated changes between sequences showed quickly that changes in the pollen records (and hence vegetation) were not synchronous, nor random, but time-transgressive across regions and continents (Davis, 1976). During the 1980s, a vigorous debate took place between palaeoecologists interpreting pollen data as controlled by ecological factors and palaeoclimatologists interpreting exactly the same data as representations of changes in past climate (e.g., Birks, 1981; Howe & Webb, 1983; Davis et al., 1986; Ritchie, 1986; Webb, 1986).

Meanwhile, in Denmark, Iversen (1941, 1949) noticed that certain mid- and late-Holocene changes in pollen records were likely to be due to the impact of prehistoric people on vegetation. Furthermore, certain abrupt changes are generally agreed to be the result of pathogen attacks on forest trees, such as the decline of *Ulmus* in the European mid-Holocene (Watts, 1961), and *Tsuga* in the eastern North American mid-Holocene (Davis, 1981). It has also been shown that the way pollen abundances change over time is similar to that expected from theoretical populations models (e.g., Tsukada & Sugita, 1982; Bennett, 1983; Walker & Chen, 1987; Magri, 1989; Fuller, 1998), introducing an ecological component to the factors bringing about palynological change. There is thus a shopping-list of about four 'explanations' of changes seen in pollen data (climate, ecology, human, pathogen) plus, of course, interactions of these.

How should palynologists proceed? Most success has been achieved by comparing records over large spatial extents (e.g., Davis, 1976), by comparing pollen data with other kinds of records, or by using selected controlled comparisons ('natural experiments'). There are now long ice-cores from Greenland and the Antarctic that provide good multi-proxy data that is informative about past climates from the last glacial to the present (Grootes et al., 1993; Petit et al., 1999), and long sequences of marine sediments that are informative about longer time periods (Martinson et al., 1987). These can be used as independent records of climate, and pollen data compared with them. Human activity can be inferred from records of microscopic charcoal, if these can be disentangled from records of natural fires. Bennett et al. (1990, 1992) used the charcoal record as part of an argument that pollen changes during the Holocene of north and west Scotland were, in part, anthropogenic rather than climatic. Tinner et al. (1999) present a detailed analysis of the interaction of fire, climate, and vegetation during the Holocene of southern Switzerland. Willis et al. (1997) used a comparison of geochemical data and pollen data from northeast Hungary to construct a detailed model of how forest and soil change interacted at the late-glacial–Holocene transition. Numerical analyses of pollen and diatom data from the same sequences have been used successfully to separate different forcing factors in the form of volcanic influence or climatic change (Lotter & Birks, 1993; Birks & Lotter, 1994). Cwynar & MacDonald (1987) used a comparison between fossil pollen data and modern genetic differentiation in an analysis of the Holocene spread of *Pinus contorta* in western North America. The pattern of post-glacial spread of *Picea glauca* in North America was investigated by Ritchie

& MacDonald (1986) using a comparison of pollen data, ice-sheet distributions and a numerical model of past climate to generate a new hypothesis about why the rate of spread of *Picea glauca* was varied spatially across North America. Many other examples include these types of data and others in comparison with pollen data.

Pollen analyses of lake sediments provide excellent measures of vegetation at specific times and places. There are many taphonomic and representation problems to be worked out (Ritchie, 1995), but modern trends towards multidisciplinary investigations are leading to major advances (Birks & Birks, 2000) where earlier unidisciplinary investigations had led to controversy. The difference is simply in the recognition that there are typically multiple possible explanations for any given event, and so a commensurate number of lines of evidence are needed to solve the puzzle.

Summary

Pollen grains are produced and dispersed as part of the plant reproductive process. They are usually spherical or elliptical and in the size range 0.01 mm to 0.1 mm. The outer walls are extremely resistant to chemical and physical attack and are ornamented and perforated in various ways. Pollen grains are found abundantly in a variety of sediment types, from which they can be concentrated and identified, using combinations of shape, size, ornamentation, and perforations. Identification is typically at the level of genus or family, but within some groups may be possible to species level.

Analysis of the pollen content of sediments, especially from peats and freshwater lakes, was introduced about a century ago, and has developed into the most widely used and applicable technique for determining past vegetation dynamics through time. Results of pollen analyses have also been used to estimate past climate and human impact on vegetation. Numerical models of pollen dispersal are becoming increasingly sophisticated and are being used, for example, to help determine the relative proportions of woodland and open vegetation in past landscapes, solely from fossil pollen data. Numerical analyses are also helping to interpret fossil pollen by identifying major directions of variation in datasets. Problems exist with taxonomy and taphonomy, but pollen analysis remains a leading technique in past environmental reconstruction.

Acknowledgements

We thank Chris Glaister for information about fine-sieving, Adam Gardner, Lindsey Gillson, Heikki Seppä and all three editors for comments on the manuscript, Simon Haberle for help with references, and Glen MacDonald for checking our insularity.

References

Andersen, S. T., 1960. Silicone oil as a mounting medium for pollen grains. Danm. Geol. Unders. Række IV 4(3): 24 pp.

Andersen, S. T., 1974. The Eemian freshwater deposit at Egernsund, South Jylland, and the Eemian landscape development in Denmark. Danm. Geol. Unders. Årbok (1974): 49–70.

Anderson, R. S. & T. R. van Devender, 1991. Comparison of pollen and macrofossils in pack-rat (*Neotoma*) middens: a chronological sequence from the Waterman Mountains of southern Arizona, U.S.A. Rev. Palaeobot. Palynol. 68: 1–28.

Armstrong, J. I., J. Calvert & C. T. Ingold, 1930. The ecology of the mountains of Mourne with special reference to Slieve Donard. Proc. R. Ir. Acad. 39B: 440–452.

Assarsson, G. & E. Granlund, 1924. En metod för pollenanalys av minerogena jordarter. Geol. För. Stockh. Förh. 46: 76–82.

Baker, C. A., P. A. Moxey & P. M. Oxford, 1978. Woodland continuity and change in Epping Forest. Field Stud. 4: 645–669.

Bates, C. D., P. Coxon & P. L. Gibbard, 1978. A new method for the preparation of clay-rich sediment samples for palynological investigations. New Phytol. 81: 459–463.

Bennett, K. D., 1983. Postglacial population expansion of forest trees in Norfolk, UK. Nature 303: 164–167.

Bennett, K. D., 1990. Pollen counting on a pocket computer. New Phytol. 114: 275–280.

Bennett, K. D., 1994. Confidence intervals for age estimates and deposition times in late-Quaternary sediment sequences. Holocene 4: 337–348.

Bennett, K. D., 1996. Determination of the number of zones in a biostratigraphical sequence. New Phytol. 132: 155–170.

Bennett, K. D., 2000. *psimpoll* and *pscomb*: computer programs for data plotting and analysis. Uppsala, Sweden: Quaternary Geology, Earth Sciences, Uppsala University. Software available on the internet at http://www.kv.geo.uu.se.

Bennett, K. D., S. Boreham, M. J. Sharp & V. R. Switsur, 1992. Holocene history of environment, vegetation and human settlement on Catta Ness, Lunnasting, Shetland. J. Ecol. 80: 241–273.

Bennett, K. D., J. A. Fossitt, M. J. Sharp & V. R. Switsur, 1990. Holocene vegetational and environmental history at Loch Lang, South Uist, Western Isles, Scotland. New Phytol. 114: 281–298.

Bennett, K. D. & R. W. Humphry, 1995. Analysis of late-glacial and Holocene rates of vegetational change at two sites in the British Isles. Rev. Palaeobot. Palynol. 85: 263–287.

Benninghoff, W. S., 1962. Calculation of pollen and spore density in sediments by addition of exotic pollen in known quantities. Pollen Spores 4: 332–333.

Berglund, B. E. & M. Ralska-Jasiewiczowa, 1986. Pollen analysis and pollen diagrams. In Berglund, B. E. (ed.) Handbook of Holocene Palaeoecology and Palaeohydrology. Wiley, Chichester: 455–484.

Birks, H. H. & H. J. B. Birks, 2000. Future uses of pollen analysis must include plant macrofossils. J. Biogeogr. 27: 31–35.

Birks, H. J. B., 1973. Past and Present Vegetation of the Isle of Skye—a Palaeoecological Study. Cambridge University Press, London, 415 pp.

Birks, H. J. B., 1981. The use of pollen analysis in the reconstruction of past climates: a review. In Wigley, T. M. L., M. J. Ingram & G. Farmer (eds.) Climate and History: Studies in Past Climates and Their Impact on Man. Cambridge University Press, Cambridge: 111–138.

Birks, H. J. B., 1986. Numerical zonation, comparison and correlation of Quaternary pollen-stratigraphical data. In Berglund, B. E. (ed.) Handbook of Holocene Palaeoecology and Palaeohydrology. Wiley, Chichester: 743–774.

Birks, H. J. B., 1995. Quantitative palaeoenvironmental reconstructions. In Maddy D. & J. S. Brew (eds.) Statistical Modelling of Quaternary Science Data, *Technical Guide* 5. Quaternary Research Association, Cambridge: 161–254.

Birks, H. J. B., 1998. Numerical tools in palaeolimnology-progress, potentialities, and problems. J. Paleolimnol. 20: 307–332.

Birks, H. J. B. & H. H. Birks, 1980. Quaternary Palaeoecology. Arnold, London, 289 pp.

Birks, H. J. B. & A. D. Gordon, 1985. Numerical Methods in Quaternary Pollen Analysis. Academic Press, London, 317 pp.

Birks, H. J. B., S. Juggins, A. Lotter & J. P. Smol (eds.), in preparation. Tracking environmental change using lake sediments: Data handling and statistical techniques. Kluwer Academic Publishers, Dordrecht.

Birks, H. J. B. & J. M. Line, 1992. The use of rarefaction analysis for estimating palynological richness from Quaternary pollen-analytical data. Holocene 2: 1–10.

Birks, H. J. B. & A. F. Lotter, 1994. The impact of the Laacher See volcano (11 000 yr B.P.) on terrestrial vegetation and diatoms. J. Paleolimnol. 11: 313–322.

Bonnefille, R. & G. Riollet, 1980. Pollens des savanes d'Afrique orientale. Centre National de la Recherche Scientifique, Paris, 140 pp.

Bonny, A. P., 1972. A method for determining absolute pollen frequencies in lake sediments. New Phytol. 71: 393–405.

Bourgeois, J. C., 1986. A pollen record from the Agassiz Ice Cap, northern Ellesmere Island, Canada. Boreas 15: 345–354.

Bradshaw, R. H. W., 1981. Modern pollen-representation factors for woods in south-east England. J. Ecol. 69: 45–70.

Bradshaw, R. H. W. & T. Webb, III, 1985. Relationships between contemporary pollen and vegetation data from Wisconsin and Michigan, USA. Ecology 66: 721–737.

Broström, A., M. J. Gaillard, M. Ihse & B. Odgaard, 1998. Pollen–landscape relationships in modern analogues of ancient cultural landscapes in southern Sweden—a first step towards quantification of vegetation openness in the past. Veget. Hist. Archaeobot. 7: 189–201.

Carrión, J. S., M. Munuera, C. Navarro, F. Burjachs, M. Dupré & M. J. Walker, 1999. The palaeoecological potential of pollen records in caves: the case of Mediterranean Spain. Quat. Sci. Rev. 18: 1061–1073.

Carrión, J. S., L. Scott & J. C. Vogel, 1999. Twentieth century changes in montane vegetation in the eastern Free State, South Africa, derived from palynology of hyrax dung middens. J. Quat. Sci. 14: 1–16.

Chitty, J., 1995. HF fatality. A.A.S.P. Newsl. 28(1): 14–15.

Clark, R. L., 1982. Point count estimation of charcoal in pollen preparations and thin sections of sediments. Pollen Spores 24: 523–535.

Coles, G. M., D. D. Gilbertson, C. O. Hunt & R. D. S. Jenkinson, 1989. Taphonomy and the palynology of cave deposits. Cave Sci. 16: 83–89.

Colinvaux, P. A., P. E. De Oliveira & E. Moreno, 1999. Amazon Pollen Manual and Atlas. Harwood Academic Publishers, Amsterdam, 332 pp.

Crane, P. R., E. M. Friis & K. R. Pedersen, 1995. The origin and early diversification of angiosperms. Nature 374: 27–33.

Cwynar, L. C., E. Burden & J. H. McAndrews, 1979. An inexpensive sieving method for concentrating pollen and spores from fine-grained sediments. Can. J. Earth Sci. 16: 1115–1120.

Cwynar, L. C. & G. M. MacDonald, 1987. Geographic variation of lodgepole pine in relation to its population history. Am. Nat. 129: 463–469.

Davis, M. B., 1965. A method for determination of absolute pollen frequency. In Kummel, B. & D. Raup (eds.) Handbook of Paleontological Techniques. W. H. Freeman, San Francisco: 674–686.

Davis, M. B., 1966. Determination of absolute pollen frequency. Ecology 47: 310–311.

Davis, M. B., 1976. Pleistocene biogeography of temperate deciduous forests. Geosci. Man 13: 13–26.

Davis, M. B., 1981. Outbreaks of forest pathogens in Quaternary history. In Bharadwaj, D. C., Vishnu-Mittre & H. K. Maheshwari (eds.) Fourth International Palynological Conference Proceedings Volume III. Birbal Sahni Institute of Palaeobotany, Lucknow: 216–228.

Davis, M. B., 1983. Quaternary history of deciduous forests of eastern North America and Europe. Ann. Missouri Bot. Gard. 70: 550–563.

Davis, M. B. & E. S. Deevey, 1964. Pollen accumulation rates: estimates from late-glacial sediment of Rogers Lake. Science 145: 1293–1295.

Davis, M. B., K. D. Woods, S. L. Webb & R. P. Futyma, 1986. Dispersal versus climate: expansion of *Fagus* and *Tsuga* into the Upper Great Lakes region. Vegetatio 67: 93–103.

Duller, A., G. Duller, I. France & H. Lamb, 1999. A pollen image database for evaluation of automated identification systems. Quat. Newsl. 89: 4–9.

Erdtman, G., 1934. Über die verwendung von essigsäureanhydrid bei pollenuntersunchungen. Svensk Bot. Tidskr. 28: 354–358.

Faegri, K. & J. Iversen, 1989. Textbook of Pollen Analysis (4th ed.). Wiley, Chichester, 328 pp.

Faegri, K. & L. van der Pijl, 1979. The Principles of Pollination Ecology (3rd ed.). Pergamon, Oxford, 244 pp.

France, I., A. W. G. Duller, G. A. T. Duller & H. F. Lamb, 2000. A new approach to automated pollen analysis. Quat. Sci. Rev. 18: 537–546.

Fuller, J. L., 1998. Ecological impact of the mid-Holocene hemlock decline in southern Ontario, Canada. Ecology 79: 2337–2351.

Funkhouser, J. W. & W. R. Evitt, 1959. Preparation techniques for acid-insoluble microfossils. Micropaleontology 5: 369–375.

Glaister, C. G., 2000. Palynology of late Pleistocene marine sediments in north Denmark. Ph. D. thesis, University of Cambridge, 210 pp.

Gordon, A. D., 1980. SLOTSEQ: a FORTRAN IV program for comparing two sequences of observations. Comput. Geosci. 6: 7–20.

Gordon, A. D. & H. J. B. Birks, 1972. Numerical methods in Quaternary palaeoecology. I. Zonation of pollen diagrams. New Phytol. 71: 961–979.

Gray, J., 1965. Extraction techniques. In Kummel, B. & D. Raup (eds.) Handbook of Paleontological Techniques. W. H. Freeman, San Francisco: 530–587.

Green, D. G., 1981. Time series and postglacial forest ecology. Quat. Res. 15: 265–277.

Green, D. G., 1995. Time and spatial analysis. In Maddy, D. & J. S. Brew (eds.) Statistical Modelling of Quaternary Science Data, *Technical Guide* 5. Quaternary Research Association, Cambridge: 65–105.

Grimm, E. C., 1991–1993. TILIA 2.0. Springfield, Illinois, USA: Illinois State Museum. Software.

Grootes, P. M., M. Stuiver, J. W. C. White, S. Johnsen & J. Jouzel, 1993. Comparison of oxygen isotope records from the GISP2 and GRIP Greenland ice cores. Nature 366: 552–554.

Havinga, A. J., 1964. An investigation into the differential corrosion susceptibility of pollen and spores. Pollen Spores 6: 621–635.

Havinga, A. J., 1984. A 20-year experimental investigation into the differential corrosion susceptibility of pollen and spores in various soil types. Pollen Spores 26: 541–558.

Hedberg, H. D. (ed.), 1976. International Stratigraphic Guide. John Wiley, New York, 200 pp.

Hooghiemstra, H. & B. van Geel, 1998. World list of Quaternary pollen and spore atlases. Rev. Palaeobot. Palynol. 104: 157–182.

Howe, S. & T. Webb, III, 1983. Calibrating pollen data in climatic terms: improving the methods. Quat. Sci. Rev. 2: 17–51.

Hughes, N. F., 1976. The enigma of angiosperm origins. Cambridge University Press, Cambridge, 242 pp.

Iversen, J., 1941. Landnam i Danmarks Stenalder. Danm. Geol. Unders. Række II 66: 68 pp.

Iversen, J., 1949. The influence of prehistoric man on vegetation. Danm. Geol. Unders. Række IV 3(6): 25 pp.

Jacobson, Jr, G. L. & E. C. Grimm, 1988. Synchrony of rapid change in late-glacial vegetation south of the Laurentide ice sheet. Bulletin of the Buffalo Society of Natural Sciences 33: 31–38.

Jacobson, Jr, G. L., T. Webb, III & E. C. Grimm, 1987. Patterns and rates of vegetation change during the deglaciation of eastern North America. In Ruddiman, W. F. & H. E. Wright, Jr (eds.) North

America and Adjacent Oceans During the Last Deglaciation, The Geology of North America Volume K-3. Geological Society of America, Boulder, CO: 277–288.

Jagudilla-Bulalacao, L., 1997. Pollen Flora of the Philippines, Volume 1. Monographs in Systematic Botany from the National Museum, Philippines. Department of Science and Technology, Special Projects Unit, Technology Application and Promotion Institute, Taguig, Metro Manila, 266 pp.

Jemmett, G. & J. A. K. Owen, 1990. Where has all the pollen gone? Rev. Palaeobot. Palynol. 64: 205–211.

Langmyhr, F. J. & K. Kringstad, 1966. An investigation of the composition of the precipitates formed by the decomposition of silicate rocks in 38–40% hydrofluoric acid. Anal. Chim. Acta 35: 131–135.

Langmyhr, F. J. & S. Sveen, 1965. Decomposability in hydrofluoric acid of the main and some minor and trace minerals of silicate rocks. Anal. Chim. Acta 32: 1–7.

Lotter, A. F. & H. J. B. Birks, 1993. The impact of the Laacher See Tephra on terrestrial and aquatic ecosystems in the Black Forest, southern Germany. J. Quat. Sci. 8: 263–276.

MacDonald, G. M., 1987. Postglacial development of the subalpine-boreal transition forest of western Canada. J. Ecol. 75: 303–320.

Maddy, D. & J. S. Brew (eds.), 1995. Statistical Modelling of Quaternary Science Data, *Technical Guide* 5. Quaternary Research Association, Cambridge, 271 pp.

Magri, D., 1989. Interpreting long-term exponential growth of plant populations in a 250000-year pollen record from Valle di Castiglione (Roma). New Phytol. 112: 123–128.

Maher, Jr, L. J., 1972. Nomograms for computing 95% limits of pollen data. Rev. Palaeobot. Palynol. 13: 85–93.

Maher, Jr, L. J., 1981. Statistics for microfossil concentration measurements employing samples spiked with marker grains. Rev. Palaeobot. Palynol. 32: 153–191.

Markgraf, V. & H. L. D'Antoni, 1980. Pollen flora of Argentina. University of Arizona Press, Tucson, AZ, 208 pp.

Martinson, D. G., N. G. Pisias, J. D. Hays, J. Imbrie, T. C. Moore, Jr & N. J. Shackleton, 1987. Age dating and the orbital theory of the ice ages: development of a high-resolution 0 to 300,000-year chronostratigraphy. Quat. Res. 27: 1–29.

Matthews, J., 1969. The assessment of a method for the determination of absolute pollen frequencies. New Phytol. 68: 161–166.

McAndrews, J. H., 1984. Pollen analyses of the 1973 ice core from Devon Island Glacier, Canada. Quat. Res. 22: 68–76.

McAndrews, J. H., A. A. Berti & G. Norris, 1973. Key to the Quaternary pollen and spores of the Great Lakes region. Life Sciences Miscellaneous Publication, Royal Ontario Museum, Toronto, 61 pp.

Moar, N. T., 1993. Pollen Grains of New Zealand Dicotyledonous Plants. Manaaki Whenua Press, Lincoln, New Zealand, 200 pp.

Moore, P. D., J. A. Webb & M. E. Collinson, 1991. Pollen Analysis. Blackwell Scientific Publications, Oxford, 216 pp.

Mosimann, J. E., 1965. Statistical methods for the pollen analyst: multinomial and negative multinomial techniques. In Kummel, B. & D. Raup (eds.) Handbook of Paleontological Techniques. W. H. Freeman, San Francisco: 636–673.

Nakagawa, T., E. Brugiapaglia, G. Digerfeldt, M. Reille, J. L. de Beaulieu & Y. Yasuda, 1998. Dense-media separation as a more efficient pollen extraction method for use with organic sediment / deposit samples: comparison with the conventional method. Boreas 27: 15–24.

Petit, J. R., J. Jouzel, D. Raynaud, N. I. Barkov, J. M. Barnola, I. Basile, M. Bender, J. Chappellaz, M. Davis, G. Delaygue, M. Delmotte, V. M. Kotlyakov, M. Legrand, V. Y. Lipenkov, C. Lorius, L. Pépin, C. Ritz, E. Saltzman & M. Stievenard, 1999. Climate and atmospheric history of the past 420,000 years from the Vostok ice core, Antarctica. Nature 399: 429–436.

Prentice, I. C., 1980. Multidimensional scaling as a research tool in Quaternary palynology: a review of theory and methods. Rev. Palaeobot. Palynol. 31: 71–104.

Prentice, I. C., 1988. Records of vegetation in time and space: the principles of pollen analysis. In Huntley, B. & T. Webb, III (eds.) Vegetation History, *Handbook of Vegetation Science* 7. Kluwer Academic, Dordrecht: 17–42.

Proctor, M., P. Yeo & A. Lack, 1996. The Natural History of Pollination. Collins, London, 479 pp.

Punt, W., 1976. The Northwest European Pollen Flora I. Elsevier Scientific, Amsterdam, 145 pp.

Punt, W. & S. Blackmore, 1991. The Northwest European Pollen Flora VI. Elsevier Scientific, Amsterdam, 276 pp.

Punt, W., S. Blackmore & G. C. S. Clarke, 1988. The Northwest European Pollen Flora V. Elsevier Scientific, Amsterdam, 154 pp.

Punt, W., S. Blackmore & P. P. Hoen, 1995. The Northwest European Pollen Flora VII. Elsevier Scientific, Amsterdam, 282 pp.

Punt, W., S. Blackmore, S. Nilsson & A. Le Thomas, 1994. Glossary of pollen and spore terminology. LPP Contributions Series No. 1. LPP Foundation, Utrecht, 71 pp.

Punt, W. & G. C. S. Clarke, 1980. The Northwest European Pollen Flora II. Elsevier Scientific, Amsterdam, 265 pp.

Punt, W. & G. C. S. Clarke, 1981. The Northwest European Pollen Flora III. Elsevier Scientific, Amsterdam, 138 pp.

Punt, W. & G. C. S. Clarke, 1984. The Northwest European Pollen Flora IV. Elsevier Scientific, Amsterdam, 369 pp.

Reille, M., 1992. Pollen et Spores d'Europe et d'Afrique du Nord. Laboratoire de Botanique Historique et Palynologie, Marseille, 520 pp.

Reille, M., 1995. Pollen et Spores d'Europe et d'Afrique du Nord: Supplement 1. Laboratoire de Botanique Historique et Palynologie, Marseille, 327 pp.

Reille, M., 1998. Pollen et Spores d'Europe et d'Afrique du Nord: Supplement 2. Laboratoire de Botanique Historique et Palynologie, Marseille, 521 pp.

Ritchie, J. C., 1986. Climate change and vegetation response. Vegetatio 67: 65–74.

Ritchie, J. C., 1995. Current trends in studies of long-term plant community dynamics. New Phytol. 130: 469–494.

Ritchie, J. C. & G. M. MacDonald, 1986. The patterns of post-glacial spread of white spruce. J. Biogeogr. 13: 527–540.

Sánchez Goñi, M. F., F. Eynaud, J. L. Turon & N. J. Shackleton, 1999. High resolution palynological record off the Iberian margin: direct land–sea correlation for the Last Interglacial complex. Earth Planet. Sci. Lett. 171: 123–137.

Southworth, D., 1990. Exine biochemistry. In Blackmore, S. & R. B. Knox (eds.) Microspores: Evolution and Ontogeny. Academic Press, London: 193–212.

Stillman, E. C. & J. R. Flenley, 1996. The needs and prospects for automation in palynology. Quat. Sci. Rev. 15: 1–7.

Stuiver, M., P. J. Reimer, E. Bard, J. W. Beck, G. S. Burr, K. A. Hughen, B. Kromer, G. McCormac, J. van der Plicht & M. Spurk, 1998. INTCAL98 radiocarbon age calibration, 24,000–0 cal BP. Radiocarbon 40: 1041–1083.

Sugita, S., 1993. A model of pollen source area for an entire lake surface. Quat. Res. 39: 239–244.

Sugita, S., 1994. Pollen representation of vegetation in Quaternary sediments: theory and method in patchy vegetation. J. Ecol. 82: 881–897.

Sugita, S., M. J. Gaillard & A. Broström, 1999. Landscape openness and pollen records: a simulation approach. Holocene 9: 409–421.

Sun, X. J., C. Q. Song, F. Y. Wang & M. R. Sun, 1997. Vegetation history of the Loess Plateau of China during the last 100,000 years based on pollen data. Quat. Int. 37: 25–36.

Thompson, R., 1980. Use of the word "influx" in paleolimnological studies. Quat. Res. 14: 269–270.

Tinner, W., P. Hubschmid, M. Wehrli, B. Ammann & M. Condera, 1999. Long-term forest fire ecology and dynamics in southern Switzerland. J. Ecol. 87: 273–289.

Traverse, A., 1988. Paleopalynology. Unwin Hyman, Boston, MA, 600 pp.

Troels-Smith, J., 1955. Karakterisering af lose jordater. Characterization of unconsolidated sediments. Danm. Geol. Unders. Række IV 3(10): 73 pp.

Tsukada, M. & S. Sugita, 1982. Late Quaternary dynamics of pollen influx at Mineral Lake, Washington. Bot. Mag., Tokyo 95: 401–418.

van der Knaap, W. O., 1989. Past vegetation and reindeer on Edgeoya (Spitsbergen) between c. 7900 and c. 3800 BP, studied by means of peat layers and reindeer faecal pellets. J. Biogeogr. 16: 379–394.

von Post, L., 1946. The prospect for pollen analysis in the study of the earth's climatic history. New Phytol. 45: 193–217.

Walanus, A. & D. Nalepka, 1997. Palynological diagram drawing in Polish POLPA1 for windows. INQUA Comm. Study Holocene: Work. gr. data-handl. meth. Newsl. 16: 3.

Walker, D. & Y. Chen, 1987. Palynological light on tropical rainforest dynamics. Quat. Sci. Rev. 6: 77–92.

Watts, W. A., 1961. Post Atlantic forests in Ireland. Proc. Linn. Soc. Lond. 172: 33–38.

Webb, III, T., 1986. Is vegetation in equilibrium with climate? How to interpret late-Quaternary pollen data. Vegetatio 67: 75–91.

Welten, M., 1944. Pollenanalytische, stratigraphische und geochronologische Untersuchungen aus dem Faulenseemoos bei Spiez. Veröff. Geobot. Inst. Rübel Zürich 21: 201 pp.

Willis, K. J., M. Braun, P. Sümegi & A. Tóth, 1997. Does soil change cause vegetation change or vice versa? A temporal perspective from Hungary. Ecology 78: 740–750.

Willis, K. J., A. Kleczkowski, K. M. Briggs & C. A. Gilligan, 1999. The role of sub-Milankovitch climatic forcing in the initiation of the Northern Hemisphere glaciation. Science 285: 568–571.

Willis, K. J., A. Kleczkowski & S. J. Crowhurst, 1999. 124,000-year periodicity in terrestrial vegetation change during the late Pliocene epoch. Nature 397: 685–688.

3. CONIFER STOMATA

GLEN M. MACDONALD (macdonal@geog.ucla.edu)
Departments of Geography and Organismic Biology
Ecology and Evolution
UCLA
Los Angeles, California
90095-1524 USA

Keywords: palynology, stomata, stomata analysis, conifers, vegetation change, treeline

Introduction

Stomata (Greek *stoma*-mouth) are minute intracellular fissures and specialized guard cells that are found on the epidermis of leaves, stems, ovules and some flowers. The stomata allow the exchange of carbon dioxide, oxygen and water vapor between the plant and the atmosphere. They are sometimes referred to in the singular as stoma, although this term is often used to describe only the intercellular opening, and not the surrounding cells. In this chapter the term stomata will be used, and be treated grammatically in the same way in which the term 'pollen' is used as a noun and adjective. Increases in turgor causes the guard cells to deform and open the stomata. In many plants, the cells surrounding the guard cells differ in form from normal epidermal cells. The surrounding structure of subsidiary cells is sometimes referred to as the Florin structure. The complete structure, including opening, guard cells and subsidiary cells, is called the stomata complex. Stomata, or stomata-like structures in the case of non-vasuclar plants, are found on a wide variety of plants ranging from non-vascular Anthocerotopsida (hornworts) and Bryophyta (mosses), to all vascular plants Psilophyta (whisk ferns), Microphyllophyta (club mosses, selaginella, quillworts), Arthrophyta (*Equisetum*), Pterophyta (ferns), Cycadophyta (cycads), Ginkophyta (ginkgoes), Gnetophyta (*Ephedra; Gnetum* and *Welwitschia*), Coniferophyta (conifers), and Anthophyta (angiosperms).

Most stomata are between 20 to 80 μm in diameter. Plant epidermal tissue may contain 10,000 to over 100,000 stomata per cm^2. The density of stomata is related to the species of the plant, but can also vary due to the environmental conditions under which the plant is growing. Changes in atmospheric CO_2 concentrations have a particularly strong impact on stomata density. Increasing CO_2 leads to significant decreases in stomata densities in many plants (Woodward, 1987).

Fossil stomata are found either as microfossils, in the form of detached individuals (Fig. 1) and small groups joined by fragmentary epidermal tissue, or preserved on macrofossil leaf and stem surfaces. The two paleoenvironmental applications of stomata analysis

33

J. P. Smol, H. J. B. Birks & W. M. Last (eds.), 2001. *Tracking Environmental Change Using Lake Sediments.*
Volume 3: Terrestrial, Algal, and Siliceous Indicators. Kluwer Academic Publishers, Dordrecht, The Netherlands.

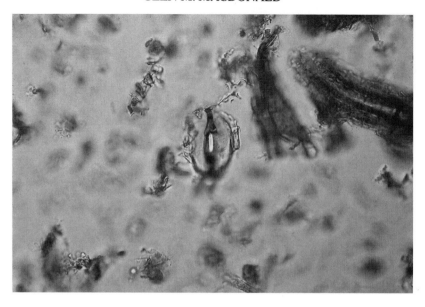

Figure 1. A *Picea* stomata in a typical palynological preparation recovered from lake sediment. The total length of a stomate along its longest axis is approximately 20 μm.

are: 1) the reconstruction of past vegetation using the presence, absence and abundance of stomata microfossils to infer the presence, absence and abundance of different plant taxa; and 2) the reconstruction of changes in atmospheric CO_2 concentrations and the impact of such changes on plant physiology through examination of changes in the density of stomata on modern and fossil leaves. This chapter will focus on the use of detached microfossil stomata in paleoenvironmental studies of lake sediments. The use of stomata density analysis is an important developing tool for paleoclimatology (Van de Water et al., 1994; Beerling & Chaloner, 1994; Poole et al., 1996). In some case, enough complete leaves and large fragments may be recovered from lake sediments to reconstruct changes in stomata density over time. However, detailed macrofossil taphonomy, and the techniques used to recover and analyze such macrofossils, are beyond the scope of this chapter (see H. H. Birks, this volume).

The usefulness of conifer stomata in paleoecology, particularly as an ancillary technique to pollen analysis (see Bennett & Willis, this volume), arises from five factors:

First, the stomata cells of conifers are lignified (Fahn, 1982). These lignified cells are more resistant to decay when deposited in lake waters than are unlignified soft tissue. Lignified stomata are also found in some cryptogams and angiosperms, but stomata from these taxa appear to be rare in lake sediments (Hansen, 1995).

Second, stomata are generally deposited in lakes on whole leaves and large leaf fragments. In this regard, their taphonomy is more similar to that of plant macrofossils than to the taphonomy of pollen grains (see Bennett & Willis, this volume). Due to their small size, pollen grains may be transported in large quantities for tens to hundreds of kilometers from the parent plant. Conifers are anemophilous and some species produce huge quantities of readily dispersed pollen grains. The propensity for pollen to be widely dispersed makes it

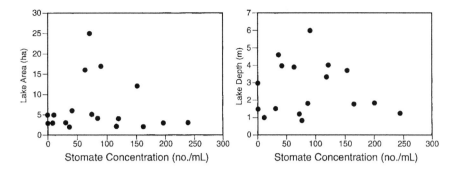

Figure 2. Relationship between lake area and depth, and stomate abundance in treeline lakes from the Kola Peninsula, Russia (data from Gervais & MacDonald, 2001).

very difficult to use the fossil pollen records from lake sediments to infer the local presence or absence of a plant species (MacDonald, 1993). In particular, it is often difficult to identify the past geographic position of either arctic or alpine conifer treelines using fossil pollen evidence (e.g., Maher, 1963; Markgraf, 1980; MacDonald & Ritchie, 1986; Hicks, 1993; H. H. Birks, this volume). The relatively large size of leaves precludes their transport by air for any great distance. Thus, for lakes which lack inflowing streams, the presence of conifer stomata indicates the local presence of coniferous plants.

Third, microfossil stomata found in lake sediments occur as detached individuals and on small fragments of epidermis. It is likely that most of the stomata recovered in lake sediment cores were released during the disintegration of leaves in shallow peripheral waters and then redeposited in deep central waters. The individual stomata and small epidermal fragments are probably resuspended and distributed throughout the lake in a manner similar to pollen (Davis, 1973). In general, terrestrial plant macrofossils are rare in sediments from the central portions of large lakes (Watts & Winter, 1967). Small deep lakes are generally thought to be the best sites to recover plant macrofossils (H. H. Birks, this volume). However, analysis of the relationship between stomata presence and lake morphometry (Fig. 2) shows no significant relationship between lake size or lake depth and the abundance of stomata (Hansen et al., 1996; Gervais & MacDonald, 2001). Thus, due to their small size and propensity for redeposition, stomata can be found in many lake sediment cores in which plant macrofossils are absent.

Fourth, many conifer stomata can be identified to species or genus using transmitted light microscopy. In some instances, genera such as *Juniperus* (juniper), *Thuja* (eastern cedar) and *Chamaecyparis* (white cedar) produce pollen which cannot be differentiated from other genera in the family, but possess stomata which are taxonomically distinctive.

Fifth, individual stomata are roughly the same size as most pollen grains. The lignified stomata can withstand the chemical treatments used in pollen preparation (see Bennett & Willis, this volume). Thus, stomata samples are prepared at the same time that pollen samples are processed. The stomata can also be counted along with the pollen using the same techniques of optical microscopy. These properties make it practical to perform stomata analysis at the same time as traditional palynological analysis.

Preparation, counting, identification and analysis

As detached stomata are roughly the same size as pollen, and have a similar density, stomata can be expected to occur in the silty lake sediments and gytjjas from which fossil pollen grains are typically recovered. Preparation of samples for stomata analysis is identical to the chemical digestion techniques used for pollen preparations (Faegri et al., 1989; Bennett & Willis, this volume). However, removal of large organic fragments using selective sieving is sometimes applied during pollen sample preparation, but can decrease stomata abundance in samples. The processing must be done in a certified fumehood, wearing impermeable gloves and apron and a face shield. The chemicals and residues are dangerous and must be stored and disposed of properly. The procedure used at the UCLA (University of California Los Angeles) lab is as follows:

1. Place 1 ml of wet sediment in an acid resistant polypropylene centrifuge tube.
2. Add a tablet or slurry containing a known quantity of an exotic palynomorph, such as *Eucalyptus* pollen or chemically treated *Lycopodium* spores. The addition of the exotic palynomorphs allows the monitoring of the effect of the processing on the fossil pollen and stomates. The exotic palynomorphs also allow the subsequent calculation of pollen and stomate concentrations in the sediment sample.
3. Add a 10% hydrochloric acid (HCl) solution. Mix thoroughly until any reaction with the exotic tablets or sediments ceases. This step removes carbonates and breaks down the matrix used to bind the exotic pollen and spore tablets. Failure to remove all carbonates at this stage will result in the formation of an insoluble precipitate when hydrofluoric acid (HF) is added to the samples later.
4. Centrifuge (all centrifuging should be at high speed for at least 5 minutes) and decant hydrochloric acid. Water wash, by filling the tube with distilled water, stirring, centrifuging and decanting.
5. Add a 10% solution of potassium hydroxide (KOH). Stir and place the centrifuge tubes in a bath of boiling water for 5 minutes. The potassium hydroxide removes humic acids.
6. Centrifuge, decant potassium hydroxide and perform distilled water washes until water is clear after centrifuging.
7. Add 50% hydrofluoric acid (HF) solution. This is an extremely dangerous acid—seek medical attention immediately if it comes in contact with skin, or if fumes are inhaled. Waste must never be stored in a glass container. Stir and place tubes in a boiling water bath for 1 hr. Stir several times during the bath. The hydrofluoric acid dissolves silicates.
8. Centrifuge and decant hydrofluoric acid. Perform a minimum of two distilled water washes. Consider the water from the washes to be highly toxic.
9. In some cases a second treatment using HCl as outlined in step 3 is required at this point to remove products of silicate digestion.
10. Add glacial acetic acid (CH_3COOH), stir well, centrifuge and decant. This step removes water from the sample.
11. Add a solution containing 10% sulfuric acid (H_2SO_4) and 90% acetic anhydride (($CH_3CO_2)_2O$) to the tubes. This mixture is called acetolysis solution. Acetolysis is highly reactive with water and should be used with caution. If acetolysis solution is added to sediment containing water, the sample will violently explode out of the centrifuge tube. Place the tubes in a boiling water bath for 5 minutes. The acetolysis hydrolyzes polysaccharides thus removing cellulose, waxes and oils from palynomorphs and stomata. Centrifuge and decant.

12. Wash and centrifuge with glacial acetic acid and decant.

13. Wash and centrifuge with distilled water and decant.

14. Add tertiary butyl alcohol ($CH_3CH_2CH_2CH_2OH$), stir, centrifuge and decant. Tertiary butyl alcohol freezes at room temperature. It can be warmed and melted in a hot water bath. It is highly flammable and care must be taken not to expose it to high heat or flame.

15. Repeat the above step. This removes water and keeps sample from clumping. Some analysts add a drop of safranin-O dye at this stage to enhance the visual contrast of surface features.

16. Use some tertiary butyl alcohol to wash the remaining sample from centrifuge tube into small sample vial. Centrifuge the samples and decant.

17. Add a small amount of 2000 cs silicone oil and stir well. Cover the sample vials with a piece of paper and allow remaining alcohol to evaporate.

As some stomata occur upon larger epidermal tissue fragments, it is desirable that the samples not be coarse-sieved to remove large organic fragments. Fine-sieving procedures (Cwynar et al., 1979) are permissible. It is possible that heavy liquid separation techniques used for concentrating pollen from lake sediments will also work for concentrating stomata.

Reference samples of stomata can be produced from dried herbarium specimens, or needles that have been collected in the field, dried and identified confidently to species. The dried needles should be ground with a mortar and pestle. The ground material is chemically processed in the manner outlined above. However, the addition of exotic pollen and spores, hydrochloric acid, and hydrofluoric acid is generally omitted.

Identification and counting of fossil stomata is conducted using transmitted light microscopy at a magnification of 400×. The suspension of pollen and stomata in silicone oil is stirred and a drop placed on a glass microscope slide. A thin cover slip is then affixed on top of the suspension and gently flattened. Small drops of diluted white glue or red metallic nail polish is used to fasten two diagonally opposite corners of the cover slip to the slide. By leaving two corners of the cover slip free, gentle tapping can cause the silicone oil to flow and thus rotate stomata and pollen grains. Stomata can be identified and counted at the same time as counting pollen on the microscope slide. However, the concentration of stomata is far less than the concentration of pollen and spores. Stomata are generally found at concentrations of <2000 per ml. The difference in the total number of stomata and pollen in a sediment sample is typically several orders of magnitude. In some cases, it is expedient to scan slides at 200× magnification and tabulate only stomata and exotic pollen or spores. Due to their relative scarcity, stomata are generally not counted to the same high numbers as fossil pollen and spores. In many cases, detecting the presence or absence of stomata, rather than their absolute or relative abundance, is the goal of the analysis.

The stomata of different plant taxa are distinctive in both plain view and cross-section. In most cases, fossil stomata are presented on the microscope slide in a polar view. Lignified subsidiary cells, and in some cases fragments of the cutinized epidermal tissue, may also be present.

The lamellae (intercellular surface) of the guard cells of conifer stomata are lignified and this preserves the shape of the guard cells in fossil specimens. Differences in the shape of the upper and lower lamellae and medial lamellae that borders the stomata opening are important in the taxonomic discrimination of fossil conifer stomata (Fig. 3). For example, *Larix* (larch) possesses lamellae that are rectangular in shape, while the guard cells and lamellae of *Thuja* are quite rounded (Fig. 3). The presence of lignified lower lamellae

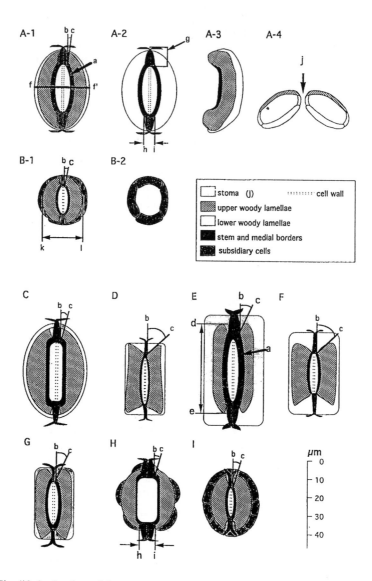

Figure 3. Simplified polar views of the stomata, lamellae, stems and subsidiary cells of: *Picea* (A-1 upper view; A-2 lower view; A-3 and A-4 cross sectional views); *Thuja* (B-1 upper view; B-2 lower view); *Picea glauca* (C upper view); *Larix laricina* (D upper view); *Pinus* (E upper view); *Abies* (F upper view); *Tsuga mertensiana* (G upper view); *Tsuga heterophylla* (H upper view); *Chamaecyparis* (I upper view). (From Hansen, 1995).

differentiates *Picea* (spruce) stomata from those of *Thuja* which possess a number of subsidiary cells below the guard cells. However, in many cases, the lower lamellae and subsidiary cells of fossil stomata either cannot be differentiated from the upper lamellae, or are missing altogether. The medial lamellae bordering the stoma are often distinctively thickened. In addition, the medial lamellae fuse to form distinctive stems at the poles of

Table I. A stomate key for some common northern conifers of North America (from Hansen, 1995).

Key to important conifer stomate types from the Hudson Bay Lowlands and southeastern Alaska (as seen in surface view; adapted from Trautmann 1953)

A. Lower woody lamellae discernible on the guard cells; the stomatal complex comprises two cells (Figs. 2A-1, A-2, and 2C–2G) ....

B. Upper woody lamellae shorter than the lower woody lamellae; lower woody lamellae distinctly visible laterally and at the poles (Figs. 2E, 2F, and 3A).......................................

C. Angle of junctions of the upper woody lamellae with the stem 20–30° (*b-c* in Fig. 2E); upper woody lamellae relatively long (*d-e* in Fig. 2E; Table 1) ....

....*Pinus* type (includes *P. contorta* and *P. banksiana*)

CC. Angle of upper woody lamellae junction with the stem 45° (*b-c* in Figs. 2F; Fig. 3*i*); stem narrow; upper woody lamellae considerably shorter than the lower woody lamellae (Fig. 2F) ....

....*Abies* type

BB. Upper woody lamellae nearly as long as the lower woody lamellae; lower woody lamellae barely visible ....

C. Outline of upper woody lamella rectangular (Figs. 2D and 2G)............................

D. Medial lamellae border narrow, <2 μm (Fig. 2D); angle of junction of upper lamellae with the stem 20–30° (*b-c* in Fig. 2D; Fig. 3B)....................................... *Larix laricina*

DD. Medial lamellae border moderately wide (Fig. 2G); angle of junction of upper lamellae with the stem < 10°; ends of the upper lamellae more rounded than angular (Fig. 3E)...............

.... *Tsuga mertensiana*

CC. Outline of upper woody lamellae oval to spherical (Figs. 2A, 2C, 3C, and 3D); angle of junction of upper lamellae with the stem less than 10° (*b-c* in Figs. 2A' and 2C); stem wide (5–6 μm at its widest part)...................... *Picea* type (includes *P. mariana, P. glauca,* and *P. sitchensis*)

AA. Lower woody lamellae not discernible; stomatal complex includes 4–9 lignified subsidiary cells below the upper woody lamellae (Figs. 2B, parts 1 and 2, and Figs. 2H, 2I, and 3G)........

B. Stem broad (about 8 μm wide; *h-i* in Fig. 2H; Fig. 3F); lower part of stomatal complex comprises 5 lignified subsidiary cells (Fig. 3G) .... *Tsuga heterophylla*

BB. Stem narrow (about 2–3 μm); lower part of the stomatal complex comprises 4–10 subsidiary cells ....

C. Length of upper woody lamellae about 32 μm; lower part of the stornatal complex comprises 6–10 subsidiary cells (Figs. 2I) .... *Chamaecyparis nootkatensis*

CC. Length of upper woody lamellae about 26 μm; lower part of the stomatal complex comprises 4–6 subsidiary cells (Figs. 2B and 3H)..*Thuja* type

the stoma (Fig. 3). The angle of attachment of the upper lamellae to the stem can also be an important characteristic for the taxonomic identification of stomata (Fig. 3).

The basis for the reliable identification of stomata taxa is a good reference collection. Keys for identification of some taxa are available from Trautmann (1953) and Hansen (1995). The stomata key provided by Hansen, and based upon the original German key produced by Trautmann, allows the identification of important northern coniferous tree taxa (Table I). In some cases, such as *Tsuga heterophylla* (western hemlock) and *Tsuga mertensiana* (mountain hemlock) the stomata can be identified to species. In the case of *Pinus* (pine) most stomata can only be identified to the genus.

Comparison of stomata and pollen from modern lake sediments

The most appealing quality of fossil stomata analysis is its ability to provide a relatively precise indicator of the past local presence or absence of coniferous plants. The potential of fossil stomata to improve upon the spatial resolution provided by pollen analysis can be assessed by comparing the relationship between the pollen and stomata content of modern lake sediments and the modern distributions of conifers. Several such studies have recently been conducted using sediment taken from latitudinal transects of lakes in the treeline zones of Canada and Russia. Lakes in the southern portions of these transects are surrounded by boreal forest dominated *Picea*, *Pinus* and *Larix*. Lakes in the north are surrounded by treeless tundra.

In north-central Canada (Fig. 4) *Pinus* is restricted to the closed forest zone, but the percentage of *Pinus* pollen (mainly from *Pinus baksiana*—jack pine) found in tundra and forest-tundra lake sediments typically exceeds 20% (Hansen et al., 1996). The combined percentages of *Picea glauca* (white spruce) and *Picea mariana* (black spruce) pollen in tundra lakes ranges from 10% to 20%. Although there is a general decline in *Pinus* and *Picea* pollen northwards from the closed forest to the tundra, it is impossible to reliably determine the geographic position of treeline (northern limits of forest-tundra) on the basis of *Pinus* and *Picea* pollen in lakes sediments. Both genera produce large amounts of anemophilous pollen which is transported far north of the treeline. In contrast, *Larix* (larch) produces small amounts of pollen. Although *Larix laricina* is locally abundant on fen and bog sites in the closed forest and forest-tundra, its pollen is very rare and is often not detected in normal 300 to 500 grain pollen counts. Pollen from *Larix* is not found in tundra lakes. However, neither is it found in all closed forest or forest-tundra lakes. Conifer stomata are found in all of the closed forest lakes and all but one of the lakes from the forest tundra (Fig. 4). The stomata include undifferentiated ones, and the taxa *Picea* cf. *mariana*, *Picea* cf. *glauca*, *Larix* and *Pinus*. Some stomata were recovered from one tundra lake, which is located extremely close to forest-tundra zone and has scattered krummholz conifers in its vicinity. Although, the absence of stomata provides a good indication of tundra sites compared to forested ones, the relative scarcity of identifiable stomata in forest and forest tundra lakes makes it difficult to infer tree abundance or relative dominance from the stomata.

A similar transect of modern lake sediment samples has been obtained from the treeline zone of western Siberia (Clayden et al., 1996). The closed forest and forest-tundra of that region (Fig. 5) are dominated by *Picea obovata* (Siberian spruce), *Larix sibirica* (Siberian larch) and *Larix dahurica* (Dahurian larch). The pollen spectra from the tundra, forest-tundra and closed forest sites all contain 10% *Pinus*, although the genus is not present in the transect region. The abundance of *Picea* pollen ranges from ~1% to ~10% in the closed forest, forest-tundra and southern tundra. The northern tundra is distinguished by *Picea* pollen percentages of <2%. *Larix* pollen occurs at frequencies of <1% to ~10% in the closed forest samples and in small amounts in the forest-tundra and tundra. As is the case in central Canada, it is not possible to identify the location of the treeline on the basis of the conifer pollen. The stomata of *Larix* and/or *Picea* were found in all of the forest and forest-tundra samples. The abundance and regularity of occurrence were greater for *Larix* stomata than for *Picea* stomata. This is attributable to the deciduous nature of *Larix* which produces a large needle-fall each autumn and generates high deposition rates of stomata into lakes. Unlike the samples from central Canada, the southern tundra samples from western Siberia did contain stomata from *Larix* and *Picea*. Some needles may have been blown across snow surfaces in winter. However, it is known that the tundra region supported extensive *Larix* and *Picea* dominated forest between 8000 and 3500 BP (radiocarbon years before A.D. 1950) (MacDonald et al., 2000). It is possible that the stomata found in the southern tundra lakes date from this period and were reworked from eroding organic shorelines found around most lakes in the region (Clayden et al., 1996).

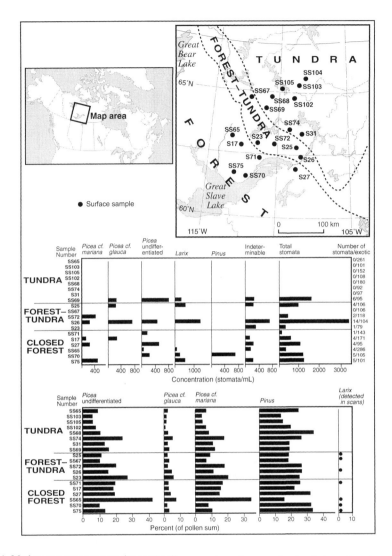

Figure 4. Modern stomate concentrations and pollen percentages from a network of lake sediment sampling sites across the treeline zone in central Canada (redrawn from Hansen et al., 1996).

Stomata and pollen records from lake sediment cores

Stomata records have been obtained from late Quaternary lake sediment cores taken in Europe, Siberia and North America. Stomata have also been recovered and analyzed from peatlands (e.g., Hansen, 1995). The salient features of stomata analysis of some selected lake core studies are outlined below and compared with pollen records from the same sites.

In studies from the Alps, fossil stomata have been recovered along with pollen from a number of sites (reviewed by Ammann & Wick, 1993). The stomata of *Picea abies*

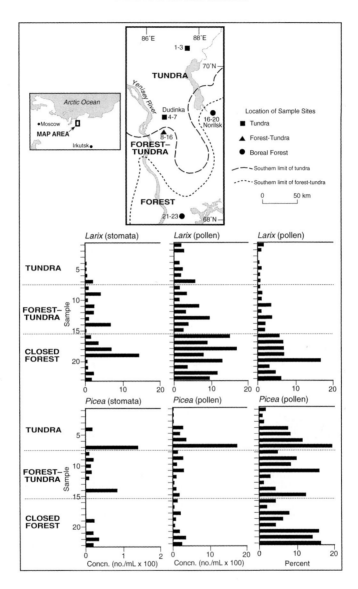

Figure 5. Modern stomate concentrations, pollen concentrations and pollen percentages from a network of lake sediment sampling sites across the treeline zone on the Taimyr Peninsula of Siberia (redrawn from Clayden et al., 1996).

(European spruce), *Pinus* and *Larix decidua* (European larch) have been used to reconstruct changes in the position of upper treeline. Stomata from *Juniperus, Pinus* and *Abies alba* (fir) have been used to reconstruct lower elevation vegetation change. Both natural vegetation change and anthropogenic changes have been detected using stomata analysis. In all cases, the abundance of stomata in the sediments is much less than that of the associated pollen

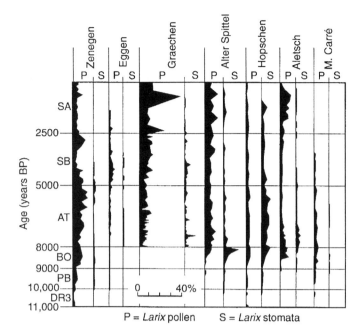

Figure 6. *Larix* pollen percentages and stomate percentages (as % of pollen sum) from a series of sediment cores from the Alps (redrawn from Amman & Wick, 1993 based upon data from Welten, 1982).

(Fig. 6). The timing of the early Holocene appearance of stomata for the different taxa is relatively consistent within different elevational bands, suggesting the stomata provide a good indicator of the presence or absence of the associated tree taxa. The appearance of stomata usually occurs higher in the stratigraphic section than does the appearance of the associated pollen type. This may reflect long-distance transport of pollen to the lake prior to the first arrival of the tree taxa in the watershed. In some cases, the stomata of *Larix decidua* appear in the same samples as the first occurrence of *Larix* pollen. This likely reflects the fact that *Larix* produces relatively small amounts of pollen, but due to its deciduous nature it disperses large amounts of needles and stomata each year. *Juniperus* stomata are generally rare in the lake sediments. It is possible that slow decomposition rates of *Juniperus* needles results in the release of less individual stomata and epidermal fragments for transport to central portions of the lakes.

The stomata and pollen record from a small lake on the Taimyr Peninsula of Russia provides a 9000 year record of vegetation change in northern Siberia (Clayden et al., 1997). The modern vegetation surrounding the lake is *Larix* forest-tundra with very few *Picea obovata*. Stomata of *Larix dahurica, Picea obovata* and *Juniperus communis* were recovered from the core (Fig. 7). *Larix* stomata are by far the most abundant. *Juniperus* stomata were only found in one sample. Both stomata and pollen evidence indicate that *Larix* and *Juniperus* were present at the site by 9000 BP. *Picea* pollen percentages, concentrations and accumulation rates all increase sharply at 6500 to 6000 BP. Such sharp increases in pollen are sometimes used to infer the timing of arrival of a plant taxa in the vicinity of a

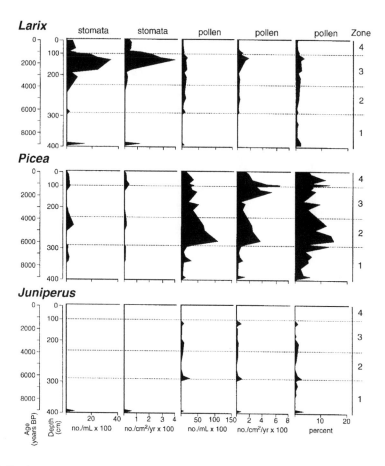

Figure 7. Fossil stomate and pollen stratigraphies from a sediment core from Derevanoi Lake on the Taimyr Peninsula, Siberia (redrawn from Clayden, 1997).

fossil pollen site (MacDonald, 1993). However, the stomata record shows that *Picea* was present at this site by 7500 BP (Fig. 7). The presence of *Picea* in the region between 8000 and 7000 BP has been corroborated by radiocarbon dates from macrofossil cones and wood recovered in the vicinity (MacDonald et al., 2000). Thus, in this case the stomata record shows that the low percentages, concentration, and accumulation rates of *Picea* pollen between at least 7500 BP and 6500 BP reflect the presence of small populations of *Picea* at the site—and not long distance transport as is often inferred from such pollen records.

An extremely detailed and valuable comparative study of stomata, macrofossils and pollen was conducted by Yu (1997) on a core from Crawford Lake, Ontario, Canada (Fig. 8). As it is usually impossible to differentiate the pollen of *Thuja* and *Juniperus*, the study of stomates and macrofossils provided the opportunity of better taxonomic resolution of Holocene vegetation changes around the lake. The stomata of *Picea, Abies, Larix, Pinus, Tsuga, Thuja* and *Juniperus* were all recovered from the core. The sediments of this small, deep lake have a particularly rich macrofossil record, including the seeds and/or needles

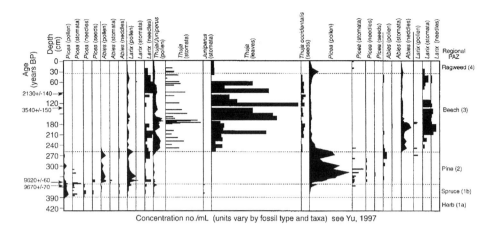

Concentration no./mL (units vary by fossil type and taxa) see Yu, 1997

Figure 8. Fossil stomate, pollen and macrofossil stratigraphies from Crawford Lake, Ontario (redrawn from Yu, 1997).

of these conifer taxa (Fig. 7). The stomata stratigraphy from Crawford Lake compares very closely to the pollen and macrofossil stratigraphies (Fig. 7). *Thuja, Picea* and *Pinus* are particularly well represented in the stomate records. The stomata indicate that *Picea, Larix* and *Juniperus* were present around the lake by 9600 BP. Although *Juniperus* stomata were recovered, macrofossils from that genus were not. Thus, even in this macrofossil-rich site, stomata analysis provides an important line of additional evidence on past vegetation. *Thuja* does not arrive at the site until the mid Holocene. The stomata and macrofossils make it clear that the majority of *Thuja/Juniperus* type pollen in the mid to late Holocene portion of the core represents *Thuja*. In contrast, the *Thuja/Juniperus* type pollen from the early Holocene sediments represents *Juniperus*. Aside from the information that the study provides on Holocene vegetation change, the Crawford Lake core demonstrates the comparability of the stomata record to macrofossil evidence.

Summary

Microfossils of conifer stomata offer considerable benefits as a paleoecological tool. Conifer stomata are frequently present in the same silty organic lake sediments from which pollen and other microfossils are recovered. Unlike pollen, stomata are not prone to long distance transport and provide a good indication of the local presence or absence of plant taxa. In this regard, stomata are similar to plant macrofossils such as seeds and leaves. However, microfossil stomata are deposited throughout lakes in a manner similar to pollen. Thus, stomata may be more frequently obtained from lake sediment cores than is the case for plant macrofossils. Stomata are recovered during the normal processing of samples for pollen analysis and can be identified and counted at the same time as pollen. Stomate identification is relatively easy and the taxonomic resolution is usually as good, or better, than for pollen.

Stomata analysis does have some limitations. The stomata are usually present in very low abundances relative to pollen concentrations. In general, stomata occur too infrequently to be used to infer absolute or relative abundance of associated plant taxa. In some cases, taxonomic resolution is worse than for the associated pollen taxa (Ammann & Wick, 1993). Finally, stomata are prone to reworking from older deposits.

Analysis of microfossil stomata has been conducted on a mere handful of lake sediment cores compared with the many cores analyzed for fossil pollen. The ease of conducting conifer stomate analysis, and the benefits of the technique in terms of spatial resolution, will generate the increasing use of the technique on new sites. In addition, many samples that have been previously prepared and analyzed for pollen likely contain conifer stomate records. It is likely that many of these older records could be easily reanalyzed to count stomata and provide additional information on past vegetation.

Acknowledgments

I am greatly indebted to Barbara Hansen for getting me interested in stomata analysis in the first place. I thank Olga Borisova, Susan Clayden, Les Cwynar, Bruce Gervais and Mike Pisaric for working with me—and keeping me interested in stomata! Keith Bennett and the volume editors provided many useful comments on an earlier version of this chapter. The writing of this chapter and the empirical results from the Kola Peninsula were supported by a NSF PALE grant (ATM 9632926). This chapter is a PACT (Paleoecological Analysis of the Circumpolar Treeline) contribution and a NSF PALE (Paleoclimatology of Arctic Lakes and Estuaries) contribution.

References

Ammann, B. & L. Wick, 1993. Analysis of fossil stomata of conifers as indicators of the alpine tree line fluctuations during the Holocene. In Frenzel, B. (ed.) Oscillations of the Alpine and Polar Tree Limits in the Holocene. Palaeoklimaforschung 9: 175–185.

Beerling, D. J. & W. G. Chaloner, 1994. Atmospheric CO_2 change since the last glacial maximum: evidence from the stomata density reords of fossil leaves. Rev. Paleobot. Palyn. 81: 11–17.

Clayden, S. L., L. C. Cwynar & G. M. MacDonald, 1996. Stomate and pollen content of lake surface sediments from across the tree line on the Taimyr Peninsula. Siberia. Can. J. Bot. 74: 1009–1015.

Clayden, S. L., L. C. Cwynar, G. M. MacDonald & A. A. Velichko, 1997. Holocene pollen and stomates from a forest-tundra site on the Taimyr Peninsula. Siberia. Arctic Alpine Res. 29: 327–333.

Cwynar, L. C., E. Burden & J. H. McAndrews, 1979. An inexpensive sieving method for concentrating pollen and spores from fine-grained sediments. Can. J. Earth Sci. 16: 1115–1120.

Davis, M. B., 1973. Redeposition of pollen grains in lake sediment. Limnol. Oceanogr. 18: 44–52.

Faegri, K., J. Iversen, P. E. Kaland & K. Krzywinsnki, 1989. Textbook of Pollen Analysis 4th Edition. Toronto, Wiley, 328 pp.

Fahn, A., 1982. Plant Anatomy. 3rd Edition. Oxford, Pergamon, 544 pp.

Gervais, B. R. & G. M. MacDonald, 2001. Modern pollen and stomate deposition in lake surface sediments from across treeline. Kola Peninsula, Russia. Rev. Palaeobot. Palyn. 114: 223–237.

Hansen, B. C. S., 1995. Conifer stomate analysis as a paleoecological tool: an example from the Hudson Bay Lowland. Can. J. Bot. 73: 244–252.

Hansen, B. C. S., G. M. MacDonald & K. A. Moser, 1996. Identifying the tundra-forest border in the stomate record: an analysis of lake surface samples from the Yellowknife area. Northwest Territories, Canada. Can. J. Bot. 74: 796–800.

Hicks, S., 1993. The use of recent pollen rain records in investigating natural and anthropogenic changes in the polar tree limit in northern Fennoscandia. In Frenzel, B. (ed.) Oscillations of the Alpine and Polar Tree Limits in the Holocene. Palaeoklimaforschung 9: 5–18.

MacDonald, G. M., 1993. Fossil pollen analysis and the reconstruction of plant invasions. Adv. Ecol. Res. 24: 67–110.

MacDonald, G. M. & J. C. Ritchie, 1986. Modern pollen spectra from the western interior of Canada and the interpretation of late Quaternary vegetation development. New Phytologist 103: 245–268.

MacDonald, G. M., A. A. Velichko, C. V. Kremenetski, O. K. Borisova, A. A. Goleva, A. A. Andreev, L. C. Cwynar, R. T. Riding, S. L. Forman, T. W. D. Edwards, R. Aravena, D. Hammarlund, J. M. Szeicz & V. N. Gattaulin, 2000. Holocene treeline history and climate change across northern Eurasia. Quat. Res. 53: 302–311.

Maher, L. J., Jr., 1963. Pollen analysis of surface materials from the southern San Juan Mountains. Colorado. Geol. Soc. Am. Bul. 74: 1485–1504.

Markgraf, V., 1980. Pollen dispersal in a mountain area. Grana 19: 127–146.

Poole, I., J. D. B. Weyers, T. Lawson & J. A. Raven, 1996. Variations in stomata density and index: implications for paleoclimatic reconstruction. Plant Cell Environ. 19: 705–712.

Trautmann, W., 1953. Zur Unterscheidung fossiler Splatoffnungen der mitteleuropaischen Conifera. Flora 140: 523–533.

Van de Water, P. K., S. W. Leavitt & J. L. Betancourt, 1994. Trends in stomata density and $^{13}C/^{12}C$ ratios of *Pinus flexilis* needles during last glacial-interglacial cycle. Science 264: 239–243.

Watts, W. A. & T. C. Winter, 1967. Plant macrofossils from Kirchner Marsh, Minnesota—a paleoecological study. Geol. Soc. Am. Bull. 77: 1339–1360.

Welten, M., 1982. Vegetationsgeschichtliche Untersuchungen in den westlichen Schweizer Alpen: Bern-Wallis. Denkschr. S.N.G. 95: 104 pp.

Woodward, F. I., 1987. Stomata numbers are sensitive to increases in CO_2 from pre-industrial levels. Nature 327: 617–618.

Yu, Z., 1997. Late Quaternary paleoecology of *Thuja* and *Juniperus* (Cupressaceae) at Crawford Lake, Ontario, Canada: pollen, stomata and macrofossils. Rev. Palaeobot. Palyn. 96: 241–254.

4. PLANT MACROFOSSILS

HILARY H. BIRKS (Hilary.Birks@bot.uib.no)

Botanical Institute
University of Bergen
Allégaten 41
N-5007 Bergen, Norway
and
Environmental Change Research Centre
University College London
26 Bedford Way, London
WC1H 0AP, UK

Keywords: plant and animal macrofossils, method of analysis, identification, ecosystem reconstruction, environment reconstruction, multi-disciplinary studies, radiocarbon dating

Introduction

A macrofossil can be considered as a fossil that is large enough to be seen by the naked eye, and that can be manipulated by hand, usually with the aid of a brush or fine forceps under a stereo-microscope. Thus a macrofossil contrasts with a microfossil. Plant macrofossils are diaspores (seeds, fruits, spores) and vegetative parts, such as leaves (including cuticles, leaf spines, etc.), buds, budscales, flowers, bulbils, rhizomes, roots, tissue fragments, bark, and wood, etc. Wood remains may range in size from small twigs or fragments to rootstocks and trunks. Large tree-remains are often termed 'megafossils' and can be found in peat deposits and exposed on river banks as well as in lakes in the boreal forest (e.g., Eronen et al., 1999). Lower plants can be found as macrofossils, most notably mosses (Dickson, 1973; 1986; Janssens, 1990) and occasionally liverworts (that are not capable of such good preservation as mosses), lichens (e.g., Sernander, 1918), and marine algae (e.g., Pedersen & Bennike, 1992). Charophytes (macroalgae) are frequently represented by their oospores, and occasionally Cyanobacteria have been found (e.g., Birks et al., 1976).

This account of plant macrofossils owes much to the earlier accounts by Dickson (1970), Birks (1980), Birks & Birks (1980), and Wasylikowa (1986) but it emphasises macrofossils in lake sediments and their use in palaeolimnology, in past vegetation, ecosystem, and climate reconstruction. It also mentions some other applications of macrofossils, in particular to radiocarbon dating.

J. P. Smol, H. J. B. Birks & W. M. Last (eds.), 2001. *Tracking Environmental Change Using Lake Sediments. Volume 3: Terrestrial, Algal, and Siliceous Indicators.* Kluwer Academic Publishers, Dordrecht, The Netherlands.

Brief history

The use of plant macrofossils for investigating vegetation history stretches back to the 19th century. Before the development of pollen analysis (see Bennett & Willis, this volume) it was the only technique available for studying floristic and vegetational history. Using plant macrofossils, a picture of vegetational development during and since the Tertiary in northwest Europe and the British Isles was built up (e.g., Reid, 1899; see Godwin, 1975). Nathorst, Andersson, Hartz and others used macrofossils in Scandinavia to reconstruct the climatic changes around the end of the last glaciation and in the post-glacial. After the development of pollen analysis in the 1920s, the use of plant macrofossils declined, leaving few people skilled in their identification, notably Knud Jessen and his students G. F. Mitchell and W. A. Watts, who continued to analyse macrofossils and used them in conjunction with pollen analysis to reconstruct floristic and vegetational history (e.g., Jessen & Milthers, 1928; Watts, 1959). Jessen used his considerable skill to reconstruct interglacial and late-glacial vegetation in Ireland (Jessen & Farrington, 1938; Jessen, 1949; Jessen et al., 1959). The early history of macrofossil analysis in North America is outlined by Warner (1988). Since the 1960s, as the limitations of pollen analysis became realised, the use of macrofossils has increased again, as the two techniques are complementary. The first modern stratigraphic macrofossil analysis was presented by West (1957) from an Ipswichian (last) interglacial sequence in eastern England. In North America, Watts developed stratigraphic macrofossil analysis, with the first diagrams by Baker (1965) and by Watts & Winter (1966), closely followed by many other studies, e.g., Watts & Bright (1968) and Wright & Watts (1969). Watts (e.g., 1979) fully integrated macrofossil techniques with pollen studies. Now the technique is fully established (e.g., Watts, 1978; Birks, 1980) and its contribution to palaeoecology validated (e.g., Wasylikowa, 1986).

Outline of methods

The stages involved in a macrofossil study are outlined in Figure 1.

1. *Site selection.* Once the problem to be addressed has been defined, suitable sites must be carefully selected. It is difficult to predict if a lake has been an effective macrofossil trap, as many apparently suitable sites contain few or no macrofossils in their sediment, but other seemingly similar sites contain an abundance of macrofossils. Because plant remains do not readily reach the centres of large, deep lakes, suitable coring sites are often in the littoral zone of small and rather shallow lakes (ca. 500 m diameter lakes or bays, up to ca. 10 m deep). Watts (1978) proposed that small lakes with steep banks overhung by forest, with minimal fringing emergent vegetation and with small or no inflows and outflows should be suitable sites in temperate regions, and this has often been found to be the case in forested environments. In treeless environments, suitable sites may be small lakes with an active inflow or slope-wash that brings terrestrial material from the catchment. Aquatic vegetation is usually poorly developed or absent in arctic or alpine lakes, although lakes in prairie or steppe environments may contain abundant aquatic vegetation. For late-glacial studies, the lake or swamp should be considered in the light of its probable late-glacial morphometry. Other suitable environments for the preservation of plant macrofossils are

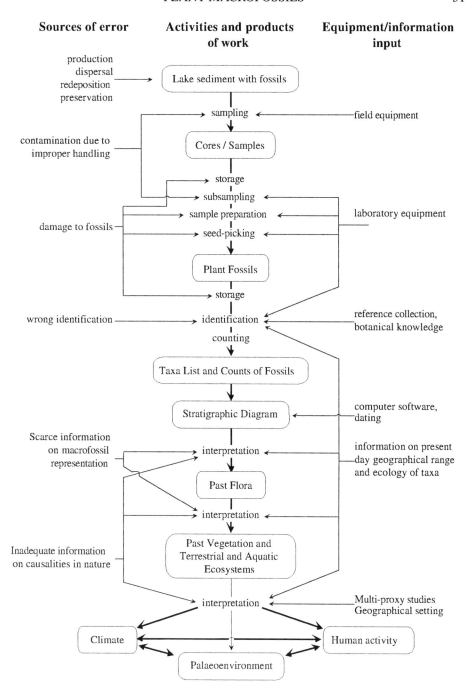

Figure 1. Flow chart showing the successive stages in a plant macrofossil study, from sampling to reconstruction of palaeoenvironments and climate, in the centre column. The sources of potential error are indicated on the left side, and the equipment and information needed during each stage are shown on the right. Modified from Wasylikowa (1986).

bends and bays in slow-moving temperate rivers (e.g., Chaney, 1924; Kelly & Osborne, 1965; Godłowśka et al., 1987; Baker & Drake, 1994), and depositional areas in arctic (glacial) rivers such as outwash plains that are preserved as peaty lenses in gravel deposits (e.g., Lea Valley near London; Reid, 1949). The processes of macrofossil deposition are illustrated in Figure 2, compiled from modern representation studies in temperate lakes (e.g., Birks, 1973; Wainman & Mathewes, 1990; McQueen, 1969; Drake & Burrows, 1980; Dunwiddie, 1987; Spicer & Wolfe, 1987; Collinson, 1983), arctic lakes (e.g., Glaser, 1981), and bogs (GreatRex, 1983). Examples of depositional and representation studies in temperate rivers are by Chaney (1924) and Spicer (1981, 1989), and in arctic rivers by Ryvarden (1971), Holyoak (1984), and West et al. (1993).

2. *Sampling.*   Lake sediments can be sampled from the ice or from open water, or occasionally from the surface of an overgrown fen. Various designs of piston corers are available. Cushing & Wright (1965) and Nesje (1992) have designed large diameter corers (ca. 10 cm) that are suitable for obtaining high-resolution (1 cm interval) samples large enough for macrofossil analysis. Occasionally sediments are available in excavations or sections, and then large-sized monoliths can be dug out in the field.

In the laboratory, the cores are subsampled at suitable resolution, taking care to use clean procedures. The subsamples can be stored in plastic bags at 4 °C until required. However, if material is required for Accelerator Mass Spectrometry (AMS) radiocarbon dating, the subsamples should preferably be deep-frozen (-20 °C) until required to prevent bacterial and fungal metabolism (see Wohlfarth et al., 1998).

3. *Sample preparation.*   A known volume of sediment (measured by displacement of water in a measuring cylinder) is taken, usually 50–100 cm^3 depending upon the type of sediment and the problem in hand. If necessary, it is soaked in sodium pyrophosphate ($Na_4P_2O_7 \cdot 10H_2O$; water softener) solution (ca. 10%) to disaggregate it. Highly humic sediments can be soaked in 10% NaOH, and calcareous sediment can be treated with 10% HCl. However, sodium pyrophosphate seems to be the most universally effective and benign sediment disperser.

The sediment is placed in a sieve with mesh diameter of ca. 125 μm. Larger meshes can be placed in series above, depending on the sediment type. It is wise to separate coarse and fine fractions, so that coarse material does not obscure tiny objects during sorting (see below). A minimum diameter of 125 μm is recommended, because many minute seeds (e.g., *Juncus, Saxifraga*, Ericaceae, Pyrolaceae, Polypodiaceae sporangia) can pass through a 250 μm mesh and be lost.

The sediment is rinsed through the sieve(s) with a gentle spray of tap water delivered through a shower head. The spray must be strong enough to wash the finest material through the sieve, but not so strong as to damage delicate fossils. The residue(s) is transferred to a lidded storage container and kept cool while awaiting examination.

4. *Seed picking.*   Small quantities of the sample residue are dispersed in 2–3 mm depth of water in a petri dish or small plate, so that small objects are separated and easily visible. The suspension is systematically examined under a stereo-microscope at a magnification of ca. ×12, until all the material has been pushed aside. Remains of interest are picked out using fine 'soft' or entomological forceps or a fine brush to minimise damage. A good maxim

Temperate lake

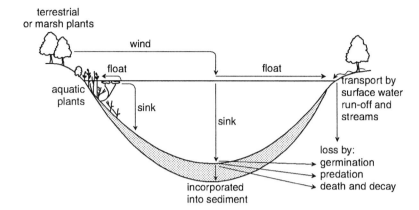

Arctic or alpine lake

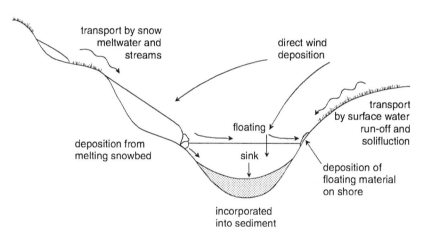

Bog / fen

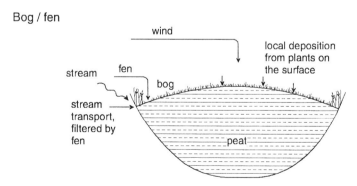

Figure 2. Illustration of the taphonomic processes leading to the preservation of a macrofossil assemblage in a temperate lake (top) (after Birks, 1980), an arctic or alpine lake (middle) (after Glaser, 1981 and Birks, 1991), and a bog or fen (bottom).

is, 'If in doubt, pick it out'. An object can always be discarded later if it is unidentifiable, but it is possible that some unlikely-looking remains may turn out to be characteristic and identifiable. The picked residue in a minimum amount of water can be stored in a plastic bag in a refrigerator in case it needs to be re-examined for different fossils or for types that were missed earlier because of inexperience.

5. *Identification and counting.* The picked-out fossils are sorted and identified.

A reference collection is indispensable for precise and dependable identifications (see Dickson, 1970). A recommended arrangement of a seed/fruit reference collection is alphabetical by family, alphabetical by genus within a family, and alphabetical by species within a genus. This enables material to be found easily and to be replaced accurately. The material is best stored on trays in small plastic boxes with transparent lids for easy preliminary scanning, and then easy removal and replacement of material during careful comparison. A reference collection can be built up from field collections, ensuring accurate identification of the specimen and/or collection of a voucher. The collection is given a number and its details recorded. Material can also be obtained from a herbarium, but seeds and fruits are not always represented on herbarium sheets, and ripe ones tend to fall off and be lost. Seeds can also be obtained through seed exchanges, but their identification and provenance should sometimes be treated with caution. Vegetative material should be included in the collection, such as flowers, buds, budscales, leaves, etc. Microscope preparations of seed coats, leaves, cuticles, petals, anthers, rhizomes, bark, etc. are useful. A collection of wood specimens should include microscope slides of transverse, longitudinal radial, and longitudinal tangential sections, as these are necessary for the identification of fossil wood. Different species converted to charcoal can also be included.

Because of the difficulties of compiling a comprehensive reference collection, seed atlases and illustrations in publications can be used as aids to identification. There is no comprehensive key to seed identification in English. However, there are several useful seed atlases, some of which include keys, and numerous articles on individual groups or genera. Some of these are listed in Appendix 1.

Plant macrofossils are usually identifiable with a stereo-microscope at high ($\times 40$) magnification. It is often helpful to place the specimen in water on a slide or piece of tissue and observe the surface as it dries out gradually, thus revealing the surface pattern. If more detail is needed, the specimen can be observed with a high-power microscope at up to $\times 400$ magnification, using either transmitted or incident light. Fluorescence lighting may also reveal features not otherwise visible, such as cuticular features on leaves. Even finer detail can be revealed by the scanning electron microscope. Extremely complex and beautiful seed-coat structures have been observed (see Appendix 1), but this method is rather impractical for routine identification, although knowledge of the fine structure helps the interpretation of observations made at a lower magnification. Modern seeds are often covered with floral or fruit structures, or have outer shiny layers that are not preserved in the fossil state. Modern seeds can be 'fossilised' by soaking or gentle heating in 10% NaOH and the soft parts and embryos removed (see Dickson, 1970, for details).

Once the fossils are identified, they can be counted. Each discrete part can be tallied. However, broken pieces are a problem. Sometimes two halves or a recognisable unique part (e.g., the lid of a *Ruppia* or *Potamogeton*, the petiole of a leaf, the tip or base of a conifer needle or a *Salix* capsule) can be counted as one, or fragments may be aggregated into whole seed amounts to provide a minimum count. Very numerous or very small fossils, such as Characeae oospores, *Juncus* seeds, and fern sporangia may be tallied without being picked out during seed picking. If a fossil is extremely numerous, it may be recorded as 'a very large number', e.g., over 500, or 1000. Vegetative parts and mosses cannot be counted as individuals. They may be estimated on a subjective abundance scale (e.g., Birks & Mathewes, 1978; Jonsgard & Birks, 1995) and Janssens (1983, 1988) has devised semi-quantitative procedures for estimating moss abundance in peats.

The counted fossils should be stored in small vials in a non-evaporative liquid, usually glycerol with the addition of a fungicide such as formalin or thymol. As experience increases it will then be possible to go back and identify unknown fossils, or to identify types to a lower taxonomic level.

6. *Presentation of results.* Results may be presented as lists, tables, or stratigraphic diagrams. For stratigraphic presentation, the numbers of fossils in a sample should be converted to numbers in a constant volume for all the samples in the sequence. The data can be entered into a spreadsheet and used to draw a concentration diagram (number cm^{-3}) (e.g., Fig. 3). The commonly used software today is the TILIA spreadsheet and the versatile TILIA-GRAPH plotting program written by E. C. Grimm (Grimm, 1990). If the sediment accumulation rate is known, a macrofossil influx diagram can be plotted (number cm^{-2} yr^{-1}) (e.g., Birks & Mathewes, 1978). Because of the great range in representation of macrofossils, from thousands of, e.g., *Najas flexilis* and Characeae oospores, to single specimens of many other species (Birks, 1973) a percentage diagram is not very useful for presenting macrofossil results (see Watts & Winter, 1966). Sources of error are considered by Wasylikowa (1986) and are outlined in Figure 1.

7. *Animal macrofossils.* During the picking out of plant remains, animal macrofossils are usually encountered. Depending on the problem in hand, these can be picked out, identified and counted at the same time as the plant macrofossils. Aquatic animal remains are common in lake sediments, and can contribute to the reconstruction of a past lake ecosystem in a multidisciplinary study (e.g., Birks et al., 1976; Birks & Wright, 2000). Commonly encountered animal macrofossils are Cladocera ephippia (e.g., *Daphnia, Simocephalus*), Bryozoa statoblasts (e.g., *Cristatella, Plumatella*), Insecta including Trichoptera (Solem & Birks, 2000) and Coleoptera (Lemdahl, 2000), Mollusca (Birks et al., 1976), Acarid mites (Solhøy & Solhøy, 2000), Ostracoda, Foraminifera, and fish scales and bones (see the relevant chapters in volume 4).

Indicator potential

Macrofossils have several advantageous features for the reconstruction of vegetation and palaeoenvironments. Many common pollen types can often be identified to family or genus only (e.g., *Betula* (tree/shrub), *Salix, Pinus, Potamogeton*, Cyperaceae, Gramineae, Caryophyllaceae, Ericaceae, Chenopodiaceae, Rosaceae, Cruciferae, etc.). Seeds and fruits can frequently be identified to genus or species level, and can thus extend the floristic detail and the ecological value of vegetational reconstructions.

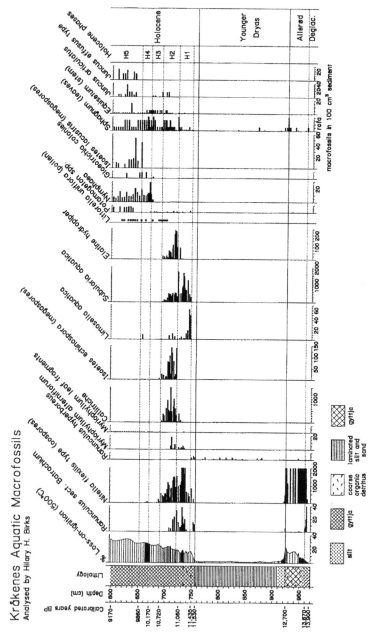

Figure 3. Selected macrofossil taxa from the late-glacial and early-Holocene sediments of Kråkenes Lake, western Norway. A range of fossils is shown; seeds/fruits, Characeae oospores, pteridophyte megaspores, pollen, Cyanobacteria colonies, moss (*Sphagnum*), vegetative parts. The shallow water isoetids are *Limosella aquatica*, *Subularia aquatica*, *Elatine hydropiper*, and *Littorella uniflora* (see text). *Isoetes* spp. usually grow in deeper water. H3 is the phase of sparse macrophytes. Drawn in TILIA.GRAPH (Grimm, 1990), and modified from Birks (2000). B.P. = before present (1950); deglac. = deglaciation phase; r o f a = rare, occasional, frequent, abundant.

Macrofossils are found of many species that are not represented by pollen or spores but that are important constituents of vegetation, such as bryophytes and Characeae. Some plants have pollen that is produced in small quantity or badly dispersed but whose macro-fossils can be readily found and identified (e.g., *Dryas, Viola,* many aquatic plants), or pollen that is poorly preserved in sediments but whose macrofossils are commonly found (e.g., *Juncus, Luzula, Najas, Zannichellia, Ceratophyllum, Callitriche,* and other aquatic taxa). Conversely, well dispersed pollen of some taxa is represented in sediments whereas macrofossils are rarely found except in favourable situations, such as many deciduous tree species, *Artemisia, Plantago* spp., and a very few pollen types can be identified to a lower taxonomic level than macrofossils, e.g., *Typha latifolia.*

Macrofossils are less readily dispersed than most types of anemophilous pollen (Birks, 1973; Birks & Birks, 1980). This confers the advantage that they tend to represent the local flora and vegetation in the lake, at its shore, and in the catchment, and thus they provide a more precise localisation and definition of the past vegetation. If a macrofossil is found of a species with abundant wind-dispersed pollen it may be concluded that the species (e.g., *Betula, Pinus, Alnus*) was growing in the vicinity and that the pollen was not derived by long-distance transport. In addition, some macrofossil types can be stratigraphic markers for different geological periods in the Tertiary, early Pleistocene, and interglacials (Godwin, 1975).

Their local dispersion confers the disadvantage to macrofossils that they are not good for regional correlation. There is no 'macrofossil rain' comparable to a regional pollen rain, and the macrofossil assemblages from different parts of the same lake may differ, depending on the local vegetation. This feature is important in the reconstruction of past lake-level fluctuations. In addition, macrofossils are produced in lower numbers than pollen and other microfossils, and therefore larger sediment samples are required for analysis. Their concentration can be low in certain types of sediment, particularly fen and bog peats where there is virtually no accumulation of fossils from neighbouring areas (Fig. 2). In full- and late-glacial minerogenic lake sediments moss-rich layers are often apparent. The species are usually aquatic, such as *Drepanocladus* and *Calliergon.* In cold, oligotrophic, clear-water lakes in polar regions and alpine areas aquatic mats of mosses may be the only macrophyte vegetation (see Birks, 1981 and references therein; Jones et al., 2000).

Uses of plant macrofossils in palaeolimnology

Limnic ecosystem reconstruction

Lake sediments potentially contain remains of plants that lived in and around the lake, and hence a record of vegetation changes over time. In a recently-formed lake, for example after deglaciation, the sediments will hold a record of the aquatic vegetation development and succession that can be linked to environmental and climatic changes, or used to reconstruct them. Such a study was done at Kråkenes Lake in western Norway (Birks, 2000). The aquatic plant macrofossil sequence is shown in Figure 3. After deglaciation ca. 14,000 cal yr B.P. (ca. 12,300 ^{14}C yr B.P.) (late-glacial interstadial, Allerød) the lake was quickly colonised by pioneer aquatic *Ranunculus* and *Nitella. Nitella* remained dominant for the next ca. 1000 years, and succession proceeded no further because of the relatively cool climate and environment that resembled low-alpine or mid-alpine conditions in the

Norwegian mountains today. The rapid change to colder conditions in the Younger Dryas almost eliminated the aquatic plants and animals, but at the opening of the Holocene, when temperatures rapidly ameliorated (Birks et al., 2000) *Ranunculus* and *Nitella* were once again the first colonisers within 1–3 decades. closely followed by other submerged species. A remarkable succession of turf-like isoetid plants took place in shallow water, particularly *Limosella aquatica, Subularia aquatica, Elatine hydropiper*, and *Littorella uniflora. Isoetes echinospora* was abundant in deeper water, and other early colonisers were *Myriophyllum alterniflorum* and *Callitriche*. Kråkenes Lake lies in acid bedrock, and nutrients and base-cations became depleted after ca. 800 years, being bound in the sediments or washed out of the lake and not replenished from the catchment. This resulted in a decline of aquatic vegetation possibly with some species surviving vegetatively (e.g., *Myriophyllum*). On the immigration of *Nymphaea* and *Potamogeton* spp., (including *P. natans*) ca. 500 years later, nutrients were recycled from the sediments by their strong roots and *Isoetes lacustris* established and became abundant. However, the shallow-water isoetid species did not return, and this habitat was occupied by *Littorella uniflora* perhaps with *I. lacustris*.

This macrofossil record demonstrated the rapidity with which macrophytes can colonise newly opened habitats, and showed the progress of succession in shallow and deeper water that was controlled by plant availability, competition between species for space and nutri-ents, accumulation of organic sediment that bound nutrients and cations and changed habitat conditions, and the influence of the catchment vegetation and soil development. Terrestrial soils rapidly became acidified and podsolised or paludified with peat development, thus providing weak acids to the lake and hastening its natural acidification (Birks et al., 2000). These successional processes, although rapid within the frame of the Holocene, take too long to be observed directly. Modern aquatic successions have only been observed over a few years or decades (see Birks, 2000). This aspect of palaeolimnology allows aquatic successions to be followed through a time perspective of centuries, and when done in conjunction with studies on other aquatic organisms (e.g., in the Kråkenes Project, Birks & Wright, 2000), an integrated reconstruction of the aquatic ecosystem and some deductions about the ecological processes and environmental impacts such as climate changes can be made (Birks et al., 2000).

Vegetation and environment reconstruction where pollen is insensitive

Ritchie (1995) reviewed past vegetation dynamics from the aspect of pollen analysis, and showed how macrofossils can aid interpretation in environments where low local pollen production is overwhelmed by long-distance dispersed pollen, for example, during the late-glacial of northwest Europe. In the species-rich humid forest in Africa, most species are insect pollinated and pollen is uninformative about forest stucture and dynamics, but these can be revealed by identification of charcoal remains from soils (Ritchie, 1995). The main environments where pollen analyses can be insensitive or misleading are all in effectively treeless situations and with a low local pollen production, such as in arctic and alpine lakes, in full- and late-glacial situations, in treeless prairie or steppe environments, or in the determination of past positions of treelines and tree-limits (Birks & Birks, 2000).

Arctic and alpine lakes: The local pollen productivity of arctic and alpine vegetation is so low that the pollen rain reaching lake sediments is dominated by long-distance transported anemophilous pollen, mostly from trees such as *Betula, Pinus,* and *Alnus* (e.g., van der Knaap, 1987). In addition the taxonomic resolution of the fossil pollen record is low. Macrofossil studies help towards resolving these difficulties, as terrestrial plant remains from the local catchment are preserved in the lake sediment, and can be identified to a low taxonomic level, thus allowing vegetation and environmental reconstructions to be made. A Holocene macrofossil record from Svalbard (Birks, 1991) showed that early Holocene summer temperatures were ca. 1.5 °C warmer before ca. 4000 B.P. with a denser vegetation containing several warmth-demanding species that do not occur in the area today.

Treeless full- and late-glacial environments: An analogous situation is found in many late-glacial sequences in northwest Europe. An interpretation of *Betula*-forest development during the interstadial is often made from relatively high percentages of *Betula* pollen (e.g., Birks, 1994a). However, macrofossil analyses from some of the sites have shown that the local vegetation was probably treeless, and a false impression is given by the percentage abundance of long-distance transported *Betula* pollen when the local pollen production is low (e.g., Birks & Mathewes, 1978; Birks, 1984, 1993; van Dinter & Birks, 1996; Birks & van Dinter, 1997). In the Younger Dryas of northwest Europe and the full-glacial of Beringia (Alaska and East Siberia) *Artemisia* pollen is common, and in combination with high Gramineae and Chenopodiaceae values, has lead to the interpretation of dry steppe-like 'no-analogue' vegetation. However, macrofossil analyses from northwest Europe (Birks & Mathewes, 1978; Birks, 1984, 1993; Jonsgard & Birks, 1995) reveal the local late-glacial vegetation to have good analogues with modern snow-bed and fell-field alpine communities. Macrofossil analyses from the Beringian full-glacial have shown the so-called 'no-analogue mammoth steppe' to be interpretable in terms of modern shrub-tundra communities (Elias et al., 1997) or of open herb-rich tundra (Goetcheus & Birks, 2000). It seems increasingly likely that the *Artemisia* and associated pollen types are long-distance dispersed, perhaps all round the Northern Hemisphere during the full-glacial and Younger Dryas times of increased wind speeds and treeless vegetation (Birks & Birks, 2000). In the North American mid-west, the late-glacial pollen assemblages are dominated by *Picea* pollen accompanied by variable amounts of pollen of deciduous trees. Baker (1965) showed from a macrofossil study that the local vegetation in northeastern Minnesota was treeless and dominated by herbs and dwarf shrubs, such as *Dryas integrifolia* and *Salix herbacea*, and that when *Picea* arrived locally as shown by the presence of its needles, the pollen assemblage changed comparatively little. *Picea* is a relatively low pollen producer, and it is probable that the pollen of deciduous thermophilous trees such as *Quercus* and *Ulmus* were long-distance transported into Minnesota during the time of the *Picea* pollen zone. The local presence of such trees in Minnesota has never been verified by macrofossils, but their macrofossils have been found to the south in northern Iowa (Baker et al., 1996) during the earliest Holocene, suggesting that their late-glacial refugia were not far distant to the south.

Prairie environments: Van Zant (1979) studied macrofossils and pollen from Lake Okoboji in northwest Iowa where terrestrial-pollen assemblages in the mid-Holocene are dominated by *Artemisia, Ambrosia*-type, Gramineae, and Chenopodiaceae. Aquatic and wetland pollen are also uninformative, but the macrofossil assemblages reveal a whole range of aquatic and wetland taxa that suggest fluctuating water-level conditions during the warm and dry mid-Holocene 'prairie period'. Baker & Drake (1994) applied macrofossil analysis to sediments laid down in bends and pools of prairie rivers where terrestrial pollen yielded little taxonomic insight into the composition of prairie vegetation in the past. The macrofossil record demonstrated distinct prairie communities and processes such as fires that affected them.

Tree-line reconstructions: Plant macrofossils aid in determining the movements of tree lines and tree limits during the Holocene. Anemophilous tree pollen travels readily over short altitudinal distances, thus blurring the reconstruction of past tree-line and tree-limit positions, particularly at the boundary to treeless vegetation where local pollen production is low. Macrofossils of tree species preserved in lake sediments indicate the local presence of the trees and can thus aid in more precise vegetational and, hence, climatic reconstructions (e.g., Schneider & Tobolski, 1985; Jackson, 1989; Wick & Tinner, 1997; Barnekow, 1999). Megafossils can also be used in this way. Tree trunks and stumps are commonly preserved in lakes at and beyond the present arctic tree-line in northern Fennoscandia and Russia, and in conjunction with radiocarbon dating and dendrochronology, have been used to reconstruct past tree-lines and species extents during the Holocene (e.g., Eronen & Zetterberg, 1996; Eronen et al., 1999; Kremenetski et al., 1998; MacDonald et al., 2000).

Historical biogeography

Macrofossils provide data on species' past distributions which can give clues to how their present distributions originated (Godwin, 1975). From Clement Reid's time, macrofossils have been used to characterise different periods in the Tertiary and Pleistocene of northwest Europe, demonstrating, in conjunction with pollen records, the increasing depauperation of the rich Tertiary flora by successive cold periods and glaciations (e.g., van der Hammen et al., 1971). Glacial-period macrofossil records have shown range expansions of arctic and alpine species and then their subsequent contractions as interglacial conditions prevailed. Holocene disjunct distributions can often be demonstrated to be relicts from previous wider distributions that have been fragmented to suitable refugia by increasing temperatures and reduction of habitats as forest spread and developed (e.g., Tralau, 1963; Godwin, 1975). Several of the species whose present distributions have been invoked as evidence for the 'nunatak hypothesis' of per-glacial survival on unglaciated mountain tops in Scandinavia, for example *Papaver* section *Scapiflora*, have now been shown to have been widespread during the glacial period, and to have readily followed the fluctuating ice-sheet margins (Birks, 1994b). The proposition of their rather unlikely mountain-top survival is unnecessary, and their modern disjunct distributions are the result of the fragmentation of their glacial ranges by habitat elimination during the Holocene. In contrast, Bennike et al. (1999) suggest that the flora and fauna of Greenland was largely exterminated during the last glacial period, and macrofossil records from lake sediments show the timing and pattern of immigration during the Holocene.

The former distribution and migration of aquatic plant species have also been shown by macrofossil records (e.g., Tralau, 1959; Godwin, 1975; Fredskild, 1992; Lang, 1992; Odgaard, 1994; Birks, 2000). At Kråkenes the rapid spreading ability of many pioneer species was documented. Other species that require some ecosystem maturity to provide a suitable habitat (e.g., *Isoetes lacustris, Littorella uniflora*) arrived later (Fig. 3) (Odgaard, 1994; Birks, 2000).

Multidisciplinary studies

Recent stratigraphic macrofossil studies have usually been made in conjunction with pollen analysis, utilising the complementarity of the two techniques. It is always preferable that the analyses are made on the same sediment core, but the pollen record is often best from the deepest part of the lake, whereas the macrofossil record is usually best in the littoral zone. If two cores are used, they can be correlated by pollen sequences on both, or, more simply by comparing their magnetic susceptibility or loss-on-ignition stratigraphies.

In recent decades, combined analyses of a variety of organisms, and also physical and chemical sedimentary features, have become more common. Early examples of multi-disciplinary studies are the Holocene environmental and climatic changes reconstructed from pollen, macrofossils, mollusca, and diatoms at Pickerel Lake, South Dakota (Watts & Bright, 1968; Haworth, 1972; Schwalb & Dean, 1998), and the documentation by Birks et al. (1976) of recent ecosystem changes in three lakes in Minnesota, USA, two of which had suffered heavy eutrophication, using macrofossils, pollen, sediment composition, diatoms, other algae, Cladocera, and some other aquatic animals. A more recent study of eutroph-ication is by Odgaard et al. (1997). As the value of chironomids as sensitive temperature indicators is being increasingly realised and refined, they are now being used in late-glacial studies also involving macrofossils (e.g., Levesque et al., 1994 in maritime Canada; Brooks & Birks, 2000 in western Norway, Brooks & Birks, 2001 in Scotland). The large late-glacial multidisciplinary study at Kråkenes, western Norway (Birks et al., 1996a; Birks & Wright, 2000) involved macrofossils, mosses, pollen, diatoms, chironomids, Cladocera, Trichoptera, oribatid mites, and Coleoptera in combination with radiocarbon dating to build up a comprehensive picture of direct aquatic ecosystem responses to the late-glacial and early Holocene climatic changes, and the importance of feedbacks and interactions within the lake ecosystem and its catchment as it developed in the early Holocene (Birks et al., 2000). Other examples of substantial multidisciplinary studies are at Lobsigensee, (A group of authors, 1985; Ammann, 1989), the Gerzensee-Leysin project (Birks & Ammann, 2000), and at Soppensee (Lotter, 1999), all from Switzerland, and of the late-glacial at Usselo, The Netherlands (van Geel et al., 1989).

The contribution of palaeolimnology to the understanding of changes in tropical lake ecosystems caused by fluctuating water levels was emphasised by the multidisciplinary study of the Kenyan Lake Oloidian (Verschuren et al., 2000). They demonstrated the im-portant role of aquatic vegetation for providing habitats for benthic Cladocera, chironomids, Ostracoda, and diatoms. This was the major control on their populations rather than the influence of changes in salinity and water depth. In a multidisciplinary study on Signy Island, Antarctica, where pollen analysis is uninformative about local conditions, Jones et al. (2000) used sedimentary records from two lakes to differentiate between local and regional influences on the aquatic ecosystems. Increased abundances of moss and animal remains of crustacea, mites, and seals were associated with increased nutrient levels during a warmer climate 3300–1200 years ago.

Radiocarbon dating

Terrestrial plant macrofossils are now widely used for AMS radiocarbon dating. High-resolution dating series through the late-glacial sediments of Rotsee, Switzerland, revealed the presence of radiocarbon plateaux (Lotter, 1991). These are stratigraphic intervals where the ^{14}C dates are the same, due to the dilution of atmospheric ^{14}C by 'old' carbon released from the ocean and the ice sheets during periods of rapid overturn and melting. High-resolution dating-series can be 'wiggle matched' against calibration date-series on annual tree-rings back to at least 11,500 years ago, and further back against series of ^{14}C dates calibrated against annual laminations in lake sediments (Kitagawa & van der Plicht, 1998) or laminated marine sediments (Hughen et al., 1998). Calibration of individual dates against the combined INTCAL98 calibration data-set can be made using the public-domain calibration software CALIB 4.0 (Stuiver et al., 1998). In this way, 'absolute' or calibrated ages for events can be estimated, and these ages closely approximate to calendar or sidereal years. For example, Gulliksen et al. (1998) estimated the calendar age of the Younger Dryas/Holocene boundary to $11, 530 \pm 50$ cal B.P. by wiggle-matching the high-resolution series of AMS dates on terrestrial macrofossils and lake gyttja from Kråkenes Lake against the Hohenheim tree-ring calibration chronology. Terrestrial plant macrofossils have also been used to radiocarbon date tephras in lake sediments (e.g., Birks et al., 1996b) and the radiocarbon and calibrated ages provide time markers wherever the tephras are found.

With the increasing emphasis on the use of macrofossils for AMS radiocarbon dating, it is important to select terrestrial macrofossils from lake sediments (e.g., Andree et al., 1986; Barnekow et al., 1998). Aquatic plant remains may show a reservoir or hard-water effect associated with the photosynthesis of $HCO_3{}^-$ derived from ground-water and ancient carbonate rocks. Törnqvist et al. (1992) showed that the expression of the hard-water effect depends on the species and whether they utilise a carbon source derived from 'old' carbon. *Potamogeton* spp. and *Nuphar lutea* showed a full hard-water effect, but *Nymphaea alba* showed the same age as terrestrial macrofossils. The isotopic composition of the source carbon is also a factor in the ^{13}C composition of aquatic plants (Keeley & Sanquist, 1992). Smits et al. (1988) demonstrated experimentally that floating and submerged leaves of *Nuphar lutea* and *Nymphaea alba* do not utilise $HCO_3{}^-$. However, *Nuphar* has more submerged leaves than *Nymphaea*, and it may utilise dissolved CO_2 that has been derived from carbonate in softwater lakes and streams (Adams, 1985). Spence & Maberly (1985) and Keeley & Sanquist (1992) surveyed the use of $HCO_3{}^-$ by aquatic macrophytes. Those that utilise $HCO_3{}^-$ are generally characteristic of hard, carbonate-rich water, and include *Ceratophyllum demersum, Myriophyllum alterniflorum, M. spicatum*, many *Potamogeton* spp., *Ranunculus aquatilis*, and *Zannichellia palustris*. In addition, Characeae algae have long been known to utilise $HCO_3{}^-$ (Smith, 1985). Macrophytes that do not utilise $HCO_3{}^-$ are generally characteristic of soft and acid water, and include *Hippuris vulgaris, Isoetes lacustris, Naias flexilis, Nuphar lutea, Potamogeton natans, P. polygonifolius, Subularia aquatica*, and many mosses (Spence & Maberly, 1985). The first group are likely to show a hard-water effect, and together with the submerged types in the second group and *Nuphar lutea*, would be unadvisable material for AMS dating. In addition, it has been shown (see Adams, 1985) that CO_2 in interstitial water of sediment is a source of carbon for some isoetid plants, interestingly those that colonise later in a succession rather than as pioneers, and also that large species with abundant intercellular gas spaces (that can include emergent species, such as *Menyanthes trifoliata, Scirpus* spp.) are adept at refixing respired CO_2 or CO_2 taken up from sediments by the roots. Therefore, AMS dates on macrofossils of aquatic plants should be interpreted with caution.

As aquatic mosses are often the only type of macrofossil in limnic sediments, it is tempting to use them for AMS dating. However, although aquatic mosses all seem to utilise

CO_2 rather than HCO_3^- (Baines & Proctor, 1980; Spence & Maberly, 1985), empirical dating experiments indicate that they show a hard-water error of the same magnitude as bulk sediment and mollusc shells (MacDonald et al., 1987; 1991) suggesting that the CO_2 is derived from dissolved inorganic carbon.

Similarly, the AMS dating of lake gyttja must be done with caution, as gyttja is derived from biological remains that could have incorporated 'old' carbon, or fluvial input, or reworked older organic debris (e.g., Andree et al., 1986; MacDonald et al., 1991; Barnekow et al., 1998). The measurements on gyttja at Kråkenes were validated empirically to show no hard-water effect by direct comparison with terrestrial macrofossil dates from the same samples (Gulliksen et al., 1998). Using annually laminated sediments as a time control, Old-field et al. (1997) demonstrated the presence of hard-water effects in all gyttja components except terrestrial macrofossils and the finest sediment fraction.

Lake-level fluctuations

Lake-level studies are important in palaeolimnology, and if they are carefully reconstructed, can provide information on climatic changes affecting the hydrological balance. Digerfeldt (1971) pioneered the use of aquatic plant macrofossils in conjunction with sediment parameters in transects of coring sites from shallow to deep water, to elegantly reconstruct lake-level changes. This is a labour-intensive method and many subsequent studies have not followed it, and thus their conclusions should be critically evaluated in the light of this. The appropriate methodology and some results are reviewed by Digerfeldt (1986) and critically assessed by Hannon & Gaillard (1997).

Environmental archaeology

Plant macrofossils play an important role in environmental archaeology. However, few studies have involved lake sediments. One of the most interesting and comprehensive is that of Godłowśka et al. (1987) who reconstructed the extent and intensity of human activities such as tree-clearance and crop cultivation during phases of the Neolithic occupation of a loess terrace above the River Vistula in Poland. The macrofossil, mollusc, and pollen evidence was collected from the sediments of an infilled oxbow lake on the floodplain of the Vistula below the terrace. Wasylikowa (1989) extended the palaeoecological interpretations with a detailed analysis using phytosociology and Ellenberg's ecological index numbers in a unique study.

Other uses of plant macrofossils

A novel use of leaf macrofossils preserved in lake sediments is the reconstruction of past atmospheric concentrations of CO_2 from their stomatal density. In several species, stomatal density is inversely proportional to CO_2 concentration, and the modern relationships can be applied to fossil leaves of known age. Beerling et al. (1995) and Rundgren & Beerling (1999) reconstructed late-glacial and Holocene changes, respectively, using fossil *Salix herbacea* leaves from lake sediments in Norway and Sweden. The reliability of the modern relationship must be carefully evaluated before reconstructions are made, or the results may be debatable (see Wagner et al. (1999) in conjunction with Birks et al. (1999)).

The stable carbon isotope ^{13}C has been measured in plant macrofossil sequences (e.g., Figge & White, 1995; Turney et al., 1997). Fluctuations in the values are hard to interpret in environmental terms and much more needs to be done to track and disentangle isotopic discrimination by different parts of the plant (e.g., Rundgren et al., 2000) and to make empirical calibrations of modern material before environmental reconstructions can be confidently made. Keeley & Sanquist (1992) discuss the ^{13}C ratios in aquatic plants, including aquatic mosses, where the relationships between physiological and biochemical processes are far more complex than for terrestrial plants.

Terrestrial plant macrofossils from shallow marine sediments have been used to quantify the so-called marine reservoir effect. The retention of CO_2 in the deep ocean for long periods of time leads to its depletion in ^{14}C. As the ocean circulates, the CO_2 is recycled and used by marine organisms. Their measured radiocarbon age can be several hundred years older than their actual age. By radiocarbon dating marine and terrestrial fossils at the same levels in shallow marine-bay sediments, the magnitude of the marine reservoir effect can be measured (Bondevik et al., 1999). During the late-glacial, this information will allow hypotheses to be tested about changes in ocean circulation during periods of rapid climatic change, but more data need to be collected to track changes in the reservoir effect during the whole glacial-Holocene transition.

Summary

This short review of plant macrofossils in lake sediments and their uses in palaeolimnology and palaeoenvironmental reconstruction illustrates the diversity and flexibility of plant macrofossils. Aided by their identification to low taxonomic levels, they provide records of past aquatic and terrestrial vegetation, from which can be deduced past ecological and environmental conditions and past climate. Their use complements pollen for this purpose (see Bennett & Willis, this volume), emphasising and defining the local vegetation rather than the regional biome and correcting false conclusions that may be drawn from pollen spectra that contain a large proportion of long-distance-transported pollen. Holocene tree-line studies are greatly enhanced by plant macrofossils, as these indicate the local presence of trees whose pollen is readily blown above or beyond the tree-line and between altitudinal and latitudinal forest belts.

Plant macrofossils play a large and important role in many multidisciplinary studies. Vegetation is basic to all ecosystems. Changes in both terrestrial and aquatic vegetation control habitat diversity and availability and soil type and nutrient status are important for many other organisms. Aquatic-plant macrofossils are essential in the reconstruction of past lake-level fluctuations.

Plant macrofossils have many other diverse use in Quaternary studies. One of the most important is the use of terrestrial macrofossils as material for AMS radiocarbon dating. Many important discoveries have been made about the global ^{14}C balance as a result, leading to hypotheses about past ocean circulation in the last glacial and Holocene periods, and how this changed at the transition. Calendar ages of marine material can be more precisely calculated as more information becomes available on the marine reservoir effect and the ages of tephras from dates on terrestrial macrofossils. AMS dates on terrestrial macrofossils from annually laminated lake sediments have been successfully used to calibrate the ^{14}C timescale back to 40, 000+ years ago (Kitagawa & van der Plicht, 1998), and to estimate the calendar age of events such as the Younger Dryas / Holocene boundary (Gulliksen et al., 1998). These discoveries are pivotal to many areas of Quaternary science.

Plant macrofossils have provided much information on the human environment, in terms of diet, clothing, equipment, agricultural practices, crops, etc. encompassed in the subject of palaeoethnobotany. More diverse uses include the reconstruction of past atmospheric CO_2 concentrations and the measurement of stable isotopes as environmental indicators.

Plant macrofossils are likely to become increasingly important in palaeolimnology and palaeoenvironment and climate reconstructions. The analyses are 'low tech' and relatively cheap, once the cores have been obtained, although the procedures of seed picking, identification, and quantification are time-consuming and demand considerable patience. The most important factor is the skill and botanical knowledge of the analyst. Another important factor is the availability of a reference collection to aid identification. Collections can be made by trained field botanists, and material may also be obtained from herbaria, although often herbarium specimens disappointingly bear only flowers and not fruits and seeds. The future of plant macrofossil analysis depends partly on the availability of reference collections, but it most acutely needs more skilled analysts to be trained in identification, plant morphology, and botanical interpretation (Birks & Birks, 2000).

Acknowledgements

Many people have stimulated and encouraged my interest in plant macrofossils over the years, in particular H. E. Wright, D. G. Wilson, W. A. Watts, K. Wasylikowa, J. Mangerud, and R. W. Battarbee. Thanks are due to all of them, and also especially to H. J. B. Birks for his continued interest and encouragement and for practical help with this chapter. Much of my own recent work quoted here was carried out with financial support from the Norwegian Research Council, NFR, for which I am grateful.

Appendix 1 Macrofossil identification and reference works

Identification Manuals

Anderberg, A.-L., 1994. Atlas of Seeds. Part 4. Resedaceae-Umbelliferae. Swedish Museum of Natural History, Stockholm, 281 pp.

Beijerinck, W., 1976. Zadenatlas der Nederlandsche Flora. Backhuys & Meesters, Amsterdam, 316 pp. In Dutch.

Berggren, G., 1964. Atlas of Seeds. Part 2. Cyperaceae. Swedish Natural Science Research Council, Stockholm, 68 pp.

Berggren, G., 1981. Atlas of Seeds. Part 3. Salicaceae-Cruciferae. Swedish Museum of Natural History, Stockholm, 259 pp.

Bertsch, K., 1941. Früchte und Samen. Ein Bestimmungsbuch zur Pflanzenkunde der vorgeschichtlichen Zeit. Ferdinand Enke, Stuttgart.

Elias, S. & O. Pollak, 1987. Photographic atlas and key to windblown seeds of alpine plants from Niwot Ridge, Front Range, Colorado, USA. INSTAAR Occasional Paper 45: 28 pp.

Jensen, H. A., 1998. Bibliography on Seed Morphology. A. A. Balkema, Rotterdam, 310 pp.

Katz, N. J., S. V. Katz & M. G. Kipiani, 1965. Atlas and keys of fruits and seeds occurring in the Quaternary deposits of the USSR. Nauka, Moscow, 365 pp. In Russian.

Martin, J. B. & W. D. Barkley, 1961. Seed Identification Manual. California University Press, Berkeley.

Montgomery, F. H., 1977. Seeds and fruits of plants in eastern Canada and northeastern United States. University of Toronto Press.

Schoch, W. H., B. Pawlik & F. H. Schweingruber, 1988. Botanische Makroreste. P. Haupt, Bern & Stuttgart. German, English, French.

U.S.D.A. 1948. Woody-Plant Seed Manual. Forest Service, US Dept. of Agriculture Misc. Publication 654.

Papers on identification

Seeds and fruits

Aalto, M., 1970. Potamogetonaceae fruits. 1. Recent and subfossil endocarps of Fennoscandian species. Acta Bot. Fenn. 88: 1–55.

Berggren, G., 1974. Seed morphology of some *Epilobium* species in Scandinavia. Svensk Bot. Tiddsk. 68: 164–169.

Bright, R. C. & R. Woo, 1969. Coating seeds with Ammonium chloride; a technique for better photographs. Turtox News 47: 226–229.

Dickson, C. A., 1970. The study of plant macrofossils in British Quaternary deposits. In Walker, D. & R. G. West (eds.) Studies in the Vegetational History of the British Isles. Cambridge University Press, Cambridge: 233–254.

Conolly, A. P., 1976. The use of the electron microscope for the identification of seeds, with special reference to *Saxifraga* and *Papaver*. Folia Quaternaria 47: 29–32.

Delcourt, P. A., O. K. Davis & R. C. Bright, 1979. Bibliography of taxonomic literature for the identification of fruits, seeds, and vegetative plant fragments. Oak Ridge National Laboratory Environmental Science Division, Publication No. 1328.

Haas, J. N., 1994. First identification key for charophyte oospores from central Europe. European Journal of Phycology 29: 227–235.

Huckerby, E., R. Marchant & F. Oldfield, 1972. Identification of fossil seeds of *Erica* and *Calluna* by scanning electron microscopy. New Phytol. 71: 387–392.

Jessen, K., 1955. Key to Subfossil *Potamogeton*. Bot. Tidsskr. 52: 1–7.

Jones, T. P. & N. P. Rowe (eds.) 1999. Fossil Plants and Spores: Modern Techniques. The Geological Society of London, 396 pp.

Kaplan, K., 1981. Embryologische, pollen- und samenmorphologische Untersuchungen zur Systematik von Saxifraga (Saxifragaceae). Bibliotheca Botanica 134: 56 pp.

Krause, W., 1986. Zur Bestimmungsmöglichkeit subfossiler Characeen-Oosporen an Beispielen aus Schweizer Seen. Vierteljahrsschrift der Naturforschenden Gesellschaft in Zürich 131: 295–313. In German.

Körber-Grohne, U., 1964. Bestimmungsschlüssel für subfossile *Juncus*-samen und Gramineen-Früchte. Probleme der Küstenforschung im Sudlichen Nordseegebiet 7: 1–47 + plates. August Lax, Hildesheim.

Kuzniewska, E., 1974. Studia systematyczne nad nasionami srodkowo—i palnocnoeuope-jskich gatunkow rodzaju *Saxifaga* L. Prace Opolskiego Towarzystwa Przyjaciol Nauk Wydzial 111: 1–142. In Polish.

Nilsson, Ö. & H. Hjelmquist, 1967. Studies on the nutlet structure of south Scandinavian species of *Carex*. Bot. Notiser 120: 460–485.

Nesbitt, M. & J. Greig, 1989. A bibliography for the archaeobotanical identification of seeds from Europe and the Near East. Circaea 7: 11–30.

van Dinter, M. & H. H. Birks, 1996. Distinguishing fossil Betula nana and B. pubescens using their wingless fruits: implications for the late-glacial vegetational history of western Norway. Veg. Hist. Archaeobot. 5: 229–240.

Mosses

Floras for identifying modern mosses are used for identifying fossil mosses.

Dickson, J. H., 1986. Bryophyte analysis. In Berglund, B. E. (ed.) Handbook of Palaeoecology and Palaeohydrology. J. Wiley & Sons Ltd. Chichester: 627–643.

Janssens, J. A., 1990. Methods in Quaternary Ecology 11. Bryophytes. Geosci. Canada 17: 13–24.

Wood

Barefoot, A. C. & Hankins, F. W. 1982. Identification of modern and Tertiary woods. Calendon Press, Oxford, 189 pp.

Grosser, D., 1977. Die Hölzer Mitteleuropas. Springer, Berlin. 208 pp.

Mork, E., 1966. Vedanatomi. With an identification key for microscopic wood sections. 2nd ed. Forlag Johan Grundt Tanum, Oslo, 69 pp + plates. In Norwegian.

Schweingruber, F. H., 1990. Anatomie europäischer Hölzer; Anatomy of european woods. Verlag Paul Haupt, Bern & Stuttgart, 800 pp. In German and English.

Cuticles-bud scales

Grosse-Brauckmann, G., 1972. Über pflanzliche Makrofossilien mitteleuropäischer Torfe-I. Gewebereste krautiger Pflanzen und ihre Merkmale. Telma 2, 19–55.

Grosse-Brauckmann, G., 1972. Über pflanzliche Makrofossilien mitteleuropäischer Torfe-II. Weitere Reste (Früchte und Samen, Moose, U.A.) und ihre Bestimmungsmöglichkeiten. Telma 4, 51–117.

Katz, N. J., S. V. Katz & E. I. Skovejeva, 1977. Atlas of Plant Remains in peat-soil. Nedra, Moscow. In Russian.

Palmer, P. G., 1976. Grass cuticles: a new palaeoecological tool for East African lake sediments. Can. J. Bot. 54: 1725–1734.

Tomlinson, P., 1985. An aid to the identification of fossil buds, bud-scales and catkin-bracts of British trees and shrubs. Circaea 3: 45–130.

Westerkamp, C. & H. Demmelmeyer, 1997. Blattoberflächen mitteleuropäischer Laubgehölze: Leaf surfaces of central European woody plants. Atlas and Keys. G. Borntraeger; Berlin, Stuttgart, 558 pp. In German and English.

References

Adams, M. S., 1985. Inorganic carbon reserves of natural waters and the ecophysiological consequences of their photosynthetic depletion: (II) macrophytes. In Lucas, W. J. & J. A. Berry (eds.) Inorganic Carbon Uptake by Aquatic Photosynthetic Organisms. American Society of Plant Physiologists: 421–435.

A group of authors, 1985. Lobsigensee - Late-glacial and Holocene environments of a lake on the Central Swiss Plateau. Dissertationes Botanicae 1985: 127–170.

Ammann, B., 1989. Late-Quaternary palynology at Lobsigensee. Regional vegetation history and local lake development. Dissertationes Botanicae 137: 1–157.

Andree, M., H. Oeschger, U. Siegenthaler, T. Riesen, M. Moell, B. Ammann & K. Tobolski, 1986. [14]C dating of plant macrofossils in lake sediment. Radiocarbon 28: 411–416.

Baines, J. T. & M. C. F. Proctor, 1980. The requirement of aquatic bryophytes for free CO_2 as an inorganic carbon source: some experimental evidence. New Phytol. 86: 393–400.

Baker, R. G., 1965. Late-glacial pollen and plant macrofossils from Spider Creek, southern St. Louis County, Minnesota. Geol. Soc. Amer. Bull. 756: 601–610.

Baker, R. G. & P. Drake, 1994. Holocene history of prairie in midwestern United States: pollen versus plant macrofossils. Ecoscience 1: 333–339.

Baker, R. G., E. A. Bettis, D. P. Schwert, D. G. Horton, C. A. Chumbley, L. A. Gonzalez & M. K. Reagan, 1996. Holocene paleoenvironments of northeast Iowa. Ecological Monographs 66: 203–234.

Barnekow, L., 1999. Holocene tree dynamics in the Abisko area, northern Sweden, based on pollen and macrofossil records, and the inferred climatic changes. The Holocene 9: 253–265.

Barnekow, L., G. Possnert & P. Sandgren, 1998. AMS ^{14}C chronologies of Holocene lake sediments in the Abisko area, northern Sweden—a comparison between dated bulk sediment and macrofossil samples. Geologiska Föreningar i Stockholm Förhandlingar 120: 59–67.

Beerling, D. J., H. H. Birks & F. I. Woodward, 1995. Rapid late-glacial atmospheric CO_2 changes reconstructed from the stomatal density of fossil leaves. J. Quat. Sci. 10: 379–384.

Bennike, O., S. Björck, J. Böcher, L. Hansen, J. Heinemeier & B. Wohlfarth, 1999. Early Holocene plant and animal remains from north-east Greenland. J. Biogeogr. 26: 667–677.

Birks, H. H., 1973. Modern macrofossil assemblages in lake sediments in Minnesota. In Birks, H. J. B. & R. G. West (eds.) Quaternary Plant Ecology. Blackwells, Oxford: 173–189.

Birks, H. H., 1980. Plant macrofossils in Quaternary lake sediments. Arch. Hydrobiol. 15: 1–60.

Birks, H. H., 1984. Late-Quaternary pollen and plant macrofossil stratigraphy at Lochan an Druim, north-west Scotland. In Haworth, E. Y. & J. W. G. Lund (eds.) Lake Sediments and Environmental History. University of Leicester Press: 377–405.

Birks, H. H., 1991. Holocene vegetational history and climatic change in west Spitsbergen-plant macrofossils from Skardtjørna, an arctic lake. The Holocene 1: 209–218.

Birks, H. H., 1993. The importance of plant macrofossils in late-glacial climatic reconstructions: an example from western Norway. Quat. Sci. Rev. 12: 719–726.

Birks, H. H., 1994a. Late-glacial vegetational ecotones and climatic patterns in western Norway. Veg. Hist. Archaeobot. 3: 107–119.

Birks, H. H., 1994b. Plant macrofossils and the Nunatak Theory of per-glacial survival. Dissertationes Botanicae 234: 129–143.

Birks, H. H., 2000. Aquatic macrophyte vegetation development in Kråkenes Lake, western Norway, during the late-glacial and early Holocene. J. Paleolim. 23: 7–19.

Birks, H. H. & B. Ammann, 2000. Two terrestrial records of rapid climate change during the glacial-Holocene transition (14,000–9,000 calendar years B.P.) from Europe. Proc. nat. Acad. Sci. 97: 1390–1394.

Birks, H. H., R. W. Battarbee & H. J. B. Birks, 2000. The development of the aquatic ecosystem in Kråkenes Lake, western Norway, during the late glacial and early Holocene—a synthesis. J. Paleolim. 23: 91–114.

Birks, H. H. & H. J. B. Birks, 2000. Future uses of pollen analysis must include plant macrofossils. J. Biogeogr. 27: 31–35.

Birks, H. H., W. Eide & H. J. B. Birks, 1999. Early Holocene atmospheric CO_2 concentrations. Science 286: 1815–1815a.

Birks, H. H., S. Gulliksen, H. Haflidason, J. Mangerud & G. Possnert, 1996b. New radiocarbon dates for the Vedde Ash and the Saksunarvatn Ash from western Norway. Quat. Res. 45: 119–127.

Birks, H. H. & R. W. Mathewes, 1978. Studies in the vegetational history of Scotland V. Late Devensian and early Flandrian pollen and macrofossil stratigraphy at Abernethy Forest, Inverness-shire. New Phytol. 80: 455–484.

Birks, H. H., M. C. Whiteside, D. Stark & R. C. Bright, 1976. Recent paleolimnology of three lakes in northwestern Minnesota. Quat. Res. 6: 249–272.

Birks, H. H. & M. van Dinter, 1997. Betula species in the west Norwegian late-glacial interstadial and early Holocene, and the reconstruction of climate gradients. Geonytt 24: 102.

Birks, H. H. & H. E. Wright, 2000. Introduction to the reconstruction of the late-glacial and early-Holocene aquatic ecosystems at Kråkenes Lake, Norway. J. Paleolim. 23: 1–5.

Birks, H. H. + 23 others, 1996a. The Kråkenes late-glacial palaeoenvironmental project. J. Paleolim. 15: 281–286.

Birks, H. J. B., 1981. Late Wisconsin vegetational and climatic history at Kylen Lake, northeastern Minnesota. Quat. Res. 16: 322–355.

Birks, H. J. B. & H. H. Birks, 1980. Chapter 5 in Quaternary Palaeoecology, 66–84. Edward Arnold, London.

Bondevik, S., H. H. Birks, S. Gulliksen & J. Mangerud, 1999. Late Weichselian marine ^{14}C reservoir ages at the western coast of Norway. Quat. Res. 52: 104–114.

Brooks, S. J. & H. J. B. Birks, 2000. Chironomid-inferred late-glacial and early Holocene mean July air temperatures for Kråkenes Lake, western Norway. J. Paleolim. 23: 77–89.

Brooks, S. J. & H. J. B. Birks, 2001. Chironomid-inferred air temperatures from Lateglacial and Holocene sites in north-west Europe: progress and problems. Quat. Sci. Rev. (in press).

Chaney, R. W., 1924. Quantitative studies of the Bridge Creek Flora. Am. J. Sci. 8: 127–144.

Collinson, M. E., 1983. Accumulations of fruits and seeds in three small sedimentary environments in southern England and their palaeoecological implications. Ann. Bot. 52: 583–592.

Cushing, E. J. & H. E. Wright, 1965. Hand-operated piston corers for lake sediments. Ecology 46: 380–384.

Dickson, C. A., 1970. The study of plant macrofossils in British Quaternary deposits. In Walker, D. & R. G. West (eds.) Studies in the Vegetational History of the British Isles. Cambridge University Press, Cambridge: 233–254.

Dickson, J. H., 1973. Bryophytes of the Pleistocene. Cambridge University Press, Cambridge, 256 pp.

Dickson, J. H., 1986. Bryophyte analysis. In Berglund, B. E. (ed.) Handbook of Palaeoecology and Palaeohydrology. J. Wiley & Sons Ltd. Chichester: 627–643.

Digerfeldt, G., 1971. The post-glacial development of the ancient lake at Torreberga, Scania, south Sweden. Geologiska Föreningar i Stockholm Förhandlingar 93: 601–624.

Digerfeldt, G., 1986. Studies on past lake-level fluctuations. In Berglund, B. E. (ed.) Handbook of Holocene Palaeoecology and Palaeohydrology. J. Wiley & Sons Ltd. Chichester: 127–143.

Drake, H. & C. J. Burrows, 1980. The influx of potential macrofossils into Lady Lake, north Westland, New Zealand. New Zealand J. Bot. 18: 257–274.

Dunwiddie, P. W., 1987. Macrofossil and pollen representation of coniferous trees in modern sediments from Washington. Ecology 68: 1–11.

Elias, S. A., S. K. Short & H. H. Birks, 1997. Late Wisconsin environments of the Bering Land Bridge. Palaeogeogr. Palaeoclim. Palaeoecol. 136: 293–308.

Eronen, M., H. Hyvärinen & P. Zetterberg, 1999. Holocene humidity changes in northern Finnish lapland inferred from lake sediments and submerged Scots pine dated by tree-rings. The Holocene 9: 569–580.

Eronen, M. & P. Zetterberg, 1996. Expanding megafossil-data on Holocene changes at the polar/alpine pine limit in northern Fennoscandia. Paläoklimaforschung 20: 127–134.

Figge, R. A. & J. W. C. White, 1995. High-resolution Holocene and late glacial atmospheric CO_2 record: variability tied to changes in thermohaline circulation. Global Biogeochemical Cycles 9: 391–403.

Fredskild, B., 1992. The Greenland limnophytes—their present distribution and Holocene history. Acta Bot. Fenn. 144: 93–113.

Glaser, P. H., 1981. Transport and deposition of leaves and seeds on tundra: a late-glacial analog. Arct. Alp. Res. 13: 173–182.

Godłowska, M., J. K. Kozłowsi, L. Starkel & K. Wasylikowa, 1987. Neolithic settlement at Pleszów and changes in the natural environment in the Vistula valley. Przegląd Archeologiczny 34: 133–159.

Godwin, H., 1975. The History of the British Flora. Cambridge University Press, Cambridge, 541 pp, 2nd edition.

Goetcheus, V. G. & H. H. Birks, 2000. Full-glacial upland tundra vegetation preserved under tephra in the Beringia National Park, Seward Peninsula, Alaska. Quat. Sci. Rev. 20: 135–147.

GreatRex, P. A., 1983. Interpretation of macrofossil assemblages from surface sampling of macroscopic plant remains in mire communities. J. Ecol. 71: 773–791.

Grimm, E. C., 1990. TILIA and TILIA.GRAPH, PC spreadsheet and graphics software for pollen data. INQUA Working Group on Data Handling Methods Newsletter 4: 5–7.

Gulliksen, S., H. H. Birks, G. Possnert & J. Mangerud, 1998. A calendar age estimate of the Younger Dryas-Holocene boundary at Kråkenes, western Norway. The Holocene 8: 249–259.

Hannon, G. E. & M.-J. Gaillard, 1997. The plant-macrofossil record of past lake-level changes. J. Paleolim. 18: 15–28.

Haworth, E. Y., 1972. Diatom succession in a core from Pickerel Lake, Northeastern South Dakota. Geol. Soc. Am. Bull. 83: 157–172.

Holyoak, D. T., 1984. Taphonomy of prospective plant macrofossils in a river catchment on Spitsbergen. New Phytol. 98: 405–423.

Hughen, K. A., J. T. Overpeck, S. J. Lehman, M. Kashgarian, J. Southon, L. C. Peterson, R. Alley & D. M. Sigman, 1998. Deglacial changes in ocean circulation from an extended radiocarbon calibration. Nature 391: 65–68.

Jackson, S. T., 1989. Postglacial vegetational changes along an elevational gradient in the Adirondack Mountains (New York). New York State Mus. Bull. 465: 29 pp.

Janssens, J. A., 1983. A quantitative method for stratigraphic analysis of bryophytes in Holocene peat. J. Ecol. 71: 189–196.

Janssens, J. A., 1988. Fossil bryophytes and paleoenvironmental reconstruction of peatlands. In Glime, J. M. (ed.) Methods in Bryology. Proc. Bryol. Meth. Workshop, Mainz: 299–306. Hattori Bot. Lab. Nichinan.

Janssens, J. A. Methods in Quaternary Ecology 11. Bryophytes. Geosci. Canada 17: 13–23.

Jessen, K., 1949. Studies in late Quaternary deposits and flora-history of Ireland. Proc. r. Irish Acad. 52: B 6, 85–290.

Jessen, K., S. T. Andersen & A. Farrington, 1959. The interglacial deposit near Gort, Co. Galway, Ireland. Proc. r. Irish Acad. 60: B 1, 1–77.

Jessen, K. & A. Farrington, 1938. The bogs at Ballybetagh, near Dublin, with remarks on late-glacial conditions in Ireland. Proc. r. Irish Acad. 44: B 10, 205–260.

Jessen, K. & V. Milthers, 1928. Stratigraphical and palaeontological studies of interglacial fresh-water deposits in Jutland and northwest Germany. Danm. geol. Unders. Række 2, 28: 1–378.

Jones, V. J., D. A. Hodgson & A. Chepstow-Lusty, 2000. Palaeolimnological evidence for marked Holocene environmental changes on Signy Island, Antarctica. The Holocene 10: 43–60.

Jonsgard, B. & H. H. Birks, 1995. Late-glacial mosses and environmental reconstructions at Kråkenes, western Norway. Lindbergia 20: 64–82.

Keeley, J. E. & D. R. Sandquist, 1992. Carbon: freshwater plants. Plant, Cell, Environment 15: 1021–1035.

Kelly, M. & P. J. Osborne, 1965. Two faunas and floras from the alluvium at Shustoke, Warwickshire. Proc. linn. Soc. Lond. 176: 37–65.

Kitagawa, H. & J. van der Plicht, 1998. Atmospheric radiocarbon calibration to 45,000 yr B.P.: Late Glacial fluctuations and cosmogenic isotope production. Science 279: 1187–1190.

Kremenetski, C. V., L. D. Sulerzhitsky & R. Hantemirov, 1998. Holocene history of the northern range limits of some trees and shrubs in Russia. Arct. Alp. Res. 30: 317–333.

Lang, G., 1992. Some aspects of European late- and post-glacial flora history. Acta Bot. Fenn. 144: 1–17.

Lemdahl, G., 2000. Late-glacial and early-Holocene Coleoptera assemblages as indicators of local environment and climate at Kråkenes Lake, western Norway. J. Paleolim. 23: 57–66.

Levesque, A., L. Cwynar & I. R. Walker, 1994. A multiproxy investigation of late-glacial climate and vegetation changes at Pine Ridge Pond, southwest New Brunswick, Canada. Quat. Res. 42: 316–327.

Lotter, A. F., 1991. Absolute dating of the late-glacial period in Switzerland using annually laminated sediments. Quat. Res. 35: 321–330.

Lotter, A. F., 1999. Late-glacial and Holocene vegetational history and dynamics as shown by pollen and plant macrofossil analyses in annually laminated sediments from Soppensee, central Switzerland. Veg. Hist. Archaeobot. 8: 165–184.

MacDonald, G. M., R. P. Beukens & W. E. Kieser, 1991. Radiocarbon dating of limnic sediments: a comparative analysis and discussion. Ecology 72: 1150–1155.

MacDonald, G. M., R. P. Beukens, W. E. Kieser & D. H. Vitt, 1987. Comparative radiocarbon dating of terrestrial plant macrofossils and aquatic moss from the "ice-free corridor" of western Canada. Geology 15: 837–840.

MacDonald, G. M., B. R. Gervais, J. A. Snyder, G. A. Tarasov & O. K. Borisova, 2000. Radiocarbon dated *Pinus sylvestris* L. wood from beyond tree-line on the Kola Peninsula, Russia. The Holocene 10: 134–147.

McQueen, D. R., 1969. Macroscopic plant remains in recent lake sediments. Tuatara 17: 13–19.

Nesje, A., 1992. A piston corer for lacustrine and marine sediments. Arct. Alp. Res. 24: 257–259.

Odgaard, B. V., 1994. The Holocene vegetation history of northern West Jutland, Denmark. Opera Botanica 123: 171 pp.

Odgaard, B., P. Rasmussen & N. J. Anderson, 1997. The macrofossil record of 20th century submerged vegetation dynamics in shallow Danish lakes. Würtzburger Geographische Manuskripte 41: 153–154.

Oldfield, F., P. R. J. Crooks, D. D. Harkness & G. Petterson, 1997. AMS radiocarbon dating of organic fractions from varved lake sediments: an empirical test of reliability. J. Paleolim. 18: 87–91.

Pedersen, P. M. & O. Bennike, 1992. Quaternary marine macroalgae from Greenland. Norw. J. Bot. 13: 221–225.

Reid, C., 1899. The origin of the British Flora. Dulau & Co., London, 191 pp.

Reid, E. M., 1949. The Late-glacial flora of the Lea Valley. New Phytol. 48: 245–252.

Ritchie, J. C., 1995. Tansley Review No. 83. Current trends in studies of long-term plant community dynamics. New Phytol. 130: 469–494.

Rundgren, M. & D. Beerling, 1999. A Holocene CO_2 record from the stomatal index of subfossil *Salix herbacea* L. leaves from northern Sweden. The Holocene 9: 509–513.

Rundgren, M., N. J. Loader & D. J. Beerling, 2000. Variations in the carbon isotope composition of late-Holocene plant macrofossils: a comparison of whole-leaf and cellulose trends. The Holocene 10: 149–154.

Ryvarden, L., 1971. Studies in seed dispersal I. Trapping of diaspores in the alpine zone at Finse, Norway. Norw. J. Bot. 18: 215–226.

Schneider, R. & K. Tobolski, 1985. Lago di Ganna - Late-glacial and Holocene environments of a lake in the Southern Alps. Dissertationes Botanicae 1985: 229–271.

Schwalb, A. & W. E. Dean, 1998. Stable isotopes and sediments from Pickerel Lake, South Dakota, USA: a 12ky record of environmental changes. J. Paleolim. 20: 15–30.

Sernander, R., 1918. Subfossile Flechten. Flora, Jena 112: 703–724.

Smith, F. A., 1985. Historical perspective on HCO_3^- assimilation. In Lucas, W. J. & J. A. Berry (eds.) Inorganic Carbon Uptake by Aquatic Photosynthetic Organisms. American Society of Plant Physiologists: 1–15.

Smits, A. J. M., M. J. H. De Lyon, G. van der Velde, P. L. M. Steentjes & J. G. M. Roelofs, 1988. Distribution of three Nymphaeid macrophytes (*Nymphaea alba* L., *Nuphar lutea* (L.) Sm. and *Nymphoides peltata* ((Gmel.) O. Kuntze) in relation to alkalinity and uptake of inorganic carbon. Aquat. Bot. 32: 45–62.

Solem, J. O. & H. H. Birks, 2000. Late-glacial and early-Holocene Trichoptera (Insecta) from Kråkenes Lake, western Norway. J. Paleolim. 23: 49–56.

Solhøy, I. W. & T. Solhøy, 2000. The fossil oribatid mite fauna (Acari: Oribatida) in late-glacial and early-Holocene sediments in Kråkenes Lake, western Norway. J. Paleolim. 23: 35–47.

Spence, D. H. N. & S. C. Maberly, 1985. Occurrence and ecological importance of HCO_3^- use among aquatic higher plants. In Lucas, W. J. & J. A. Berry (eds.) Inorganic Carbon Uptake by Aquatic Photosynthetic Organisms. American Society of Plant Physiologists: 125–143.

Spicer, R. A., 1981. The sorting and deposition of allochthonous plant material in a modern environment at Silwood Lake, Silwood Park, Berkshire, England. U.S. Geological Survey Professional Paper 1143: 77 pp.

Spicer, R. A., 1989. The formation and interpretation of plant fossil assemblages. Adv. Botan. Res. 16: 95–191.

Spicer, R. A. & J. A. Wolfe, 1987. Plant taphonomy of late Holocene deposits in Trinity (Clair Engle) Lake, northern California. Paleobiology 13: 227–245.

Stuiver, M., P. J. Reimer, E. Bard, J. W. Beck, G. S. Burr, K. A. Hughen, B. Kromer, G. McCormac, J. van der Plicht & M. Spurk, 1998. INTCAL98 radiocarbon age calibration, 24,000–0 cal B.P. Radiocarbon 40: 1041–1083.

Törnqvist, T. E., A. F. M. de Jong, W. A. Oosterbaan & K. van der Borg, 1992. Accurate dating of organic deposits by AMS ^{14}C measurement of macrofossils. Radiocarbon 34: 566–577.

Tralau, H., 1959. Extinct aquatic plants of Europe. Bot. Notiser 112: 385–406.

Tralau, H., 1963. The recent and fossil distribution of some boreal and arctic montane plants in Europe. Ark. Bot. Ser.2, 5: 533–582.

Turney, C. S. M., D. J. Beerling, D. D. Harkness, J. J. Lowe & E. M. Scott, 1997. Stable carbon isotope variations in northwest Europe during the last glacial-interglacial transition. J. Quat. Sci. 12: 339–344.

van der Hammen, T., T. A. Wijmstra & W. H. Zagwijn, 1971. The floral record of the Late Cenozoic of Europe. In Turekian, K. K. (ed.) Late Cenozoic Glacial Ages. Yale University Press, New Haven and London: 391–424.

van der Knaap, W. O., 1987. Long-distance transported pollen and spores on Spitsbergen and Jan Mayen. Pollen Spores 24: 449–453.

van Dinter, M. & H. H. Birks, 1996. Distinguishing fossil Betula nana and B. pubescens using their wingless fruits: implications for the late-glacial vegetational history of western Norway. Veg. Hist. Archaeobot. 5: 229–240.

van Geel, B., G. R. Coope & T. van der Hammen, 1989. Palaeoecology and stratigraphy of the lateglacial type section at Usselo (The Netherlands). Rev. Palaeobot. Palynol. 60: 25–129.

van Zant, K., 1979. Late glacial and postglacial pollen and plant macrofossils from Lake West Okoboji, Northwestern Iowa. Quat. Res. 12: 358–380.

Verschuren, D., J. Tibby, K. Sabbe & N. Roberts, 2000. Effects of depth, salinity, and substrate on the invertebrate community of a fluctuating tropical lake. Ecology 81: 164–182.

Wagner, F., S. J. P. Bohncke, D. L. Dilcher, W. M. Kürschner, B. van Geel & H. Visscher, 1999. Century-scale shifts in early Holocene atmospheric CO_2 concentration. Science 284: 1971–1973.

Wainman, N. & R. W. Mathewes, 1990. Distribution of plant macroremains in surface sediments of Marion Lake, southwestern British Columbia. Can. J. Bot. 68: 364–373.

Warner, B. G., 1988. Methods in Quaternary Ecology #3. Plant macrofossils. Geosci. Canada 15: 121–129.

Wasylikowa, K., 1986. Analysis of fossil fruits and seeds. In Berglund, B. E. (ed.) Handbook of Palaeoecology and Palaeohydrology. J. Wiley & Sons Ltd. Chichester: 571–590.

Wasylikowa, K., 1989. Paleoecological characteristics of the settlement periods of the Linear Pottery and Lengyel Cultures at Cracow-Nowa Huta (on the basis of plant material). Przegląd Archeologiczny 36: 57–87.

Watts, W. A., 1959. Interglacial deposits at Kilbeg and Newtown, Co. Waterford. Proc. Roy. Irish Acad. 60: B 79–134.

Watts, W. A., 1978. Plant Macrofossils and Quaternary Paleoecology. In Walker, D. & J. C. Guppy (eds.) Biology and Quaternary Environments. Australian Academy of Science, Canberra: 53–67.

Watts, W. A., 1979. Late Quaternary vegetation of central Appalachia and the New Jersey coastal plain. Ecol. Monographs 49: 427–469.

Watts, W. A. & R. C. Bright, 1968. Pollen, seed, and mollusk analysis of a sediment core from Pickerel Lake, Northeastern South Dakota. Geol. Soc. am. Bull. 79: 855–876.

Watts, W. A. & T. C. Winter, 1966. Plant macrofossils from Kirchner Marsh, Minnesota - a paleoecological study. Geol. Soc. am. Bull. 77: 1339–1360.

West, R. G., 1957. Interglacial deposits at Bobbitshole, Ipswich. Phil. Trans. r. Soc., Lond. B, 241: 1–31.

West, R. G., R. Andrew & M. Pettit, 1993. Taphonomy of plant remains on floodplains of tundra rivers, present and Pleistocene. New Phytol. 123: 203–231.

Wick, L. & W. Tinner, 1997. Vegetation changes and timberline fluctuations in the Central Alps as indicators of Holocene climatic oscillations. Arct. Alp. Res. 29: 445–458.

Wohlfarth, B., G. Skog, G. Possnert & B. Holmqvist, 1998. Pitfalls in the AMS radiocarbon dating of terrestrial macrofossils. J. Quat. Sci. 13: 137–145.

Wright, H. E. & W. A. Watts, 1969. Glacial and vegetational history of northeastern Minnesota. Minnesota Geological Survey, University of Minnesota, SP-11. 59 pp.

5. CHARCOAL AS A FIRE PROXY

CATHY WHITLOCK (whitlock@oregon.uoregon.edu)
Department of Geography
University of Oregon
Eugene
OR 97403-1251 USA

CHRIS LARSEN
Department of Geography
University of Buffalo, SUNY
Buffalo
NY 14261-0023 USA

Keywords: charcoal analysis, fire history, lake-sediment records

Introduction

Charcoal analysis of lake sediments is used to reconstruct long-term variations in fire occurrence that can complement and extend reconstructions provided by dendrochronological and historical records. In the last 15 years, several papers have reviewed the methods for charcoal analysis of lake-sediment cores and its use as a tool for studying fire history (e.g., Tolonen, 1986; Patterson et al., 1987; MacDonald et al., 1991; Clark, 1988a; Clark et al., 1998; Long et al., 1998; Whitlock & Anderson, in press). In most cases, pollen and charcoal data from the same cores are used to examine the linkages among climate, vegetation, fire, and sometimes anthropogenic activities in the past. The growing use of charcoal analysis reflects a heightened interest within the paleoecological community to consider fire as an ecosystem process operating on long and short time scales, as well as an increasing need on the part of forest managers to understand prehistoric fire regimes. In this chapter, we discuss issues of site selection, chronology, and methodology in charcoal analysis, based on recent advances in the discipline. We also review the theoretical and empirical basis for charcoal analysis, including assumptions about the charcoal source area and the processes that transport and deposit charcoal into lakes.

Fire reconstructions based on lake-sediment records are derived from three primary data sources: particulate charcoal that provides direct evidence of burning; pollen evidence of fluctuations in vegetation that can be tied to disturbance; and lithologic evidence of watershed adjustments to fire, such as erosion or the formation of fire-altered minerals. Charcoal analysis quantifies the accumulation of charred particles in sediments during and following a fire event. Stratigraphic levels with abundant charcoal (so-called charcoal

J. P. Smol, H. J. B. Birks & W. M. Last (eds.), 2001. *Tracking Environmental Change Using Lake Sediments.*
Volume 3: Terrestrial, Algal, and Siliceous Indicators. Kluwer Academic Publishers, Dordrecht, The Netherlands.

peaks) are inferred to be evidence of past fires. Pollen analysis is used to detect past fires on the assumption that fire and post-fire succession will alter somewhat the local plant community and its pollen representation in the sediments. Lithologic analyses supplement charcoal data by detecting changes in the input of allochthonous sediment and evidence of soil mineral alteration due to heating. The lithologic record has been used to deduce the location of a fire within a watershed and also fire intensity.

Charcoal production, transport, and deposition

Charcoal is produced when a fire incompletely combusts organic matter. The rate at which charcoal accumulates in a lake depends on the characteristics of the fire (e.g., how much charcoal is produced) and the processes that transport and deliver charcoal to the lake (Fig. 1). *Primary* charcoal refers to the material introduced during or shortly after a fire event. *Secondary* charcoal is introduced during non-fire years, as a result of surface run-off and lake-sediment mixing. The relationships between fire characteristics and the accumulation of primary charcoal and between taphonomic processes and the deposition of secondary charcoal are discussed separately, but it is important to remember that both sources comprise the sedimentary charcoal record.

Fire size, intensity, and severity all affect charcoal production and aerial transport, although little information is known about these relationships. Because charcoal particles can be carried aloft to great heights and transported great distances (Radtke et al., 1991; Andreae, 1991), the source of the charcoal may be from regional (distant) fires, extralocal (nearby but not within the watershed) fires, or local (within the watershed) fires. The distance that charcoal is carried during a fire has been discussed in several papers, including Swain (1978), Tolonen (1986), Patterson et al. (1987), Clark (1988a), Whitlock & Millspaugh (1996); Clark & Royall (1995, 1996), Clark et al. (1998), and Gardner & Whitlock (2001). Simple Gaussian plume models suggest that particles $>1000 \mu m$ diameter are released relatively close to the ground and deposited near a fire (Clark & Patterson, 1997). These models predict that particles $<100 \mu m$ in size travel well beyond 100 m, and very small particles are lofted to great heights and travel long distances. Theoretical models also suggest a "skip distance" between the base of the convective column and the site of deposition. In principle, few charcoal particles smaller than $200 \mu m$ in diameter should be deposited within 6 km of the convection column (Fig. 2).

Four studies following modern fires confirm model predictions by showing a decrease in charcoal abundance away from the source. In one study, charcoal accumulation in small lakes following the 1988 fires in Yellowstone National Park indicated that charcoal particles $>125 \mu m$ diameter were abundant in sites <7 km from the fire (Whitlock & Millspaugh, 1996); beyond that distance the accumulation of such particles declined sharply. A more comprehensive study of the upper sediment of 35 lakes followed a 1996 fire in the Cascade Range of Oregon (Gardner & Whitlock, 2001). Levels of $>125 \mu m$-sized charcoal were compared for the upper two core samples (0–2 cm and 2–4 cm depth) in burned sites and sites located within few kilometers upwind and downwind of the fire. Cores from the burned sites had statistically greater charcoal abundance in the top sample than those from unburned sites, and the peaks (i.e., difference between the top and second sample) were better defined than in unburned sites. Sites downwind of the fires had more charcoal in

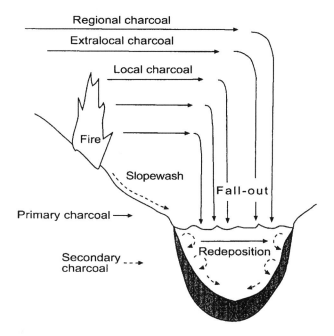

Regional charcoal

Extralocal charcoal

Local charcoal

Fire

Slopewash

Fall-out

Primary charcoal →

Secondary charcoal

Redeposition

Figure 1. Schematic diagram of illustrating the sources of primary and secondary charcoal in a watershed.

the top sample than did sites upwind. The results suggest that lakes that received highest charcoal inputs lay inside the burned perimeter or just downwind of the site.

In a third study, Clark et al. (1998) described the abundance and particle size of charcoal collected in a series of traps during a prescribed fire in Siberia in 1993. Charcoal abundance dropped off sharply at the edge of the burned margin. In both the Siberian and Cascade studies, the observed sharp decline was not consistent with the presence of a skip distance (i.e., a zone of no charcoal deposition at the base of the convective column), although charcoal accumulation at great distances was not evaluated. In the Siberian study, particle size distributions were the same in traps from the burned area as they were for those located 80 m beyond the burn.

The fourth study examined 704 charcoal traps distributed within, and up to 100 meters outside, of three separate experimental fires in boreal Scandinavia (Ohlson & Tryterud, 2000). Traps within the burned area contained 56× more large particles (i.e., >0.5 mm diameter) than traps outside the fire perimeter. Moreover, large particles were found in about 80% of the traps inside the fire perimeter, in about 25% of traps located 0.1 to 0.9 meters outside the perimeter, and in <5% of the traps located 1 to 100 meters of the fire. These results confirm that macroscopic charcoal is not transported far from the fire margin.

Emissions of particulate matter vary depending on the fire and fuel conditions that affect combustion efficiency. Fires of low intensity (i.e., low heat release per unit time) are known to produce high emissions of particulate matter, because of their low combustion efficiency (Pyne et al., 1996). However, large particles are often associated with high-

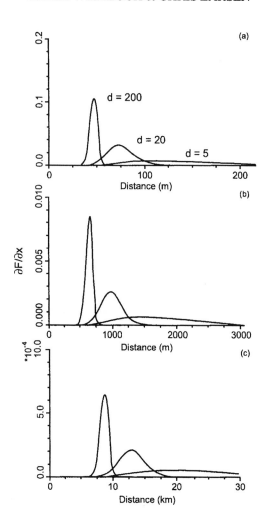

Figure 2. Relationship between distance from the base of a fire's convective column and the amount of charcoal deposited as determined by theoretical models for charcoal particles with diameters of 200, 20 and 5 μm and convective columns with a height of (a) 10 m, (b) 100 m, and (c) 1000 m. (after Clark, 1988a). Notice a theoretical "skip distance" between the fire and the first deposition of charcoal.

intensity fires, because turbulent winds move such particles beyond the combustion zone (Ward & Hardy, 1991). As a result, fires of high intensity (with long flame length) often produce proportionately larger particles than do low intensity, smoldering combustion fires (Ward & Hardy, 1991). The composition of the charcoal also changes with temperature. In experiments run at high temperatures ($>500\,^{\circ}$C), early combustion of grass and leaf material resulted in a high representation of wood charcoal (Umbanhowar & McGrath, 1998). Wood particles also become denser and more fractured at high temperatures, which makes them more prone to waterlogging and settling (Vaughn & Nichols,

1995). In forests in the western United States, particulate charcoal in lake-sediment records consists of predominately wood particles, suggesting a bias towards preserving the record of high-intensity convection-driven fires. Fire regimes characterized by frequent and efficient ground fires do not produce much charcoal, partly because such fires are often small, and charred particulates are not carried aloft. Prairie fires, which are generally cool and fast, produce significant amounts of charred herbaceous material (Umbanhowar, 1996; Pearl, 1999).

If fire processes were the only factors involved, sedimentary charcoal would all be primary and thus a direct measure of biomass burning. However, because the record is composed of both primary and secondary sources, estimating fire size, severity, or intensity is possible only in the most general terms. Studies of modern charcoal accumulation in lakes indicate that charcoal deposition can take place several years after the actual fire. For example, Whitlock & Millspaugh (1996) observed that lakes in both burned and unburned watersheds in Yellowstone received charcoal during the 1988 fires, but the amounts continued to increase significantly for five years in burned watersheds. Anderson et al. (1986) described accrual of charcoal into a lake in Maine for several decades following a 1910 fire. Patterson et al. (1987) report steady increases in microscopic charcoal for several decades after a watershed fire in 1947. The secondary charcoal in these cases may have been introduced from standing burned snags and downfallen trees along the lake margin. Surface run-off may also have delivered charcoal in the few years following a fire, but the importance of this process diminishes as the watersheds became revegetated.

Another source of secondary charcoal, noted in the Yellowstone study (Whitlock & Millspaugh, 1996), was the accumulation of particles that landed on the lake during the fire and were blown to the shore and deposited in the littoral zone. In the years after the fire, this material was refocused to deep water. Bradbury (1996) documented similar movement of littoral charcoal in Elk Lake, a 1.01 km^2 lake in north-central Minnesota. By associating the charcoal peaks in the deep-water core with changes in the diatom record, Bradbury (1996) demonstrated that shallow-water charcoal was mobilized in the lake during spring circulation. In both the Yellowstone and Elk Lake studies, the focusing of charcoal to deep water occurred within a few years of the fire event. In the case of most sites, this refocused material would be part of the charcoal peak. Thus, it is important to note that a charcoal peak in the stratigraphic record is probably composed of particles deposited during and after a fire. For this reason, it may be difficult to infer levels of fire intensity or fire size in the past based on charcoal abundance in lakes.

Site selection

Fire history reconstructions, like most paleoecological procedures, are time consuming, and it is important that sites are chosen carefully. The characteristics of both the watershed and the lake should be considered. Large watersheds provide a large source area for charcoal, because fires can occur over a large area, both near and close to the lake. A site with a large watershed relative to the lake size will magnify the allochthonous inputs (Birks, 1997). For example, in Whitlock & Millspaugh (1996), lakes with large watersheds relative to their size had higher amounts of macroscopic charcoal after a fire than did lakes with small watersheds. Rhodes & Davis (1995) chose a lake in Maine specifically because it had a small surface area and a 50× larger watershed. The large ratio between watershed size and lake surface area magnified the limnological signal of each disturbance event. Such sites, however, also increase the introduction of secondary charcoal, which might distort fire history interpretations.

Steep slopes may increase the introduction of secondary charcoal through erosion (Swanson 1981, Meyer et al., 1995). High rates of erosion following fire are generally attributed to unvegetated ground, hydrophobic soils, and reduced infiltration, but the effects last only a few years. A study from the Colorado Rockies showed a 1000-fold increase in surface soil movement following a stand-replacement fire. Erosion rates remained ten times greater than pre-fire rates for up to four years (Morris & Moses, 1987). Meyer et al. (1995) presented evidence from Yellowstone National Park to suggest that significant amounts of charcoal were transported soon after a fire by large mass-wasting events triggered by intense rains; surface run-off contributed very little to the charcoal record. The presence of riparian vegetation at the lake margin may trap some of this material and thus limit the input of secondary charcoal (Whitlock & Millspaugh, 1996; Terasmae & Weeks, 1979). In this way, a riparian fringe may enhance the resolution of the primary fire signal, particularly if it has existed through the duration of the record. Lakes chosen for fire-history studies should also have small or no inflowing streams that could transport secondary charcoal.

Chronology issues

Adequate chronological control is necessary for any high-resolution time series. Varved sediment records provide the option of seasonal to coarser temporal resolution, and thus they are preferred for fire-history reconstructions. In sites with non-varved sediments, the chronology for the fire reconstruction is based on ^{210}Pb dating of sediments that span the last 200 years and AMS ^{14}C dating of charcoal and terrestrial macrofossils from the remainder of the core. Radiocarbon years should be converted to calendar years using standard calibration programs (e.g., Stuiver et al., 1998) in order to calculate true charcoal accumulation rates. In developing an age-depth model for homogeneous sediment types, it is important to use as smooth a regression curve as possible to avoid sharp discontinuities in deposition time that will influence the charcoal accumulation rates. Of course, sharp changes in sediment type suggest discontinuities in deposition and may justify changing the age-depth model.

Variations in sedimentation rate usually make it difficult to sample a core at equally spaced time intervals. Such changes are not a problem when annually laminated sediments are used; however, in non-varved records, variations in sedimentation rate affect the calculation of charcoal accumulation rates and fire frequency. For this reason, charcoal records from nonlaminated sediments should be converted to intervals that are regularly spaced in time. Because direct interpolation of charcoal data to a constant time interval may not conserve the quantity of charcoal within the intervals, concentration values and deposition times should be interpolated to pseudo-annual intervals. Those values may be integrated over broader intervals (e.g., ten years, but ideally that of the temporally longest subsample) and then divided by the average deposition time over those intervals to produce a series of charcoal accumulation rates (number, area, or mass of charcoal cm^{-2} yr^{-1} = CHAR) spaced at broader (i.e., decadal) intervals (see Long et al., 1998).

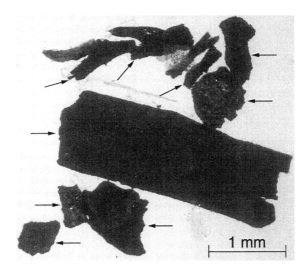

Figure 3. Macroscopic charcoal particles (arrow) left after washing sediment through a 250 μm screen.

Methods

Charcoal is produced between temperatures of 280 and 500 °C (Chandler et al., 1983). Higher temperatures convert the material to ash through glowing combustion and lower temperatures may lightly scorch the material, but not char it. Charcoal particles are visually recognizable as opaque, angular and usually planar, black fragments (Fig. 3). Other black particles in sediments, such as minerals, plant fragments, and insect cuticles, may sometimes be confused with charcoal. Minerals are, however, distinguishable by their crystalline form, such as the octahedral or cubic shape of pyrite, or by their birefringence in polarized light (Clark, 1984). Insect cuticles are thinner than charcoal. Dark plant fragments can be distinguished from charcoal by applying pressure to the particles using a dissecting needle. Charcoal particles fracture under pressure into smaller angular fragments, whereas plant fragments impale or compress.

The visual and physical characteristics of charcoal may be learned by looking at and breaking experimentally created charcoal (Umbanhower & McGrath, 1998) and by examining published photographs (Clark, 1984; Sander & Gee, 1990). Burning of plant material at 350 °C appears to provide the greatest amount of charcoal. The created charcoal should be processed using the same steps employed for the fossil charcoal (see procedures described below). Even so, experimentally produced particles will not have undergone the same taphonomic processes as the sedimentary material and will have slightly different shape characteristics (Umbanhower & McGrath, 1998).

One issue in fire-history studies has been the lack of a standardized methodology. Several methods have been proposed for processing charcoal samples and quantifying the results (Table I). Methods concerned with fire occurrence in a general way have focused on the analysis of microscopic charcoal (with size fractions < 100 μm size) on pollen slides (e.g., Swain, 1973; Cwynar, 1978; Clark, 1982). In this approach, the number or area of charcoal particles is calculated along a series of traverses or on a grid, and the data are expressed as charcoal accumulation rates, a percentage of the pollen sum, or as a ratio of the pollen sum. Because small particles can travel great distances, the source area is poorly defined but probably regional in extent. Another method, which has gained widespread favor, has been the analysis of macroscopic charcoal ($>100 \mu$m size) to reconstruct local fires (e.g., Millspaugh & Whitlock, 1995; Long et al., 1998; Mohr et al., 2000; Hallett & Walker, 2000). Again, the data are presented as accumulation rates of area or particle number, and when contiguous samples are analyzed, they have been used to calculate fire frequency. A third method has been a chemical digestion to calculate charcoal abundance by weight (Winkler, 1985; Laird & Campbell, 2000). This approach avoids assessment of particle sizes, because in principle all charcoal in a subsample is analyzed. The procedure, although simple, seems to produce unreliable results, probably because of inaccuracies in measuring small charcoal quantities and weight-losses associated with the decomposition of clay minerals upon ignition (e.g., MacDonald et al., 1991). The method will not be discussed further.

Microscopic charcoal

Iversen (1941) was the first to recognize that pollen-slide charcoal could be used as a fire proxy, and today most sedimentary charcoal studies are based on an analysis of particles contained in pollen preparations. The method has intrinsic limitations: (1) samples in most Holocene studies are spaced centimeters apart in a core, and gaps of decades to centuries exist in the record; (2) charcoal particles are broken during pollen preparation, thus creating an artificially high abundance of microscopic particles (<100 um); and (3) the exact source area of microscopic charcoal is generally vague—somewhere in the region, but often not the immediate watershed (Table I). Fire frequency per se cannot be calculated from pollen-slide charcoal, because the source area is ambiguous and the records are discontinuous. Despite these caveats, the data are useful in that they disclose periods of burning in the past, and often the paleoclimatic inferences are consistent with those based on the pollen record (perhaps because the source areas of pollen and microscopic charcoal are similar). For example, a common conclusion from studies that combine pollen and microscopic charcoal analysis is that many fires occurred during periods when disturbance-adapted species were more prevalent; thus both charcoal and pollen suggest climate conditions suitable for fires (e.g., Cwynar, 1987; MacDonald, 1989; Horn, 1993; Sarmaja-Korjonen, 1998).

Samples for microscopic charcoal analysis are prepared as part of routine pollen analysis (see Bennett & Willis, this volume). Because charcoal area on pollen slides decreases with increased numbers of steps in pollen processing (Clark, 1984), samples should receive similar treatments. The data are presented as abundance of charcoal particles or charcoal area. Both measurements are often converted to accumulation rates by dividing charcoal concentration, typically assessed by the use of an exotic tracer (e.g., Stockmarr, 1971), by the deposition rate (yr cm^{-1}). Charcoal area is calculated from size-classes, point-counts, or computerized imaging techniques (described below). The size-class method (Waddington, 1969) involves measuring the area of each particle by use of a gridded eyepiece in the microscope. The size of each piece is recorded or placed into a size class. Geometric size classes are usually used because more small particles are present than large. Particles <50–90μm^2 in size are usually not measured because of their great abundance and minimal

Table I. Methods of charcoal analysis from lake sediments.

Method	Procedure (P) and Quantification (Q)	Objective	Advantages (Adv) and Disadvantages (Dis)	References
Pollen Slide	P- Standard pollen-preparation methods. Q-A grid (in microscope eyepiece) is moved on traverses across pollen slide. Number or area of charcoal is expressed as an accumulation rate by division with ratio of counted to added marker grains or as a relative measure as a ratio of total pollen count. Q-A grid is moved step-by-step across a pollen slide and only charcoal particles that intersect a grid line are counted. Area of charcoal particles is estimated.	To determine the importance of fire in a region on centennial or millennial time scales.	Adv: charcoal is counted on pollen slides without additional preparation Dis: spatial and temporal resolution of fire reconstruction is poor; difficult to identify breakage; problems calculating concentration or accumulation rates	Swain, 1973; Cwynar, 1978; Clark, 1982
Thin-section	P- Varved sediments are dehydrated with acetone, impregnated with epoxy, cured, and then thin sectioned. Q-Measurements are based on size classes. A grid is moved on traverses across each varve. Number and area of macroscopic charcoal ($>50\,\mu m$) are recorded.	To reconstruct history of local and extralocal fires on annual to millennial time scales.	Adv: provides record with annual resolution Dis: expensive, varved-sediment lakes are rare	Clark, 1988b; Rhodes & Davis, 1995
Macroscopic Sieving	P-Contiguous 1 cm core intervals are gently washed through analytical sieves (mesh sizes >0.100 mm). Sieved samples put in gridded petri dish. Q-Macroscopic charcoal ($>100\,\mu m$) are counted under stereomicroscope. Recorded as charcoal per volume.	To reconstruct history of local & extralocal fires on decadal to millennial time scales.	Adv: easy, can be used for non-varved lake sediments, preserves macrofossils for AMS-dating Dis: nonarboreal, difficult to disaggregate	Millspaugh & Whitlock, 1995; Long, et al., 1998
Chemical Extraction	P-Sediment is digested in nitric acid, then weighed. Sample is ignited at 500 °C then weighed again. Q-To calculate % charcoal: weight after nitric digestion is subtracted from weight after ignition. Results are multiplied by 100, then divided by weight of sample.	To determine the importance of fire on millennial time scales	Adv: analyzes all particle size ranges Dis: method considered unreliable	Winkler, 1985
Image Analysis	P-A video camera is mounted on a microscope to scan preparation for charcoal particles. Q-Scanner recognizes charcoal based on optical density and records number, area, and size-class distributions of charcoal.	To quantify charcoal area for different size ranges	Adv: use of scanner is less time consuming than visual counting. Dis: scanner mis-identifies other types of dark particles, underrepresents counts	MacDonald et al., 1991; Horn et al., 1992; Earle et al., 1996; Clark & Hussey, 1996

contribution to total area; particles $>2000 \mu m^2$ are recorded individually because of their great contribution to total area (Patterson et al., 1987; Pitkänen & Huttunen, 1999). The number of particles in each size class is multiplied by its midpoint size and these are summed across the classes.

The point count method (Clark, 1982) involves selecting random points on the pollen slide and determining the percentage of points that overlie charcoal. This method tends to produce values of zero in cases where the surface-area method indicates low values, and it is not faster than the size class method when charcoal content is low (Patterson et al., 1987). Both methods typically add 5–10 minutes to the time required to count a pollen slide. The ratio of charcoal-to-pollen accumulation rates was introduced by Swain (1973) to better identify a fire event by integrating an increase in charcoal with a presumed decrease in pollen as a result of burned vegetation. The ratio, however, appears to broaden and dampen the charcoal peaks based on charcoal accumulation rates (e.g., Swain, 1973; Cwynar, 1978) and does not register some fires (e.g., MacDonald et al., 1991).

Comparison with historic fire records points to the regional nature of the fire reconstructions provided by pollen-slide charcoal. For example, peaks in charcoal accumulation rates were matched with fires occurring within a 120 km radius of a lake in the Canadian boreal forest (MacDonald et al., 1991). Similarly, charcoal peaks in a lake from the mixed deciduous forest of Switzerland corresponded with the dates of fires that occurred 20–50 km away (Tinner et al., 1998). Other studies have used peaks in microscopic charcoal to reconstruct local fire history (e.g., Swain, 1973; Tolonen, 1978; Cwynar, 1978; MacDonald et al., 1991; Larsen & MacDonald, 1998a). Microscopic charcoal abundance increases during local fires, but other proxy records, such as macroscopic charcoal or lithologic changes, are needed to confirm if the fire is local. Rhodes & Davis (1995) found that peaks in the charcoal-to-pollen ratio coincided with 8 of 9 fires inferred from pollen, sedimentological, and paleolimnological data. Larsen & MacDonald (1998b) observed that peaks in pollen-slide charcoal coincided with 10 of 16 fires inferred from pollen and macroscopic charcoal, but at least 15 other charcoal peaks did not match other proxy data.

Macroscopic charcoal

A convincing demonstration that large particles provide a record of local fires comes from comparing macroscopic charcoal from varved lake sediments with known watershed fires (e.g., Clark, 1990). Similarly, peaks in macroscopic charcoal in nonlaminated sediments match times of local fires, although with less temporal precision (Millspaugh & Whitlock, 1995; Long et al., 1998; Mohr et al., 2000) (Fig. 4). Clark & Hussey (1996) compared macroscopic charcoal measurements based on area, volume and mass, and concentration in several lakes in northeastern North America. Although different methods produced different peak magnitudes, the records all showed a similar time-series of peaks.

Macroscopic charcoal is generally quantified from petrographic thin sections or in sieved sediment fractions. The thin-section method is desirable for varved-sediment records, because it permits fire history reconstructions with annual precision (Clark, 1988b). Anderson & Smith (1997) also used the thin-section method to analyze eight wet-meadow cores from widely separated sites in the Sierra Nevada, California. The use of petrographic thin sections enabled them to tally charcoal particles at 1-mm intervals, thus increasing the temporal resolution.

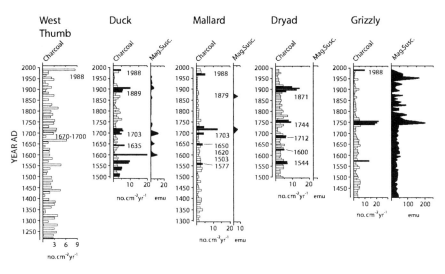

Figure 4. Charcoal accumulation rates (CHAR) and magnetic susceptibility (emu) from sediment cores from a large lake (>4250 ha) and four smaller (14–47 ha) lakes in Yellowstone National Park (after Millspaugh & Whitlock, 1995). Chronology is based on an age model extrapolated from a series of lead-210 dates. Black bars indicate charcoal peaks inferred to represent a local fire event. Dates of known local fires, based on tree-ring studies, are shown next to appropriate peaks. The stratigraphic record extends the fire history beyond the tree-ring record.

The sieving method is used to reconstruct local fire frequency in lakes with nonlaminated sediments. Compared with the thin-section method, sieving is inexpensive and relatively fast. Enumeration is based on simple counts of particles of different size (Millspaugh & Whitlock, 1995; Mehringer et al., 1977) or area measures (MacDonald et al., 1991; Horn et al., 1992; Earle et al., 1996) (Table I). Charcoal is generally analyzed in contiguous samples, usually 1-cm-thick. In most lakes from temperate North America, a single centimeter represents about 5–20 years, depending on the sedimentation rate. Where fires are infrequent, this time span is short enough to discriminate particular fire events, but in regions of frequent burning, a single sample may represent one or more fires occurring years apart. For that reason, the term "fire event" (*sensu* Agee, 1993), rather than "fire", is more appropriate for the information provided by the sieving method. Although sub-samples at 0.25-cm intervals have provided distinct peaks in boreal lakes (Larsen & MacDonald 1998b), sampling at intervals of <1 cm did not improve the temporal resolution in temperate lakes probably because bioturbation blurs the charcoal signal at a finer scale.

The sieving method uses between 1 and 5 cm^3 of wet sediment from each 1-cm interval, depending on the charcoal concentration. Each sample is soaked in a deflocculant (e.g., solution of 5% sodium hexametaphosphate) for a few days and then gently washed through a series of nested sieves (we use mesh sizes of 250, 125, and 63 μm). As a first step, charcoal in the different size fractions is tallied for several samples to assure that the three fractions show similar trends. Most studies use the 125–250 μm fraction or the >100 μm fraction as the most practical size range for analysis. In our experience, a fire event is typically represented by >50 particles cm^{-3} and a nonfire event by substantially fewer or no particles. The resulting data set is converted to charcoal concentration (number of charcoal particles cm^{-3}) and then to charcoal accumulation rates by dividing by the deposition time (yr cm^{-1}).

Several studies have quantified charcoal using computerized image analysis (e.g., MacDonald et al., 1991, Szeicz & MacDonald 1991; Horn et al., 1992; Earle et al., 1996; Clark & Hussey, 1996). Image analysis estimates are often lower than those determined by eye (MacDonald et al., 1991; Horn et al., 1992), because the particle edge has a lower optical density than the center, resulting in small particles not being observed and large particles appearing smaller than they are. If the software is set to characterize the low-density edges as charcoal, then it also falsely characterizes many non-charred objects as charcoal. In more recent procedures (Clark & Hussey, 1996), charcoal particles are first identified using optical microscopy, and then measured using image analysis. The image is captured by video camera and analyzed while the sample is still on the microscope so particles can be compared with those on the enhanced image. A threshold density is set on the image for optimal differentiation of charcoal from optically dense organic and mineral matter. Other dark objects "misidentified" by this density slice are dismissed prior to analysis. The criteria are similar to those used without image analysis, but the approach allows particle dimensions and area to be calculated.

Interpretation of charcoal records

Interpretation of the charcoal time series rests on the ability to calibrate charcoal peaks with known fire events. Dendrochronology and historical documents provide information on historic fires. If the charcoal peaks in the upper sediments match poorly with known fires, the ability of that site to accurately depict older fires is suspect, and another lake should be considered (see site selection section).

Dendrochronological reconstructions of past fires are based on an analysis of fire-scarred tree rings and stand ages (see Arno & Sneck, 1977; Agee, 1993; Johnson & Gutsell, 1994 for a discussion of these methods). Fire scars on trees disclose the exact year of a fire, and fire-history reconstructions based on this method are spatially specific. However, since scars typically form during low-severity ground fires, they reflect incomplete stand destruction and thus may be from fires that did not produce much charcoal (Mohr et al., 2000). Stand-age analysis is used in regions of severe fires, where the forest structure provides an age on past disturbance events. The accuracy of the fire reconstruction fades with time (the so-called telescoping effect) as younger fires destroy the evidence of older events (Agee, 1993; Larsen, 1996; Kipfmueller & Baker, 1998).

Dendrochronological data have been used in a number of studies to calibrate charcoal data with fire age, size, and proximity (Swain, 1973, 1978; Cwynar, 1978; Clark, 1990; MacDonald et al., 1991, Millspaugh & Whitlock, 1995, Larsen & MacDonald, 1998a,b). In principle, a threshold value based on modern calibration should provide a tool for identifying significant charcoal peaks down core. However, local fires located downwind of lakes are often not recorded as charcoal peaks, and, conversely, some charcoal peaks may correspond with extralocal events (Fig. 4; Millspaugh & Whitlock, 1995).

Decomposition of the charcoal record

Clark & Royall (1996) and Long et al. (1998) outline methods for decomposing charcoal records into separate time series that describe different aspects of the fire history.

Their motivation is based on the fact that most time series of charcoal accumulation rates (CHAR) display a low-frequency or slowly varying component, called the *background component*, and a higher frequency or rapidly varying component, called the *peaks component*. Several sources may contribute to the background component or general trends in the data, but they are often difficult to separate. For example, a general time-varying level of background CHAR may be the result of changes in fuel accumulation and its influence on charcoal production. Millspaugh et al. (2000) argue that an increase in background CHAR in a Yellowstone lake ca. 11,000 years ago occurred as a result of changes in fuel during the transition from open meadow to forest vegetation. Background CHAR has also been attributed to secondary charcoal, i.e., material stored in the watershed that is delivered to the lake over a long period. In this case, the background component is not directly related to the fire regime. An increase in charcoal in late-Holocene lake sediments in the Oregon Coast Range was attributed to increased mass movements brought about by the onset of a wetter climate (Long et al., 1998). This hypothesis was supported by the high magnetic susceptibility of late-Holocene sediments. A third contributor of background charcoal may be extra-local or regional fires. This possibility has been proposed by Clark & Royall (1996), although we know of no studies that compare the background component of the macroscopic charcoal record with peaks of a pollen-slide charcoal record to see if they both record the same regional events. If the background component reflects variations in charcoal production and secondary charcoal delivery, these, in turn, are affected by changes in vegetation, climate, and fire weather, and possibly also by changes in hydrology, fluvial geomorphology, and lake characteristics.

A charcoal peak represents the contribution of charcoal from a fire event. As discussed above, this component probably has its source area within the watershed if small basins are chosen, but sometimes fires from adjacent upwind basins can also be recorded. In addition to a particular fire event, peaks may also represent "noise" from analytical error (Whitlock & Millspaugh, 1996) and natural random variations in CHAR. In practice, the largest variations in the peaks component are attributed to fire events, and the minor "noise" component is disregarded.

Peaks of significance are identified by assigning a threshold value, such that CHAR higher than that value is assumed to signal a fire event. Depending on the deposition time, an event may represent one or more fires occurring during the time span represented by the peak. In sites with fast deposition times, a peak is generally less than 20 years (one or two centimeters thick) (Millspaugh, 1997; Long et al., 1998), whereas in sites with slow sedimentation, a comparable size peak may span several decades (Mohr et al., 2000; Anderson & Smith, 1997).

Values for the window width to infer background levels and the threshold-ratio are selected by (1) examining the CHAR from the short core relative to the record of recent fires near the site, and (2) using a variety of values of the two parameters to decompose the long record. The results of the decomposition are compared with information on present-day fire regimes in the region. This iterative approach helps assess the robustness of the method and the sensitivity of the outcomes to the choice of parameter values (Fig. 5).

To detect individual fires or calculate the mean fire interval (MFI; Romme, 1980), the sample interval must be significantly shorter than the average time between fires. Suppose for a given period that the MFI is described by a negative exponential distribution

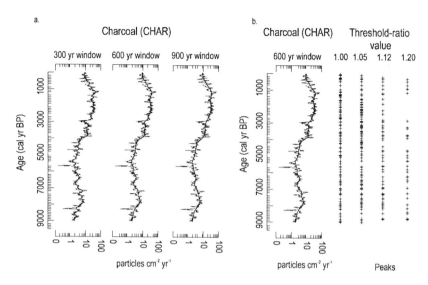

Figure 5. Comparison of (a) different window widths and (b) threshold-ratio values for the decomposition of CHAR at Little Lake in the Oregon Coast Range (after Long et al., 1998). CHAR values were log-transformed and interpolated to a constant time step. Different window-widths were considered to define background levels. The comparison of threshold ratios was based on a background window width of 600 years. The study used a threshold ratio of 1.12 for the fire history reconstruction because it correctly identified eight fire events in the last 1500 years and no additional events.

of fire-intervals, i.e., short intervals are more frequent than long ones. The cumulative proportion of all intervals up to a given length x can be calculated as:

$$\sum f(x) = 1 - e^{-px},$$ (1)

where

$f(x) =$ frequency of a fire

$e =$ base of natural logarithms

$p =$ probability of fire in any year (inverse of the MFI).

(Van Wagner, 1978; Agee, 1993). This equation can be used to examine the influence of different sampling resolutions and the relation between actual MFI and the lowest possible MFI that can be estimated. The shortest interval that can be detected between fires is twice the sampling interval. The equation is used to first calculate the expected fire-interval distribution for a given MFI. Then, the portion of the distribution that was twice a particular sampling resolution is used to estimate the shortest possible MFI that this resolution can detect. The estimated shortest-possible MFI in all cases is longer than the actual MFI, an observation also made by Green (1983) for pollen data. The ratio between the actual and shortest-possible estimated MFI increases as the ratio between the estimated MFI and the sample resolution approaches two (i.e., every other sample is inferred to be a significant charcoal peak). When the actual MFI is 4× the sample resolution, the estimated shortest possible MFI is ca. 1.7× the actual MFI; when the actual MFI is 8× the sample resolution, the estimated shortest-possible MFI is ca. 1.3× the actual MFI. Although these results are based on a simple calculation that does not characterize the changing nature of fire regimes on long time scales, it does point to the importance of sample interval in estimating MFI from nonlaminated sediment records.

Use of other data for verification

Evidence of fire-related erosion has been used to help constrain the charcoal source area, inasmuch as the co-occurrence of a charcoal peak and evidence of erosion provides confirmation that the fire event occurred within the watershed. Selecting a site suitable for lithologic analyses is not straightforward, because changes in lithology or geochemistry may be unrelated to fire. To maximize the input of the allochthonous component following fire, Birks (1997) and Rhodes & Davis (1995) suggest selecting a site with a large watershed relative to the lake.

The magnetic properties of lake sediments have been used to trace the input of allochthonous clastic material (Thompson & Oldfield, 1986; Gedye et al., 2000). The usefulness of such measurements depends on fire location, fire type and intensity, and soils and substrate type. In Millspaugh & Whitlock (1995), lakes that recorded the highest sediment magnetism were located in steep-sided watersheds, where the potential for post-fire erosion was greatest. Low-gradient watersheds, in comparison, showed no signal. Long et al. (1998) found that magnetic susceptibility increased dramatically in the late Holocene but peaks of magnetic susceptibility did not match charcoal peaks. Fire-induced erosion has also been inferred from increases in the content of aluminum, vanadium, and silt in sediments associated with charcoal peaks (Cwynar, 1978) and from an increase in varve thickness (Tolonen, 1978; Larsen & MacDonald, 1998a).

The decomposition approach described above for charcoal has also been applied to magnetic susceptibility data. Background levels of magnetic minerals provide information on pedologic and geomorphic processes that operate within the basin over the long term. Peaks in magnetic susceptibility measurements indicate individual geomorphic events, such as landslides, similar to the CHAR peaks. In Yellowstone, such peaks corresponded well with charcoal peaks, suggesting that they were from fire-related erosion events (Millspaugh & Whitlock, 1995). In other studies in the western United States, no direct relation between CHAR peaks and magnetic susceptibility peaks was noted, even when the possibility of a time lag was considered (Millspaugh, 1997; Long et al., 1998; Mohr et al., 2000; Brunelle & Anderson, in press).

The pollen record often complements the reconstruction provided by charcoal data by suggesting the proximity and size of the fire through changes in the composition of the vegetation. A number of studies have noted the correspondence between charcoal peaks and changes in key pollen taxa (Swain, 1973; Tolonen, 1978; Patterson & Backman, 1988; Pitkänen & Huttunen, 1999) or assemblages of pollen taxa that represent different stages of forest succession (Swain, 1978, 1980; Patterson & Backman, 1988; Rhodes & Davis,

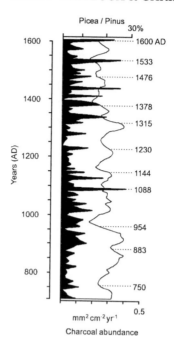

Figure 6. Comparison of *Picea/Pinus* ratio and CHAR (black) for a Lake Pönttölampi in eastern Finland from 700–1600 AD (after Pitkänen & Huttunen, 1999). Eleven local fires are identified by the peak in CHAR and associated decline in the *Picea/Pinus* ratio.

1995; Larsen & MacDonald, 1998a,b; Tinner et al., 1999) (Fig. 6). Cross-correlations between pollen records and either charcoal or a fire-sensitive pollen taxon have been used to identify pollen taxa with repeated sequences of peaks and troughs relative to a fire record (e.g., Green, 1981; Clark et al., 1989; Larsen & MacDonald, 1998a,b; Tinner et al., 1999). For example, cross-correlation results for four pollen types from a site in northern Alberta that show peaks at different lengths of time after a peak in pollen-slide charcoal. These relations were used to identify fires in an 840-year record (Larsen & MacDonald, 1998a) (Fig. 7).

An interesting approach in fire reconstructions is the use of computer models to simulate the pollen source area (Sugita et al., 1997) and then estimate fire size and proximity based on pollen changes within that area. The method assumes that local fires lead to a decrease in pollen abundance. Model results for a site in the boreal forest of Canada suggest that a small lake (3 ha surface area) would register a 10% decline in local pollen from a 4-ha fire at the lake shore, a 100- ha fire on one side of the lake, or a 2500- ha fire within 100 m from one side of the lake. A large lake (314 ha surface area) exposed to fires of the same size and proximity would record decreases in local pollen of approximately 0, 1 and 2%. Decreases in the local pollen of 30% would be observed in the 100-m-radius lake if a 100 ha fire burned around the lake shore, and in the 1000-m-radius lake if a 2500 ha fire burned around the lake shore. The patchiness of the forest and the pollen productivity of the locally dominant species (Sugita, 1994) also affected the simulated relationships. If the modeled relations are correct, it is not surprising that the pollen record

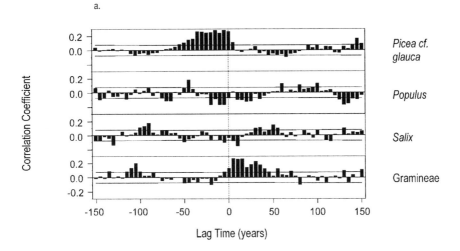

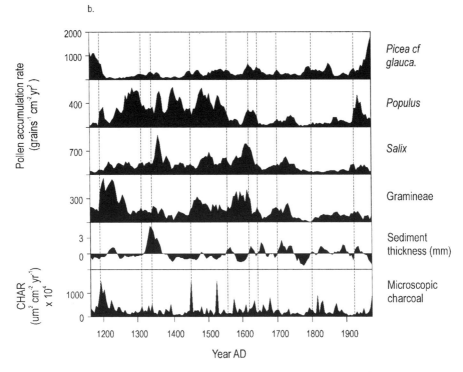

Figure 7. Analysis of pollen assemblages as a fire proxy at Rainbow Lake (59°51'N, 112°15'W) in the Canadian boreal forest (after Larsen & MacDonald, 1998a). (a) Cross-correlograms developed through cross-correlation analysis between the 840-year record of particular pollen taxa and pollen-slide charcoal accumulation rate. The solid bars indicate the cross-correlations at each time lag, and the horizontal lines indicate the 95% confidence limits. Positive cross-correlations after the peak in charcoal at year zero indicate a record whose peak values follow that of charcoal. (b) The 840-year records of *Picea* cf. *glauca, Populus, Salix,* Gramineae, pollen slide charcoal, and the detrended sediment thickness in each 5-year sample. Vertical dashed lines indicated inferred local fire events. All but the charcoal record was smoothed using a 3-sample mathematical average.

does not respond to every fire detected by the charcoal record. However, the pollen record may be sensitive enough to detect the large fire events that result in major changes in vegetation composition.

Conclusions

To realize the potential of charcoal data as a paleoenvironmental proxy requires standardization of both the techniques and assumptions used to interpret such data. Too many charcoal studies are based on imprecise or unsubstantiated assumptions and analytical approaches. Studies of modern charcoal transport and deposition are rare. Information on modern taphonomic processes is needed to calibrate charcoal data and refine the interpretation of the stratigraphic record. Similarly, additional modeling efforts that focus on the relationship between fire and charcoal production and transport are needed to verify the assumptions developed from empirical studies.

We offer some recommendations to improve fire-history reconstructions based on lake-sediment records:

1. Charcoal studies should routinely examine macroscopic charcoal in order to get a local fire reconstruction. The source area of macroscopic charcoal is better known than that of microscopic charcoal, and fire location is an essential part of any fire reconstruction.

2. Contiguous sampling at a fine interval is critical to calculate fire-event frequency; discontinous sampling misses charcoal peaks and often background trends are interpreted as fire events. The sample resolution ideally should be ca. 1/8th the estimated MFI to differentiate closely recurring fires.

3. An adequate chronology is essential, as is some method of calibration to identify a significant threshold level. Thus, charcoal studies require varved-sediments or a chronology based on a suite of calibrated AMS ^{14}C and ^{210}Pb dates.

4. The choice of a specific method of charcoal enumeration, whether charcoal counts through a sieving procedure, charcoal area measurements from image analysis, or charcoal abundance by sediment weight or volume, is less important than the decision to undertake high-resolution sampling and careful calibration. Most high-resolution methods seem to produce similar trends, although further comparison of the results derived from different methods is needed.

5. In analyzing the data, it is important to recognize that the time series consists of at least two components, a slowly varying background component, superimposed upon which is a peaks component. The information contained in these two components is different and should be interpreted separately. Periods with abundant charcoal may not necessarily represent times of more fires; they could be periods of high background charcoal as a result of a shift in fire severity or the introduction of secondary charcoal.

6. Each macroscopic charcoal record is a local reconstruction; to infer landscape, regional, or larger-scale patterns requires a network of sites, done to a similar high standard.

7. In addition to charcoal data, other fire proxy are worth considering, to supplement the fire reconstructions. The sensitivity of pollen and lithologic records to a particular fire event should to be carefully tested in each locality.

Summary

Charcoal particles preserved in lake sediments provide a means of reconstructing fire history beyond documentary and dendrochrological records. Recent refinements in charcoal analysis and interpretation have greatly improved our ability to use charcoal records as proxy of past fire events and to calculate long-term variations in fire frequency. Standardization has also facilitated synthesis of different researchers' data. Interpretating charcoal records in terms of the fire location, size, and intensity requires an understanding of the processes that influence charcoal production, transport, and deposition. Studies of charcoal deposition following modern fires, as well as theoretical models of charcoal particle transport, suggest that macroscopic particles (>100 microns in size) are not transported far from source before settling. They become entrapped in lake sediments within a few years of the fire event through airborne fall-out and secondary reworking. Microscopic charcoal particles (<100 microns in size), in contrast, are able to be carried aloft during a fire and can travel long distances before settling. A record of these small particles provides a reconstruction of regional or extralocal fires. Macroscopic charcoal is tallied or measured in petrographic thin sections or in sieved residues and used to calculate charcoal accumulation rates. Microscopic charcoal is usually counted as a part of routine pollen analysis and its abundance is often presented as a ratio of the pollen sum. The choice of particle size dictates whether regional or local fire events are reconstructed, and whether calculation of fire frequency is possible. Interpretation of the charcoal record requires a well-constrained chronology, in order to analyze charcoal samples taken at a finer time interval than the mean fire return interval inferred from ecological data. In most cases, it is necessary to distinguish between background charcoal in the stratigraphic record, which may be introduced through secondary processes like erosion, and the primary charcoal signal of peaks that represents fire events. Calibration of the charcoal record in terms of background and peaks is also provided by comparing the uppermost stratigraphy with known fire events, inferred from documentary or dendrochronological evidence. Current efforts to be rigorous in methodology and explicit in assumption promise to produce a network of high-resolution charcoal records that can be more easily compared and interpreted.

Acknowledgments

Much of this material arises from research supported by grants. Whitlock was funded by National Science Foundation (SBR-9616951, EAR-9906100) and the U.S.D.A. Forest Service (USFS PSW-95-0022CA, USFS PNW-98-5122-1CA). Larsen received support from NSERC, Northern Training grants and the Science Council of British Columbia. Helpful comments were provided by the editors, J. A. Mohr, and an anonymous reviewer.

References

Agee, J. K., 1993. Fire ecology of Pacific Northwest forests. Island Press. Washington, DC, 493 pp.

Anderson, R. S., R. B. Davis, N. G. Miller & R. Stuckenrath, 1986. History of late- and post-glacial vegetation and disturbance around Upper South Branch Pone. northern Maine. Can. J. Bot. 64: 1977–1986.

Anderson, R. S. & S. J. Smith, 1997. The sedimentary record of fire in montane meadows. Sierra Nevada, California, USA: a preliminary assessment. In Clark, J. S., H. Cachier, J. G. Goldammer, B. Stocks (eds.) Sediment Records of Biomass Burning and Global Change. NATO ASI Series 1: Global Environmental Change, vol. 51, Springer (Berlin): 313–328.

Andreae, M. O., 1991. Biomass burning: its history, use, and distribution and its impact on environmental quality and global climate. In Levin, J. (ed.) Global Biomass Burning: Atmospheric, Climatic, and Biospheric Implications. MIT Press, Cambridge (MA): 3–21.

Arno, S. F. & K. M. Sneck, 1977. A method for determining fire history in coniferous forests in the mountain west. U.S.D.A. Forest Service General Technical Report, INT-42, 28 pp.

Birks, H. J. B., 1997. Reconstructing environmental impacts of fire from the Holocene sedimentary record. In Clark, J. S., H. Cachier, J. G. Goldammer & B. Stocks (eds.) Sediment Records of Biomass Burning and Global Change. NATO ASI Series 1: Global Environmental Change, vol. 51, Springer (Berlin): 295–312.

Bradbury, J. P., 1996. Charcoal deposition and redeposition in Elk Lake. Minnesota, USA. The Holocene 6: 339–344.

Brunelle, A. & R. S. Anderson, in press. Sedimentary charcoal as an indicator of late Holocene drought in the Sierra Nevada, California and its relevance to the future. The Holocene.

Chandler, C., P. Cheney, P. Thomas, L. Trabaud & D. Williams, 1983. Fire in forestry: volume I: forest fire behaviour and effects. John Wiley and Sons, New York.

Clark, J. S., 1988a. Particle motion and the theory of stratigraphic charcoal analysis: source area, transport, deposition, and sampling. Quat. Res. 30: 67–80.

Clark, J. S., 1988b. Stratigraphic charcoal analysis on petrographic thin sections: applications to fire history in northwestern Minnesota. Quat. Res. 30: 81–91.

Clark, J. S., 1990. Fire and climate change during the last 750 years in northwestern Minnesota. Ecol. Mon. 60: 135–159.

Clark, J. S. & T. C. Hussey, 1996. Estimating the mass flux of charcoal from sedimentary records: effects of particle size, morphology, and orientation. The Holocene 6: 129–145.

Clark, J. S. & W. A. Patterson, III, 1997. Background and local charcoal in sediments: scales of fire evidence in the paleorecord. In Clark, J. S., H. Cachier, J. G. Goldammer & B. Stocks (eds.) Sediment Records of Biomass Burning and Global Change. NATO ASI Series 1: Global Environmental Change, vol. 51, Springer (Berlin): 23–48.

Clark, J. S. & P. D. Royall, 1995. Particle size evidence for source areas of charcoal accumulation in late Holocene sediments of eastern North American lakes. Quat. Res. 43: 80–89.

Clark, J. S. & P. D. Royall, 1996. Local and regional sediment charcoal evidence for fire regimes in presettlement northeastern North America. J. Ecol. 84: 365–382.

Clark, J. S., J. Lynch, J. B. Stocks & J. Goldammer, 1998. Relationships between charcoal particles in air and sediments in West-central Siberia. The Holocene 8: 19–29.

Clark, J. S., J. Merk & H. Muller, 1989. Post glacial fire, vegetation and human history of the northern Alpine forelands, south-western Germany. J. Ecol. 77: 897–925.

Clark, R. L., 1982. Point count estimation of charcoal in pollen preparations and thin sections of sediment. Pollen Spores 24: 523–535.

Clark, R. L., 1984. Effects on charcoal of pollen preparation procedures. Pollen Spores 26: 559–576.

Cwynar, L. C., 1978. Recent history of fire and vegetation from annually laminated sediment of Greenleaf Lake, Algonquin Park, Ontario. Can. J. Bot. 56: 10–12.

Cwynar, L. C., 1987. Fire and the forest history of the north Cascade Range. Ecol. 68: 791–802.

Earle, C. J., L. B. Brubaker & P. M. Anderson, 1996. Charcoal in northcentral Alaskan lake sediments: relationships to fire and late-Quaternary vegetation history. Rev. Palaeobot. Palynol. 92: 83–95.

Gardner, J. J. & C. Whitlock, 2001. Charcoal accumulation following a recent fire in the Cascade Range, northwestern USA, and its relevance for fire-history studies. The Holocene. 11: 541–549.

Gedye, S. J., R. T. Jones, W. Tinner, B. Ammann & F. Oldfield, 2000. The use of mineral magnetism in the reconstruction of fire history: a case study from Lago di Origlio, Swiss Alps. Palaeogeog., Palaeoclimatol., Palaeoecol. 164: 101–110.

Green, D. G., 1981. Time series and postglacial forest ecology. Quat. Res. 15: 265–277.

Green, D. G., 1983. The ecological interpretation of fine resolution pollen records. New Phytol. 94: 459–477.

Hallett, D. J. & R. C. Walker, 2000. Paleoecology and its application to fire and vegetation management in Kootenay National Park, British Colombia. J. Paleolimnology 24: 401–414.

Hilton, J., 1985. A conceptual framework for predicting the occurrence of sediment focusing and sediment redistribution in small lakes. Limnol. Oceanogr. 30: 1131–1143.

Horn, S. P., 1993. Postglacial vegetation and fire history in the Chirripa Paramo of Costa Rica. Quat. Res. 40: 107–116.

Horn, S. P., R. D. Horn & R. Byrne, 1992. An automated charcoal scanner for paleoecological studies. Palynology. 16: 7–12.

Iversen, J., 1941. Land occupation in Denmark's Stone Age. Danmarks Geologiske Forenhandlungen II 66.

Johnson, E. A. & S. L. Gutsell, 1994. Fire frequency models, methods and interpretations. Adv. Ecol. Res. 25: 239–287.

Kipfmueller, K. F. & W. L. Baker, 1998. A comparison of three techniques to date stand-replacing fires in lodgepole pine forests. Forest Ecol. & Manage. 104: 171–177.

Laird, L. D. & I. D. Campbell, 2000. High resolution palaeofire signals from Christina lake, Alberta: a comparison of the charcoal signals extracted by two different methods. Palaeogeog., Palaeoclimatol., Palaeoecol. 164: 111–123.

Larsen, C. P. S., 1996. Fire and climate dynamics in the boreal forest of northern Alberta. Canada, from AD 1850 to 1989. The Holocene 6: 449–456.

Larsen, C. P. S. & G. M. MacDonald, 1998a. An 840-year record of fire and vegetation in a boreal white spruce forest. Ecology 79: 106–118.

Larsen, C. P. S. & G. M. MacDonald, 1998b. Fire and vegetation dynamics in a jack pine and black spruce forest reconstructed using fossil pollen and charcoal. J. Ecol. 86: 815–828.

Long, C. J., C. Whitlock, P. J. Bartlein & S. H. Millspaugh, 1998. A 9000-year fire history from the Oregon Coast Range, based on a high-resolution charcoal study. Can. J. For. Res. 28: 774–787.

MacDonald, G. M., 1989. Postglacial palaeoecology of the subalpine forest-grassland ecotone of southwestern Alberta: new insights on vegetation and climate change in the Canadian Rocky Mountains and adjacent foothills. Palaeogeog., Palaeoclimatol., Palaeoecol. 73: 155–173.

MacDonald, G. M., C. P. S. Larsen, J. M. Szeicz & K. A. Moser, 1991. The reconstruction of boreal forest fire history from lake sediments: a comparison of charcoal, pollen, sedimentological, and geochemical indices. Quat. Sci. Rev. 10: 53–71.

Mehringer, P. J., S. F. Arno & K. L. Petersen, 1977. Postglacial history of Lost Trail Pass Bog, Bitterroot Mountains, Montana. Arct. Alp. Res. 9: 345–368.

Meyer, G. A., S. G. Wells & A. J. T. Jull, 1995. Fire and alluvial chronology in Yellowstone National Park: climatic and intrinsic controls on Holocene geomorphic processes. Geol. Soc Amer. Bull. 107: 1211–1230.

Millspaugh, S. H., 1997. Late-glacial and Holocene variations in fire frequency in the Central Plateau and Yellowstone-Lamar Provinces of Yellowstone National Park. Ph.D. dissertation, University of Oregon, Eugene, OR.

Millspaugh, S. H. & C. Whitlock, 1995. A 750-year fire history based on lake sediment records in central Yellowstone National Park. USA. The Holocene 5: 283–292.

Millspaugh, S. H., C. Whitlock & P. J. Bartlein, 2000. Variations in fire frequency and climate over the last 17,000 years in central Yellowstone National Park. Geology 28: 211–214.

Mohr, J. A., C. Whitlock & C. J. Skinner, 2000. Postglacial vegetation and fire history, eastern Klamath Mountains. California. The Holocene 10: 587–601.

Morris, S. E. & T. A. Moses, 1987. Forest fire and the natural soil erosion regime in the Colorado Front Range. Ann. Assoc. Amer. Geog. 77: 245–254.

Ohlson, M. & E. Tryterud, 2000. Interpretation of the charcoal record in forest soils: forest fires and their production and deposition of macroscopic charcoal. The Holocene 10: 519–525.

Patterson, W. A. III & A. E. Backman, 1988. Fire and disease history of forests. In Huntley, B. & T. Webb III (eds.) Vegetation History. Kluwer Academic Publishers, Dordrecht, p. 603–632.

Patterson, W. A., III, K. J. Edwards & D. J. MacGuire, 1987. Microscopic charcoal as a fossil indicator of fire. Quat. Sci. Rev. 6: 3–23.

Pearl, C. A., 1999. A Holocene environmental history of the Willamette Valley, Oregon: insights from an 11,000-year-record from Beaver lake. M.S. thesis, University of Oregon, Eugene, OR.

Pitkänen, A. & P. Huttunen. 1999. A 1300-year forest-fire history at a site in eastern Finland based on charcoal and pollen records in laminated lake sediment. The Holocene 9: 311–320.

Pyne, S. J., P. L. Andrews & R. D. Laven, 1996. Introduction to Wildland Fire. John Wiley & Sons, Inc., New York, 769 pp.

Radtke, L. F., D. A. Hegg, P. V. Hobbs, J. D. Nance, J. H. Lyons, K. K. Laursen, R. E. Weiss, P. J. Riggan & D. E. Ward, 1991. Particulate and trace gass emissions from large biomass fires in North America. In Levine, J. S. (ed.) Global Biomass Burning: Atmospheric, Climatic, and Biospheric Implications. MIT Press, Cambridge (MA): 209–224.

Rhodes, T. E. & R. B. Davis, 1995. Effects of late Holocene forest disturbance and vegetation change on acidic Mud Pond. Maine, USA. Ecology 76: 734–746.

Romme, W. H., 1980. Fire history terminology: report of the ad hoc committee. In: Proceedings of the Fire History Workshop. Tucson, Arizona, p. 135–137. USDA For. Serv. Gen. Tech. Rep. RM-81.

Sander, P. M. & C. T. Gee, 1990. Fossil charcoal: techniques and applications. Rev. Palaeobot. Palynol. 63: 269–279.

Sarmaja-Korjonen, K., 1998. Latitudinal differences in the influx of microscopic charred particles to lake sediments in Finland. The Holocene 8: 589–597.

Stockmarr, J., 1971. Tablets with spores used in absolute pollen analysis. Pollen Spores 13: 615–621.

Stuiver, M., P. J. Reimer, E. Bard, J. W. Beck, G. S. Burr, K. A. Hughen, B. Kromer, G. McCormac, J. Van der Plicht & M. Spurk, 1998. INTCAL 89 radiocarbon age calibration, 24,000-0 cal B.P. Radiocarbon 40: 1041–1083.

Sugita, S., 1994. Pollen representation of vegetation in Quaternary sediments: I. Theory and methods in patchy vegetation. J. Ecol. 82: 881–897.

Sugita, S., G. M. MacDonald & C. P. S. Larsen, 1997. Reconstruction of fire disturbance and forest succession from fossil pollen in lake sediments: potential and limitations. In Clark, J. S., H. Cachier, J. G. Goldammer & B. Stocks (eds.) Sediment Records of Biomass Burning and Global Change. NATO ASI Series 1: Global Environmental Change, vol. 51, Springer (Berlin): 387–412.

Swain, A. M., 1973. A history of fire and vegetation in northeastern Minnesota as recorded in lake sediments. Quat. Res. 3: 383–396.

Swain, A. M., 1978. Environmental changes during the past 2000 yr in north-central Wisconsin: analysis of pollen, charcoal and seeds from varved lake sediments. Quat. Res. 10: 55–68.

Swanson, F. J., 1981. Fire and geomorphic processes. In Mooney, H. A., T. M. Bonnicksen, N. L. Christensen, J. E. Lotan & W. A. Reiners (eds.) Proceedings, Fire Regimes and Ecosystem Properties. USDA For. Serv. Gen. Tech. Rep. WO-28: 401–420.

Szeicz, J. M. & G. M. MacDonald, 1991. Postglacial vegetation of oak savanna in southern Ontario. Can. J. Bot. 69: 1507–1519.

Terasmae, J. & N. C. Weeks, 1979. Natural fires as an index of paleoclimate. Can. Field Naturalist 93: 116–125.

Thompson, R. & F. Oldfield, 1986. Environmental Magnetism. Allen and Unwin Ltd., London, England, 227 pp.

Tinner, W., M. Conedera, B. Ammann, H. W. Gaggeler, S. Gedye, R. Jones & B. Sagesser, 1998. Pollen and charcoal in lake sediments compared with historically documented forest fires in southern Switzerland since AD 1920. The Holocene 8: 31–42.

Tinner, W., P. Hubschmid, M. Wehrli, B. Ammann & M. Conedera, 1999. Long-term forest fire ecology and dynamics in southern Switzerland. J. Ecol. 87: 273–289.

Tolonen, M., 1978. Palaeoecology of annually laminated sediments in Lake Ahvenainen, S. Finland. I. Pollen and charcoal analyses and their relation to human impact. Ann. Bot. Fenn. 15: 177–208.

Tolonen, K., 1986. Charred particle analysis. In Berglund, B. E. (ed.) Handbook of Holocene Palaeoecology and Palaeohydrology. John Wiley and Sons, Ltd., New York: 485–496.

Umbanhowar, C. E., Jr., 1996. Recent fire history of the northern Great Plains. Amer. Midl. Nat. 135: 115–121.

Umbanhowar, C. E., Jr. & M. J. McGrath, 1998. Experimental production and analysis of microscopic charcoal from wood, leaves, and grasses. The Holocene 8: 341–346.

Van Wagner, C. E., 1978. Age-class distribution and the forest fire cycle. Can. J. For. Res. 8: 220–227.

Vaughan, A. & G. Nichols, 1995. Controls on the deposition of charcoal: implications for sedimentary accumulations of fusain. J. Sed. Res. A65: 129–135.

Waddington, J. C. B., 1969. A stratigraphic record of the pollen influx to a lake in the Big Woods of Minnesota. Geol. Soc. Amer., Spec. Pap. 123: 263–283.

Ward, D. E. & C. C. Hardy, 1991. Smoke emissions from wildland fires. Env. Intl 17: 117–134.

Whitlock, C. & R. S. Anderson, in press. Fire history reconstructions based on sediment records from lakes and wetlands. In Veblen, T. T., W. L. Baker, G. Montenegro & T. W. Swetnam (eds.) Fire and Climate Change in the Americas. Springer-Verlag, Berlin.

Whitlock, C. & S. H. Millspaugh, 1996. Testing assumptions of fire history studies: an examination of modern charcoal accumulation in Yellowstone National Park. The Holocene 6: 7–15.

Winkler, M. G., 1985. Charcoal analysis for paleoenvironmental interpretation: a chemical assay. Quat. Res. 23: 313–326.

6. NON-POLLEN PALYNOMORPHS

BAS VAN GEEL (vanGeel@science.uva.nl)
Institute for Biodiversity and Ecosystem Dynamics
Kruislaan 318
1098 SM Amsterdam
The Netherlands

Keywords: non-pollen palynomorphs, palaeoecology, palaeoenvironmental indicators, zygnemataceous spores, *Pediastrum, Botryococcus*, cyanobacteria, fungal spores, rotifer eggs

Introduction

In pollen preparations other microfossils of various origin are often preserved. Among the microfossils that attracted the attention of palynologists are a variety of organisms, e.g., algae (Korde, 1966; Jankovská & Komárek, 1992, 2000), eggs produced by aquatic flatworms (Haas, 1996), eggs of Tardigrada (Jankovská (1990, 1991), bacteria (Nilsson & Renberg, 1990) and stomates (MacDonald, this volume). For a period of more than 30 years, deposits of Eemian, Weichselian and Holocene age have been studied palynologically at the University of Amsterdam. The analysis of pollen was combined with the study of all 'extra' microfossils (non-pollen palynomorphs: NPP) with a characteristic morphology. The aim was to discern still unexplored fossils, and this strategy resulted in an increase in the number of palaeoenvironmental indicators. Among the extra fossils were spores of fungi, remains of algae, cyanobacteria (formerly known as blue-green algae) and invertebrates. Most of the deposits studied were situated in Northwest Europe, but some cores from Colombia were also studied. In a series of papers the descriptions and illustrations of the NPP ('Types') were published and their indicator value was discussed. Morphological descriptions were always combined with stratigraphic information, often in the form of pollen and macrofossil diagrams. Several hundred Types have now been distinguished, each one with a Type-number (Bakker & van Smeerdijk (1982), Batten & van Geel (1985), Carrion & van Geel (1999), Ellis-Adam & van Geel (1978), Kuhry (1985, 1997), López-Sáez et al. (1998), Pals et al. (1980), van Dam et al. (1988), van Geel (1976, 1978, 1986, 1998), van Geel & Grenfell (1996), van Geel et al. (1981, 1983, 1986, 1989, 1994, 1995). In most cases there was initially no, or hardly any taxonomic/ecological knowledge about the distinguished Types. The identification of the fossils was attempted with the aid of literature and by consulting colleagues in invertebrate zoology, phycology, mycology and plant anatomy. Among the NPP there still are many taxa which are not properly identified, but some of them nevertheless can be used as palaeoenvironmental indicators. In such cases, the ecological information is inferred from the co-occurrence (curve matching) with

99

identified taxa (e.g., pollen, seeds). Most fungal and algal taxa and cyanobacteria have a much longer geological history than angiosperms and gymnosperms, and, therefore, some of the NPP-studies are also interesting for palynologists specialised in the analysis of pre-Quaternary deposits.

In this chapter only a small selection of NPP taxa is shown. For more information and for morphological descriptions and illustrations, reference is made to the above-mentioned publications. The palaeoecological exploration of NPP in lake deposits is certainly not yet finished. Some egg types of rotifers are shown here as an example of a taxonomic group which still has to be studied in detail (compare Merkt & Müller, 1999), so that the full palaeoenvironmental information can be used in future studies.

Methods used

All the recorded non-pollen palynomorphs were found in pollen samples that had been treated according to the pollen preparation method as described below (compare Bennett & Willis, this volume, Faegri & Iversen, 1989 and Moore et al., 1991):

1) Boil the sample (0.5–3 cc) in 10% KOH (or in 10% Na-pyrophosphate for clayey sediments).

2) sieve (meshes 215 μm); pour into centrifuge tubes and centrifuge until a speed of at least 4500 rpm.

3) Wash the material with water and centrifuge until the supernatant is clear. NB: the following steps (4–7; 10–11) are undertaken in the fume cupboard.

4) Dehydrate with 96% acetic acid and centrifuge.

5) Prepare an acetolysis mixture by slowly adding 1 part H_2SO_4 to 9 parts acetic anhydride (stirring and cooling in a water bath is essential; no contact between acetolysis mixture and water!; safety glasses and gloves required).

6) Acetolyse the material by heating the sample in the acetolysis mixture to 100 °C for ca 10 minutes in a water bath.

7) Cool the sample tubes in a water bath and then centrifuge.

8) Wash with distilled water and centrifuge twice.

9) Wash with 96% alcohol and centrifuge twice.

10) The separation of organic material from sand and clay (if any) is done by using a heavy liquid (bromoform-alcohol mixture, specific gravity 2; gloves!). The procedure for this treatment, if required, is:

11) Add the bromoform-alcohol mixture to the sample and centrifuge at 1500 rpm for 10 minutes. Do not use the centrifuge brake in this case and pour the material floating above the bromoform mixture into a tube that is half-filled with 96% alcohol.

12) Centrifuge again (until 4500 rpm). Decant and wash the sample into a residue tube, using 96% alcohol. Centrifuge (until 4500 rpm) the residue tubes. Decant and add a drop/drops of glycerine to the residue (dependent upon the residue size).

13) Put the residue tube for a night in the oven at 40 °C.

14) Prepare microscope slides.

In studies by the author, HF treatment in order to remove minerogenic material was never applied. It is not known if HF would destroy any of the NPP occurring in samples which are not treated with HF. Clarke (1994) discussed differential recovery of palynomorphs by three processing techniques. None of the techniques seemed to cause notable deterioration to fungal palynomorphs.

Indicator potential and applications of a selection of non-pollen palynomorphs

Zygnemataceae

Zygnemataceae are unbranched filamentous green algae, inhabiting shallow, stagnant, oxygen-rich freshwater lakes, ponds, small pools or wet soils. The cell walls of the filaments do not fossilize, but the morphologically characteristic and resistent spore walls preserve. Within the family, twelve genera are distinguished, among which are *Mougeotia, Zygnema, Spirogyra* and *Debarya* (for some examples of fossil spore types see Figure 1(1–10)). Zygnemataceae reproduce sexually by conjugation. During the conjugation process, two filaments become closely aligned and adjoining vegetative cells become connected by a conjugation tube. Conjugation results in the thick-walled zygospores. Apart from zygospores, asexual resting spores (aplanospores) can also be formed. In the temperate climatic zone Zygnemataceae produce their spores during spring in shallow (often less than 0.5 m deep), relatively warm water. Dormant spores of Zygnemataceae may be exposed to (summer) desiccation without damage to the living contents. Spores germinate in the next year in early spring. With the taxonomy of the extant Zygnemataceae, the morphology and number of chloroplasts is important but in most cases the morphological characteristics of the spores (which we find as microfossils) are necessary for identification to species level.

For general information on extant Zygnemataceae, reference is made to Transeau (1951), Randhawa (1959) and Kadlubowska (1984). A research review concerning the morphology, distribution, ecology, reproduction, physiology, biochemistry, cytology, genetics, systematics and phylogeny of the extant Zygnemataceae is given by Hoshaw & McCourt (1988).

The recognition of the zygnemataceous origin of various morphological spore types in palynological slides started with a study by van Geel (1976). An overview of information about Zygnemataceae of relevance for palynologists and palaeolimnologists is given by van Geel & Grenfell (1994). Fossil spores of Zygnemataceae can be very common in pollen preparations and such spores have often been described—especially in pre-Quaternary deposits—as form-taxa of unknown taxonomic affinity. Based on the zygospore record, Zygnemataceae range from the Carboniferous to the present time. The morphological differentiation of the various spore types (*Mougeotia, Zygnema, Spirogyra, Debarya*) may have happened during the Early Carboniferous, or even earlier (van Geel, 1979).

Figure 1. 1–3: *Mougeotia*, zygospores (×1000). 4 and 5: *Zygnema*-type, zygospores or aplanospores (×1000). 6–8: *Spirogyra*, zygospores or aplanospores (×1000, ×500, ×750). 9 and 10: *Debarya*, zygospores (×1000). 11 and 12: *Aphanizomenon*, akinetes (×1000). 13–16: *Anabaena*, akinetes (×1000).

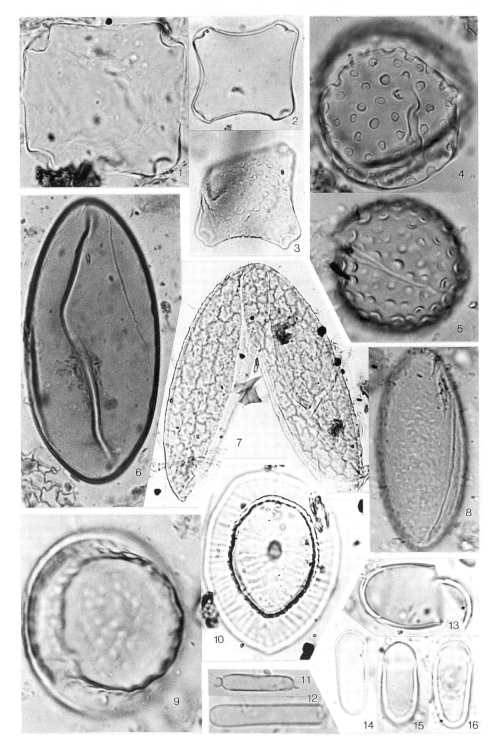

Fossil spores of Zygnemataceae from Quaternary and older sediments have been described and used as palaeoenvironmental indicators by van Geel (1976), van Geel & van der Hammen (1978), Ellis-Adam & van Geel (1978), Rich et al. (1982), van Geel et al. (1989), and Head (1993), as well as others. Within the representatives of extant Zygnemataceae few ecological differences are known. However, different Types of fossil spores often characterise different habitats (e.g., specific sediment types, different trophic conditions). Such differences in ecological amplitudes for specific spore types become evident when the complete spectrum of micro- and macrofossils from closely spaced samples is considered. Van Geel (1978) observed a succession of three different zygnemataceous spore types at the base of the Holocene raised bog Engbertsdijksveen in The Netherlands. The succession was presumably caused by changing trophic conditions (oligotrophication) under influence of the formation of a peaty layer of increasing thickness. At particular sites, conditions can remain favourable to Zygnemataceae for longer periods (millennia in lakes, or near lake margins; see van Geel & van der Hammen, 1978), or only during very short transitional intervals where Zygnemataceae played a role in the vegetation succession. Van Geel & Grenfell (1994) showed an example of such a short successional phase in which *Mougeotia* zygospores played a pioneer role after a local rise of the water table in the Holocene raised bog Engbertsdijksveen.

Pediastrum

Pediastrum species (Fig. 5(52)) are radially-symmetrical colonial green algae (Batten, 1996; Nielsen & Sørensen, 1992; Jankovská & Komárek, 1982, 1995). The outermost cells each show one or two horns. Records of *Pediastrum* species present in pollen slides seem to indicate a wide range of environmental responses (Batten, 1996). Factors mentioned as causative for the occurrence of *Pediastrum* species in sediments include changes in erosion in the catchment, turbidity, water chemistry, nutrient status, and pH. The range of responses may be due to the fact that in the palynological record often different *Pediastrum* species are lumped together. Further knowledge on the present-day ecology of *Pediastrum* species is needed. According to Crisman (1978) *Pediastrum* species are common in hard-water eutrophic lakes. Cronberg (1982) studied a late Holocene lake deposit in S. Sweden. She showed differences in the respons of various *Pediastrum* species and used those species as indicators of the changing trophic status of the lake. Jankovská & Komárek (2000) and Komárek & Jankovská (2001) illustrated *Pediastrum* species and other coccal green algae and reviewed their palaeoecological indicator value.

Botryococcus

Botryococcus species (Fig. 5(53)) are colonial green algae with densely-packed conical cells radiating and branching from the center of the roughly-spherical colony. Modern *Botryococcus* is widely dispersed in temperate and tropical regions, and is known to tolerate seasonally cold climates. It generally lives in freshwater fens, temporary pools, ponds and lakes, where it may form a thick surface scum, but considerable abundances in variable salinity habitats are also known. For morphological details and palaeoenvironmental significance, reference is made to Guy-Ohlson (1992), Komárek & Marvan (1992), Batten & Grenfell (1996), and Jankovská & Komárek (2000).

Cyanobacteria

The analysis of cyanobacteria allows a better understanding of the changing local environments in lakes and pools in the past. Van Geel et al. (1989) recorded *Gloeotrichia* (Fig. 2(17–20)) as an aquatic pioneer during the early part of a Late-Glacial deposit in The Netherlands. The records were of crucial importance for understanding the local environmental development during the early Late-Glacial, because *Gloeotrichia* played a pioneer role in nutrient (nitrogen) poor conditions, thanks to its ability to fix nitrogen and thus making conditions suitable for other aquatic plants.

Akinetes of *Aphanizomenon* (Figs. 1(11) and 1(12)) and *Anabaena* (Fig. 1(13–16)) were recorded in the laminated sediments of Lake Gościąż, Poland (van Geel et al., 1994, 1996; see also Findlay et al., 1998), and the record showed that an increasing human impact in the catchment area (evident from the pollen record) had an effect on trophic conditions in the lake water. From ca 1000 AD onwards, *Anabaena* and *Aphanizomenon* were present in enormous quantities. The increases of these cyanobacteria could be interpreted as the effect of the intensification of farming and land fertilization in the area around Lake Gościąż, causing eutrophication of the lake. Phosphorus enrichment in the catchment area of the lake became so high that N-limited growth conditions occurred. In such conditions cyanobacteria could bloom, since they are capable of nitrogen fixation.

Fungi

Most of the fossil fungi appeared to be ascospores, conidia and chlamydospores (produced by, respectively, Ascomycetes and Dematiaceae). Many of the recorded fungal Types were found in peat deposits, especially in peat layers which were formed under relatively dry conditions. In lake deposits, however, fungal remains normally are of rare occurrence (in open water there is no strictly local production of fungal spores which do preserve as fossils). Another factor which influences the fossil record is the fact that only relatively big (heavy) fungal spores with thick walls are normally preserved. Most of the thin walled spores, which disperse better and which are known from the records of spores in the present atmosphere, obviously do not fossilize. From the various studies of fossil fungal spores, it became clear that the recorded spores in most cases were of strictly local occurrence. They were fossilised at, or near, the place were they had been produced, or the spores were deposited at only a short distance from the place where sporulation took place. Among the analysed fossil fungi were, among others: (1) parasitic fungi and saprophytes, which were always found in combination with certain host plants or on their remains; (2) fungi which were only under certain conditions or incidentally present on special hosts, or on their remains; (3) fungi growing on animal dung; and (4) fungi occurring on burnt plant remains. A selection of examples of fossil fungi is given here:

Amphisphaerella amphisphaerioides, ascospores (Fig. 2(21)):   This fungus with its typical ascospores (3–6 pores in the equatorial plane, wall thickened around these pores, especially at the apical sides) is a parasitic species occurring on *Populus*. Van Geel et al. (1981) recorded it in low frequencies in the early Holocene in a deposit with relatively high frequencies of *Populus* pollen. Additional observations in early Holocene deposits by the author (still unpublished) also support the idea of a host-parasite relationship between *Populus* and *A. amphisphaerioides*. In these cases the presence of the ascospores is a valuable extra indication for the presence of *Populus* (the pollen of which is not always easy to recognize; Bennett & Willis, this volume).

Ustulina deusta, ascospores (Fig. 2(22–24)):   This ascomycete is a mild parasite, causing soft-rot of wood, on several tree species (in NW-Europe: *Abies, Acer, Aesculus, Alnus, Betula, Carpinus, Castanea, Fagus, Fraxinus, Populus, Quercus, Salix, Taxus, Tilia* and *Ulmus*). The pollen of these trees is transported over relatively long distances from the source and can be analysed in sediment samples from lakes and in peat deposits. The ascospores of *U. deusta*, however, are common at a short distance (several metres) from the trees, but the spores are scarce or almost absent in samples from relatively large lakes and bogs (van Geel & Andersen, 1988).

Some coprophilous fungi: Cercophora-type, Podospora-type and Sporormiella-type: Fossil ascospores of *Cercospora*-type (Fig. 2(26)) were first recognized by van Geel (1978) and records were made by van Geel et al. (1981), Bakker & van Smeerdijk (1982) and by Witte & van Geel (1985). According to Lundqvist (1972), representatives of the sordariaceous genus *Cercophora* are coprophilous or occur on decaying wood and on herbaceous stems and leaves. The fossil record thus far (e.g., van Geel et al., 1981; Witte & van Geel, 1985; Mateus, 1992; Buurman et al., 1994) shows circumstantial evidence that the presence of ascospores of the *Cercophora*-type can often be used as an indicator for dung in the surroundings of the sample site.

Although always present in low frequencies, ascospores of the *Podospora*-type (Fig. 2(25)) are regularly recorded in archaeological samples (van Geel et al., 1981; 1983; Buurman et al., 1994; van Geel, unpublished). A relation with the presence of humans or cattle (providing dung as a substrate) for this sordariaceous genus (see Lundqvist, 1972) seems probable. Further observations of ascospores of the *Podospora*-type were made (by the author) in deposits containing remains of *Mammuthus* in The Netherlands (Moershoofd Interstadial; Cappers et al., 1993), in Norwegian *Calluna-Sphagnum* peat deposits with palynological evidence for grazing (project P. E. Kaland, Bergen University), in a Roman Iron Age site in The Netherlands (van Geel, unpublished) and in late Holocene deposits from Mexico (van Geel, unpublished).

Ascospores of extant *Sporormiella* species are three- to many-septate. Every ascospore cell shows a germ slit, extending the entire length of the cell. The ascospores easily split up in separate cells, and as a consequence, in the fossil state, no observations of complete ascospores can be made (Fig. 2(27–30)). Species identification of *Sporormiella*-type is not possible because fruitbodies, asci and complete ascospores are not available (compare Ahmed & Cain, 1972). The representatives of the related, coprophilous genus *Sporormia* are without germ slits, but as the descriptions of Ahmed & Cain (1972) are based on non-germinated spores, a slit may appear after germination of *Sporormia*-spores, and thus we cannot exclude that *Sporormia* is also among our fossil spores. Therefore we refer to the fossil spores as *Sporormiella*-type instead of *Sporormiella*. Fossil spores of *Sporormiella*-type were distinguished by Davis et al. (1977), Davis & Turner (1986), Davis (1987) and by van Geel (archaeological sites; unpublished) who concluded that increased quantities of the spores can be used as indicators of dung, produced by relatively high population densities of domestic herbivores. Samples from the N-Siberian site of the Jarkov mammoth (van Geel, unpublished) are rich in spores of the *Sporormiella*-type.

Chaetomium spec., ascospores (Fig. 2(31)): *Chaetomium* species are strong decomposers of cellulose and occur wherever this substrate is abundant, such as on plant remains and dung. Apart from the occurrence of the lemon-shaped ascospores in peat deposits representing strictly natural habitats, *Chaetomium* species also appeared to be indicators for human impact in the past (Buurman et al., 1995; van Geel, unpublished). In settlements there will have been extra dung, damp straw, clothes, leather and other suitable substrates.

Neurospora spec., ascospores (Figs. 2(32) and 2(33)): The dark ascospores of *Neurospora* with their longitudinal grooves are rare, but their morphology is so characteristic and the indicator value so evident that it is worth while recording their presence. Shear & Dodge (1927) and Dennis (1968) mention *Neurospora* on vegetable matter, developing often after it has been charred by fire. Van Geel (1978) found *Neurospora* ascospores in a layer of charred *Molinia* remains in the Holocene bog Engbertsdijksveen, The Netherlands. *Neurospora* is an indicator for local fires (see also Bakker & van Smeerdijk, 1982, p. 134), but it is also evident that often *Neurospora* ascospores are absent in the charred layers that represent former fires in fens and bogs. The occurrence of ascospores in lake deposits is possible, but no records are available yet.

Glomus spec. (Fig. 2(34)): The globose chlamydospores of *Glomus* (extremely variable in size: 18–138 μm in diameter, exclusive of the hyphal attachment) are of regular occurrence on pollen slides. *Glomus* spores recorded at the Late-Glacial site Usselo (van Geel et al., 1989) closely resembled those of *Glomus fasciculatum*; a relationship of this mycorrhizal fungus (occurs on a variety of host plants) with the roots of local stands of *Betula* at Usselo was probable. Anderson et al. (1984) identified *G. fasciculatum* in post-glacial lake sediments in Maine (U.S.A.). The fungus became established with tundra vegetation on newly developing soils soon after the melting of Wisconsin ice. It was postulated by Anderson et al. (1984) that erosion in the area around the lake accounted for the abundance of *Glomus* in Late-Glacial sediments, and the reduced abundance in Holocene sediments was attributed to a decrease in the rate of soil erosion (and related sedimentation in the lake) after the establishment of forest.

Figure 2. 17: cluster of sheaths of *Gloeotrichia*-type (×400). 18 and 19: sheaths of *Gloeotrichia*-type (×500).
20: colony of *Gloeotrichia*-type (×85; recorded during macrofossil analysis). 21: *Amphisphaerella amphisphaerioides*, ascospore (×1000). 22–24: *Ustulina deusta*, ascospores (×1000). 25: *Podospora*-type, ascospore
(×1000). 26: *Cercophora*-type, ascospore (×1000). 27–30: *Sporormiella*-type, ascospore cells (×1000). 31:
Chaetomium, ascospore (×1000). 32 and 33: *Neurospora*, ascospore, in high and middle focus (×1000). 34:
Glomus, chlamydospore with hyphal attachment (×500). 35–37: *Gaeumannomyces*, hyphopodia (×1000).

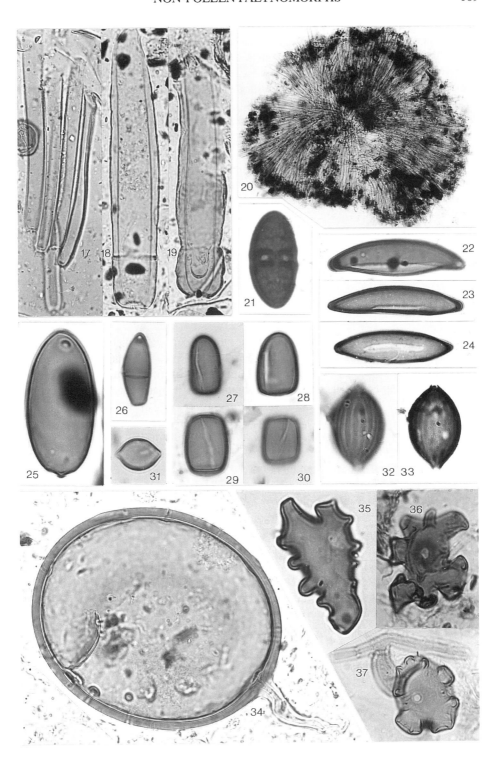

Gaeumannomyces spec. (Fig. 2(35–37)):   The lobed hyphopodia of *Gaeumannomyces* appeared to be characteristic for the local occurrence of *Carex* species. As *Carex* species often play a role in lake margins, and shallow phases of lakes, records of the hyphopodia could be useful in palaeolimnological studies. A comparison of the curve of fossil hyphopodia of *Gaeumannomyces* with the curves of macrofossils and the Cyperaceae pollen curve in a Holocene deposit was made by Pals et al. (1980). A correspondence in the representation of *Gaeumannomyces* and local *Carex* species in the core was obvious. Van Geel et al. (1983) observed and illustrated fossil hyphopodia of *Gaeumannomyces*, still in contact with the epidermis of *Carex*. The corresponding presence of *Gaeumannomyces* and *Carex* species has interesting implications for the palynological analysis of, e.g., Late-Glacial material (van Geel et al., 1989). Since Cyperaceae are usually included in the pollen sum (used as a basis for calculations of pollen percentages) in Late-Glacial deposits, the answer to the question whether Cyperaceae constituted an element of the local vegetation is a crucial piece of information (Bennett & Willis, this volume). The local presence/absence, of *Carex* species becomes evident from the analysis of hyphopodia of *Gaeumannomyces* during microfossil analyses.

Rotifers

Resting eggs of rotifers are of regular occurrence in lake deposits, but normally no records are made. In his study of the Otterstedter See, Müller (1970) recorded resting eggs of rotifers. Müller tried to explain their changing frequencies by a comparison with records of other taxa which indicated changes in lake conditions and human impact in the landscape.

Among the taxa that can be identified are *Anuraeopsis fissa* (Fig. 3(38–40)), *Brachionus* (Figs. 3(41) and 3(42)), cf. *Conochilus hippocrepis* (Figs. 3(43) and 3(44)), *Conochilus natans*-type (Fig. 4(45)), *Filinia longiseta*-type (Fig. 4(46)), *Keratella* (Fig. 4(47)), *Hexarthra mira* (Fig. 4(48)), cf. *Polyarthra dolichoptera* (Fig. 4(49)) and *Trichocerca cylindrica* (Fig. 4(50)). For relevant literature on extant and fossil rotifers see Bogoslovsky (1967, 1969), Pontin (1978), Ruttner-Kolisko (1972), Voigt (1956–1957), Voigt & Koste (1978), Müller (1970), Ralska-Jasiewiczowa & van Geel (1992) and van Geel (1998).

Many rotifer taxa pass the winter as resting eggs. Some rotifers live in shallow puddles. For such species the resting egg is a stage which can survive when the puddle dries up. Van Geel (1998) focussed on the species *Hexartra mira*. It is a cosmopolitan warmth-demanding planktonic organism. In Europe the species has its optimum habitat in the temperature range of 13 to 28 °C, when the pH of its environment is higher than 7. It occurs in lakes and pools where it feeds on detritus. Koste (1979) showed the distribution of *H. mira* in Holocene deposits of two lakes in NW-Germany.

The resting eggs of the planktonic species *Trichocerca cylindrica* showed a sharp decline and disappeared in the Otterstedter See in historical time when several pollen curves showed that human impact in the surroundings increased considerably (Müller, 1970). Also in Lake Gościąż *T. cylindrica* decreased in historical time and this is probably also related to human-induced changes of environmental conditions in the lake (van Geel, unpublished).

Filinia longiseta-type comprises the eggs of *F. longiseta* and *F. passa* (H. Müller, Hannover, pers. comm., 1990). Detailed drawings of the eggs of both species are given by Bogoslovsky (1967). *Filinia longiseta* and related varieties are planktonic taxa (Voigt

& Koste, 1978) and, according to Ruttner-Kolisko (1972), *Filinia* also occurs in brackish water. The eggs of *Filinia* are of regular occurrence in lake deposits. In the Otterstedter Lake (Müller, 1970), eggs of *Filinia* showed the highest representation during the Atlantic and early Subboreal. Increased human impact in the surroundings of the lake apparently had a negative effect on *Filinia*, at least on its egg production. Merkt & Müller (1999) showed that Late-Glacial climate changes, as evident from the pollen record of Hämelsee (Germany), were also reflected in the oscillating frequencies of rotifer remains.

Vegetative remains of Nymphaeaceae

Nymphaeaceae have mucilaginous hairs. The suberized basal cells of these hairs (Fig. 4(51)) with their central pore and concentric rings are very common in pollen slides from deposits of lakes and pools where *Nuphar* and/or *Nymphaea* played a role in the local vegetation (Pals et al., 1980). Also the trichoscilereids are characteristic, but those are less frequent. The high frequency of the suberized basal cells is in strong contrast with the often rare pollen of (entomophilous) Nymphaeaceae. Ralska-Jasiewiczowa et al. (1992) showed that the rise of the suberized cells at the Late-Glacial/Holocene transition in the Polish Lake Gościąż is a better indication for the increase of the thermophilous Nymphaeaceae than pollen of *Nuphar* and *Nymphaea*.

Conclusions and future directions

Palynologists often restrict themselves to the study of fossil pollen only. A full or selected analysis of non-pollen palynomorphs (which have demonstrable indicator value) on the same pollen slides may result in useful additional palaeoenvironmental information.

Summary

A selection of non-pollen palynomorphs is shown and discussed. Spores of Zygnemataceae are common in shallow water deposits, and based on the records of fossil spores, it is evident that the various spore types occur under different environmental conditions. Remains of cyanobacteria appear to be valuable indicators of nitrogen-poor conditions. Such conditions can occur when there is a general low level of nutrients. An extra input of phosphorus-components in lakes (as a consequence of human impact in the catchment area) can also result in the blooming of cyanobacteria. Among the fossil fungi there are (1) parasites indicating the presence of their host plants, (2) indicators for dung, (3) indicators for fires, (4) indicators for the local terrestrialization by *Carex*, (5) soil inhabiting taxa whose presence in lake deposits points to erosion. Eggs of rotifers, which may be present in pollen slides, are also important because of their potential palaeoenvironmental indicator value.

Acknowledgements

The author would like to thank Dr Helmut Müller (Hannover) for help with the identification of eggs of rotifers. The author thanks Dr Dmitri Mauquoy for help with improvements to the text.

Figure 3. 38–40: *Anuraeopsis fissa*, eggs (×1000). 41 and 42: *Brachionus*, eggs (×750 and ×500). 43 and 44: cf. *Conochilus hippocrepis*, egg, in high and middle focus (×750).

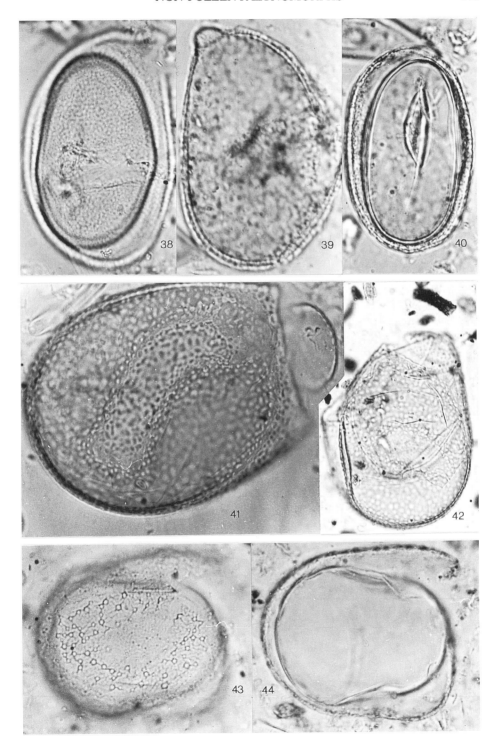

Figure 4. 45: *Conochilus natans*-type, egg (×750). 46: *Filinia longiseta*-type, egg (×750). 47: *Keratella*, egg, in high and middle focus (×750). 48: *Hexarthra mira*, egg (×500). 49: cf. *Polyarthra dolichoptera*, egg, in high and middle focus (×750). 50: *Trichocerca cylindrica*, egg, in high and middle focus (×750). 51: Nymphaeaceae, suberized basal cells of mucilaginous hairs (×1200).

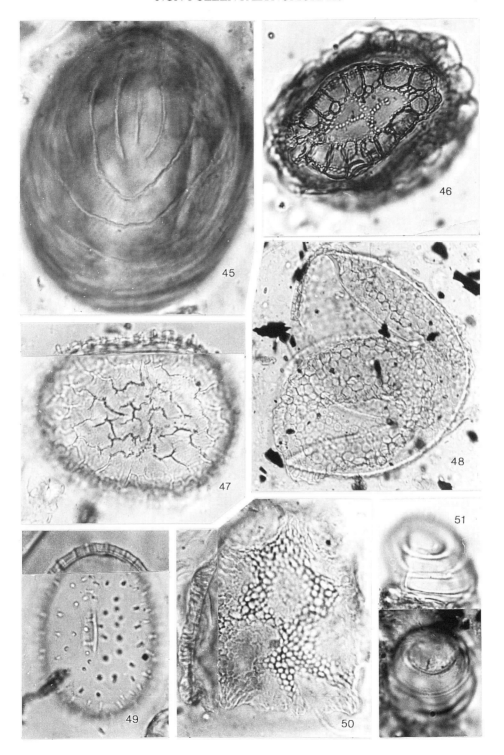

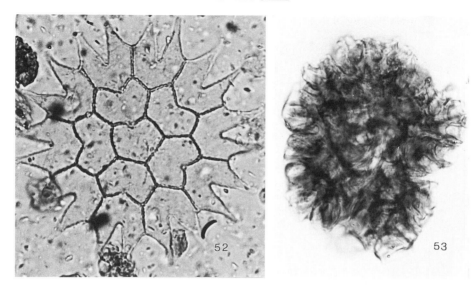

Figure 5. 52: *Pediastrum* (×750). 53: *Botryococcus* (×1000).

References

Ahmed, S. I. & R. F. Cain, 1972. Revision of the genera *Sporormia* and *Sporormiella*. Can. J. Bot. 50: 419–477.

Anderson, R. S., R. L. Homola, R. B. Davis & G. L. Jacobson Jr., 1984. Fossil remains of the mycorrhizal fungal *Glomus fasciculatum* complex in postglacial lake sediments from Maine. Can. J. Bot. 62: 2325–2328.

Bakker, M. & D. G. van Smeerdijk, 1982. A palaeoecological study of a Late Holocene section from "Het Ilperveld", western Netherlands. Rev. Palaeobot. Palynol. 36: 95–163.

Batten, D. J., 1996. Colonial Chlorococcales. In Jansonius, J. & D. C. McGregor (eds.) Palynology: Principles and Applications. American Association of Stratigraphic Palynologists Foundation, Vol. 1: 191–203.

Batten, D. J. & H. R. Grenfell, 1996. *Botryococcus.* In Jansonius, J. & D. C. McGregor (eds.) Palynology: Principles and Applications. American Association of Stratigraphic Palynologists Foundation, Vol. 1: 205–214.

Batten, D. J. & B. van Geel, 1985. *Celyphus rallus*, probable Early Cretaceous rivulariacean blue-green alga. Rev. Palaeobot. Palynol. 44: 233–241.

Bogoslovsky, A. S., 1967. Materials to the study of the resting eggs of rotifers. Communication 2. Bjulleten Moscovskogo obscestva ispytatelej prirody. Otdel Biologiceskij 72: 46–67 (in Russian with English summary).

Bogoslovsky, A. S., 1969. Materials to the study of the resting eggs of rotifers. Communication 3. Bjulleten Moscovskogo obscestva ispytatelej prirody. Otdel Biologiceskij 74: 60–79 (in Russian with English summary).

Buurman, J., B. van Geel & G. B. A. van Reenen, 1995. Palaeoecological investigations of a Late Bronze Age watering-place at Bovenkarspel, The Netherlands. In Herngreen, G. F. W. & L. van der Valk (eds.) Neogene and Quaternary Geology of North-West Europe. Meded. Rijks Geol. Dienst 52: 249–270.

Cappers, R. T. J., J. H. A. Bosch, S. Bottema, G. R. Coope, B. van Geel, B. Mook-Kamps & H. Woldring, 1993. De reconstructie van het landschap. Hoofdstuk 3 (p. 27–41) in van der Sanden, W. A. B., R. T. J. Cappers, J. R. Beuker & D. Mol (eds.) Mens en Mammoet. De mammoeten van Orvelte en de vroegste bewoning van Noord-Nederland. Archeol. Monogr. Drents Museum Deel 5. Assen, 62 pp.

Carrión, J. S. & B. van Geel, 1999. Fine-resolution Upper Weichselian and Holocene palynological record from Navarrés (Valencia, Spain) and a discussion about factors of Mediterranean forest succession. Rev. Palaeobot. Palynol. 106: 209–236.

Clarke, C. M., 1994. Differential recovery of fungal and algal palynomorphs versus embryophyte pollen and spores by three processing techniques. In Davis, O. K. (ed.) Aspects of Archaeological Palynology: Methodology and Applications. AASP Contributions Series 29: 53–62.

Crisman, T. L., 1978. Algal remains in Minnesota lake types: a comparison of modern and late-glacial distributions. Verh. Internat. Verein. Limnol. 20: 445–451.

Cronberg, G., 1982. *Pediastrum* and *Scenedesmus* (Chlorococcales) in sediments from Lake Växjösjön, Sweden. Arch. Hydrobiol. Suppl. 60, 4: 500–507.

Davis, O. K., D. A. Kolva & P. J. Mehringer, 1977. Pollen analysis of Wildcat Lake, Whitman County, Washington: the last 1000 years. Northwest. Sci. 51: 13–30.

Davis, O. K. & R. M. Turner, 1986. Palynological evidence for the historic expansion of Juniper and desert shrubs in Arizona, U.S.A. Rev. Palaeobot. Palynol. 49: 177–193.

Davis, O. K., 1987. Spores of the dung fungus *Sporormiella*: increased abundance in historic sediments and before Pleistocene megafaunal extinction. Quat. Res. 28: 290–294.

Dennis, R. W. G., 1968. British Ascomycetes. Cramer, Lehre, 585 pp.

Ellis-Adam, A. C. & B. van Geel, 1978. Fossil zygospores of *Debarya glyptosperma* (De Bary) Wittr. (Zygnemataceae) in Holocene sandy soils. Acta Bot. Neerl. 27: 389–396.

Faegri, K. & J. Iversen (IV edition: Faegri, K., P. E. Kaland & K. Krzywinski), 1989. Textbook of pollen analysis. Wiley, Chichester, 328 pp.

Findlay, D. L., H. J. Kling, H. Rönicke & W. J. Findlay, 1998. A paleolimnological study of eutrophied Lake Arendsee (Germany). J. Paleolim. 19: 41–54.

Guy-Ohlson, D., 1992. *Botryococcus* as an aid in the interpretation of palaeoenvironment and depositional processes. Rev. Palaeobot. Palynol. 71: 1–15.

Haas, J. N., 1996. Neorhabdocoela oocytes—palaeoecological indicators found in pollen preparations from holocene freshwater lake sediments. Rev. Palaeobot. Palynol. 91: 371–382.

Head, M. J., 1993. Dinoflagellates, sporomorphs, and other palynomorphs from the Upper Pliocene St. Erth Beds of Cornwall, Southwestern England. J. Paleont. 67, Suppl. to no. 3, 62 pp.

Hoshaw, R. W. & R. W. McCourt, 1988. The Zygnemataceae (Chlorophyta): a twenty-year update of research. Phycologia 27: 511–548.

Jankovska, V., 1990. The evolution of late-glacial and holocene vegetation in the vicinity of Svetlá nad Sázavou (in the western forland of the Bohemian-Moravian uplands). Folia Geobot. Phytotax. 25: 1–25.

Jankovska, V., 1991. Unbekannte Objekte in Pollenpräparaten—Tardigrada. In Kovar-Eder, J. (ed.) Palaeovegetational Development in Europe and Regions Relevant to its Palaeofloristic Evolution. Proceedings of the Pan-European Palaeobotanical Conference, Vienna, 13–19 September 1991, Museum of Natural History, Vienna: 19–23.

Jankovska, V. & J. Komárek, 1982. Das Vorkommen einiger Chlorokokkalalgen im böhmischen Spätglazial und Postglazial. Folia Geobot. Phytotax. 17: 165–195.

Jankovska, V. & J. Komárek, 1995. *Pediastrum orientale* from subfossil layers. Folia Geobot. Phytotax. 30: 319–329.

Jankovska, V. & J. Komárek, 2000. Indicative value of *Pediastrum* and other coccal green algae in palaeoecology. Folia Geobot. 35: 59–82.

Kadlubowska, J. Z., 1984. Conjugatophyceae. I. Zygnemales. Süsswasserflora von Mitteleuropa. Band 16; Fischer, New York, 532 pp.

Komárek, J. & P. Marvan, 1992. Morphological differences in natural populations of the genus *Botryococcus* (Chlorophyceae). Arch. Protistenk. 141: 65–100.

Komárek, J. & V. Jankovská, (2001). Review of the green algal genus *Pediastrum*; implication for pollen-analytical research. Biblioth. Phycol. Band 108; J. Cramer, Berlin, 127 pp.

Korde, N. W., 1966. Algenreste in Seesedimenten. Zur Entwicklungsgeschichte der Seen und umliegenden Landschaften. Arch. Hydrobiol. Beih. (Ergebn. Limnol.) 3: 1–38.

Koste, W., 1979. Das Rädertier-Porträt. *Hexarthra mira*, ein sechsarmiges Planktonrädertier. Mikrokosmos 5: 134–139.

Kuhry, P., 1985. Transgressions of a raised bog across a coversand ridge originally covered with an oak-lime forest. Rev. Palaeobot. Palynol. 44: 313–353.

Kuhry, P., 1997. The palaeoecology of a treed bog in western boreal Canada: a study based on microfossils, macrofossils and physico-chemical properties. Rev. Palaeobot. Palynol. 96: 183–224.

López-Sáez, J. A., B. van Geel, S. Farbos-Texier & M. F. Diot, 1998. Remarques paléoécologiques à propos de quelques palynomorphes non-polliniques provenant de sédiments quaternaires en France. Rev. Paléobiol. 17: 445–459.

Lundqvist, N., 1972. Nordic Sordariaceae s. lat. Symb. Bot. Ups. 20: 1–374.

Mateus, J., 1992. Holocene and present-day ecosystems of the Carvalhal Region, Southwest Portugal. PhD Thesis, Utrecht University, 184 pp.

Merkt, J. & H. Müller, 1999. Varve chronology and palynology of the Lateglacial in Northwest Germany from lacustrine sediments of Hämelsee in Lower Saxony. Quaternary International 61: 41–59.

Moore, P. D., J. A. Webb & M. E. Collinson, 1991. Pollen Analysis. Blackwell, Oxford, 216 pp.

Müller, H., 1970. Ökologische Veränderungen im Otterstedter See im Laufe der Nacheiszeit. Ber. Naturhist. Ges. Hannover 114: 33–46.

Nielssen, H. & I. Sørensen, 1992. Taxonomy and stratigraphy of late-glacial *Pediastrum* taxa from Lysmosen, Denmark—a preliminary study. Rev. Palaeobot. Palynol. 74: 55–75.

Nilsson, M. & I. Renberg, 1990. Viable endospores of *Thermoactinomyces vulgaris* in lake sediments as indicators of agricultural history. Appl. envir. Microbiol. 56: 2025–2028.

Pals, J. P., B. van Geel & A. Delfos, 1980. Paleoecological studies in the Klokkeweel bog near Hoogkarspel (prov. of Noord Holland). Rev. Palaeobot. Palynol. 30: 371–418.

Pontin, R. M., 1978. A key to the freshwater planktonic and semi-planktonic Rotifera of the British Isles. Freshwater Biological Association Scientific Publication No. 38, 178 pp.

Ralska-Jasiewiczowa, M. & B. van Geel, 1992. Early human disturbance of the natural environment recorded in annually laminated sediments of Lake Gościąż, Central Poland. Veget. Hist. Archaeobot. 1: 33–42.

Ralska-Jasiewiczowa, M., B. van Geel, T. Goslar & T. Kuc, 1992. The record of the Late Glacial/Holocene transition in the varved sediments of lake Gosciaz, central Poland. Sveriges Geologiska Undersökning, Ser. Ca 81: 257–268.

Randhawa, M. S., 1959. Zygnemaceae. Indian Council of Agricultural Research, New Delhi, 478 pp.

Rich, F. J., D. Kuehn & T. D. Davies, 1982. The palaeoecological significance of *Ovoidites*. Palynology 6: 19–28.

Ruttner-Kolisko, A., 1972. Rotatoria. In Die Binnengewässer, Band 26, Das Zooplankton der Binnengewässer, 1. Teil, pp. 99–234, Schweizerbart'sche Verlagsbuchhandlung, Stuttgart.

Shear, C. L. & B. O. Dodge, 1927. Life histories and heterothallism of the red bread-mold fungi of the *Monilia sitophila* group. J. Agric. Res. 34: 1019–1042.

Transeau, E. N., 1951. The Zygnemataceae. Columbus Graduate School Monographs, Contributions in Botany, 1. Columbus, Ohio, 327 pp.

van Dam, H., B. van Geel, A. van der Wijk, J. F. M. Geelen, R. van der Heijden & M. D. Dickman, 1988. Palaeolimnological and documented evidence for alkalization and acidification of two moorland pools (The Netherlands). Rev. Palaeobot. Palynol. 55: 273–316.

van Geel, B., 1976. Fossil spores of Zygnemataceae in ditches of a prehistoric settlement in Hoogkarspel (The Netherlands). Rev. Palaeobot. Palynol. 22: 337–344.

van Geel, B., 1978. A palaeoecological study of Holocene peat bog sections in Germany and The Netherlands. Rev. Palaeobot. Palynol. 25: 1–120.

van Geel, B., 1979. Preliminary report on the history of Zygnemataceae and the use of their spores as ecological markers. Proc. IVth Int. Palynol. Conf. Lucknow (1976–1977) 1: 467–469.

van Geel, B., 1986. Application of fungal and algal remains and other microfossils in palynological analyses. In Berglund, B. E. (ed.) Handbook of Holocene Palaeoecology and Palaeohydrology. Wiley, Chichester. pp. 497–505.

van Geel, B., 1998. Are the resting eggs of the rotifer *Hexarthra mira* (Hudson 1871) the modern analogs of *Schizosporis reticulatus* Cookson and Dettman 1959? Palynology 22: 83–87.

van Geel, B. & T. van der Hammen, 1978. Zygnemataceae in Quaternary Colombian sediments. Rev. Palaeobot. Palynol. 25: 377–392.

van Geel, B., S. J. P. Bohncke & H. Dee, 1981. A palaeoecological study of an upper Late Glacial and Holocene sequence from "De Borchert", The Netherlands. Rev. Palaeobot. Palynol. 31: 367–448.

van Geel, B., D. P. Hallewas & J. P. Pals, 1983. A Late Holocene deposit under the Westfriese Zeedijk near Enkhuizen (Prov. of N-Holland, The Netherlands): palaeoecological and archaeological aspects. Rev. Palaeobot. Palynol. 38: 269–335.

van Geel, B., A. G. Klink, J. P. Pals & J. Wiegers, 1986. An Upper Eemian lake deposit from Twente, eastern Netherlands. Rev. Palaeobot. Palynol. 47: 31–61.

van Geel, B. & S. T. Andersen, 1988. Fossil ascospores of the parasitic fungus *Ustulina deusta* in Eemian deposits in Denmark. Rev. Palaeobot. Palynol. 56: 89–93.

van Geel, B., G. R. Coope & T. van der Hammen, 1989. Palaeoecology and stratigraphy of the Lateglacial type section at Usselo (The Netherlands). Rev. Palaeobot. Palynol. 60: 25–129.

van Geel, B., L. R. Mur, M. Ralska-Jasiewiczowa & T. Goslar, 1994. Fossil akinetes of *Aphanizomenon* and *Anabaena* as indicators for medieval phosphate-eutrophocation of Lake Gościąż (Central Poland). Rev. Palaeobot. Palynol. 83: 97–105.

van Geel, B., J. P. Pals, G. B. A. van Reenen & J. van Huissteden, 1995. The indicator value of fossil fungal remains, illustrated by a palaeoecological record of a Late Eemian/Early Weichselian deposit in The Netherlands. In Herngreen, G. F. W. & L. van der Valk (eds.) Neogene and Quaternary Geology of North-West Europe. Meded. Rijks Geol. Dienst 52: 297–315.

van Geel, B. & H. R. Grenfell, 1996. Spores of Zygnemataceae. In Jansonius, J. & D. C. McGregor (eds.) Palynology: Principles and Applications. Am. Ass. Strat. Palynol. Found., Vol. 1: 173–179.

van Geel, B., B. V. Odgaard & M. Ralska-Jasiewiczowa, 1996. Cyanobacteria as indicators of phosphate-eutrophication of lakes and pools in the past. PACT 50: 399–415.

Voigt, A., 1956–1957. Rotatoria. Die Rädertiere Mitteleuropas. I Textband, II Tafelband. Borntraeger, Berlin, 508 pp.

Voigt, A. & W. Koste, 1978. Rotatoria. Die Rädertiere Mitteleuropas. I Textband, II Tafelband. Borntraeger, Berlin, 673 pp.

Witte, H. J. L. & B. van Geel, 1985. Vegetational and environmental succession and net organic production between 4500 and 800 B.P. reconstructed from a peat deposit in the western Dutch coastal area (Assendelver Polder). Rev. Palaeobot. Palynol. 45: 239–300.

7. PROTOZOA: TESTATE AMOEBAE

LOUIS BEYENS (lobe@ruca.ua.ac.be)
Departement Biologie
Sectie Polaire Ecologie, Limnologie en Paleobiologie
Universiteit Antwerpen (UA)
Groenenborgerlaan 171, B-2020 Antwerpen
Belgium

RALF MEISTERFELD (meisterfeld@rwth-aachen.de)
Institut für Biologie II (Zoologie)
Rheinisch Westfälische Technische Hochschule (RWTH) Aachen
Kopernikusstrasse 16, D-52056 Aachen
Germany

Keywords: protozoa, protists, rhizopods, testate amoebae, preparation methods, paleolimnology, pH, water table, eutrophication, catchment processes, climatic change, surface moisture, sea level changes

Introduction

Testate amoebae are a polyphyletic group of protozoans that includes lobose and filose amoebae. They reproduce mainly by binary fission but meiosis is known (Mignot & Raikov, 1992). About 1900 species and subspecies have been described (Decloitre, 1986), many of them inadequately, and critical revision will certainly reduce this figure significantly. Many regions of the world, however, are incompletely sampled and numerous new taxa remain to be found (e.g., Beyens & Chardez, 1997).

The lobose order Arcellinida is the largest group. They contain about three-quarters of all known species of testate amoebae. All of them have a test or tectum. This key character that binds the taxon Arcellinida together may not be homologous within the entire group, but at the moment the ultrastructural and molecular data do not allow us to clarify the relations within this order or to other Sarcodina.

Shell morphology

The architecture and composition of the shell have been central in the taxonomy of all shelled amoebae, whether lobose or filose. Two principal types are common: proteinaceous and agglutinate. Following the concepts of Ogden (1990) and Meisterfeld (2001 a, b) those with a proteinaceous test can be divided into:

121

a) Those with a rigid sheet of fibrous material (Hyalospheniidae, Amphitremidae).

b) Those with a test constructed of regularly arranged hollow building units to form an areolate surface (Microchlamyiidae, Arcellidae). These shell types have not been described from the filose testate amoebae.

c) Families that have a more or less flexible membrane that encloses the cytoplasm (Microcoryciidae or the filose Chlamydophryidae). These forms can occur in sediments but are normally not preserved.

Agglutinate shells of lobose testate amoebae have either a cement matrix of often perforated building units, or a sheet-like cement in which foreign material (xenosomes) such as mineral particles (small sand grains) or diatoms is incorporated (suborder Difflugina). The morphology of this cement is of increasing importance for the classification of lobose testate amoebae (Ogden & Ellison, 1988; Ogden, 1990). In filose amoebae, the amount of organic cement varies and may dominate a shell, or it may be limited to small cement strands to glue the mineral particles together. In all genera studied it is sheet-like.

A few lobose genera (of the Lesquereusiidae) have siliceous tests, composed of endogenous rods, nails or rectangular plates which are produced by the organisms themselves (idiosomes). In filose testate amoebae a siliceous test composed of secreted plates is characteristic for the order Euglyphida. Here the plates differ in shape, size and arrangement from genus to genus and also from one species to another and are often essential for the identification (Fig. 1). The tests may disarticulate and the single plates may be studied alongside diatoms (see Douglas & Smol, this volume).

Calcareous shells are characteristic for the Paraquadrulidae, and the genus *Cryptodifflugia*. The former has rectangular calcite plates bound by an internal sheet of cement, while the latter has a thick layer of calcium phosphate which is deposited within an organic template.

The testate amoebae display a great diversity of shell shapes. Especially in drier habitats such as mosses and soils, the apertural apparatus has evolved to a slit-like structure often covered by the anterior lip (*Plagiopyxis*) or a visor (e.g., *Planhoogenraadia*). The aperture can be protected by teeth or located at the end of a tube (tubular section of test leading inwards to the aperture). Bonnet (1975) has classified the different types of shell shape and apertural morphology and used these types to suggest their phylogenetic relationship.

Biogeography

Testate amoebae as a group are found worldwide and inhabit almost all substrates from marine sandy beaches over all freshwater biotopes, as well as terrestrial habitats like mosses and soils. Many species are ubiquistic with little habitat specificity while others live exclusively in one habitat type. Many species have a cosmopolitan distribution, but others have a restricted geographical distribution. For instance several species from the Nebelidae, Distomatopyxidae and Lamtopyxidae occur only in the tropics or the Southern Hemisphere. Some of these species are quite common there, thus this characteristic distribution is certainly not a result of uneven sampling effort (e.g., Smith & Wilkinson, 1986; Meisterfeld & Tan, 1998).

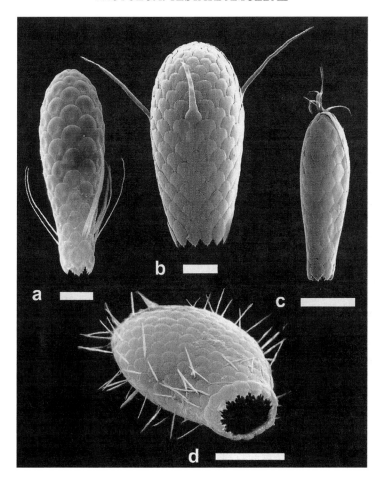

Figure 1. Morphological diversity in spined *Euglypha* species. a) *E. brachiata*; b) *E. crenulata*; c) *E. cristata*; d) *E. strigosa*. Scale bar 20 μm. The spines are either prolonged body plates as in *E. crenulata* and *E. cristata* or distinct elements (*E. brachiata* & *E. strigosa*). Important diagnostic characters are the morphology, number and arrangement of body as well as apertural plates.

Fossil record

Their empty test remains intact after the death of the amoebae and can be determined in most cases to the species level. Although tests fossilize and have been found in deposits as old as the Carboniferous (Loeblich & Tappan, 1964; Wightman et al., 1994), Triassic (Poinar et al., 1993) or Cretaceous (Medioli et al., 1990), most records are from Quaternary deposits.

Phylogeny and classification

The evolutionary origins of the Arcellinida are uncertain. Page (1987) and Schönborn (1967, 1989) believe the family Cochliopodiidae are ancestors of the Arcellinida. The phylogenetic position of the suborder Phryganellina presents some difficulties because it is not clear how taxonomically important their distinctive reticulolobopodia are. Griffin (1972), Hedley et al. (1977) and Ogden & Pitta (1990) have shown that the ultrastructure of these pseudopodia, with their system of pointed pseudopodia and a connecting network of thin pseudopodial strands, is quite different from normal lobopodia. But pseudopodial characters are variable to some degree depending on the task performed and the physiological status of the cell. Molecular data will help to solve these problems and to reconstruct the true phylogeny of this group. Considering these uncertainties, the classification of the order Arcellinida remains provisional. Conventionally (Bovee, 1985; Meisterfeld, 2001a) three suborders are recognized: Arcellina, Difflugina and Phryganellina (with reticulolobopodia).

Species with filopodia are traditionally grouped in the subclass Testaceafilosia De Saedeleer, 1934 or the order Gromiida Claparède & Lachmann, 1858 (e.g., Lee et al., 1985). It is now generally called into question whether they form a natural assemblage (Page, 1987; Meisterfeld, 2001b). Only a few testate amoebae with filopodia have been studied with modern methods. There is a recent study of the ssuRNA gene (Bhattacharya et al., 1995) which places two members of the Euglyphida (*Euglypha rotunda* and *Paulinella chromatophora*) and the chlorarachniophyte algae together. Our own sequence data of several genera (Wylezich et al., 1999) show that the Euglyphida are monophyletic and support the view of Cavalier-Smith (1997) and Cavalier-Smith & Chao (1997) that the euglyphid testate amoebae form a sister group to flagellates like *Cercomonas*. The siliceous body-plates are considered to be homologous in all Euglyphida.

Classification within the species group differs between different traditions and schools of splitters and lumpers (e.g., Decloitre, 1977a, 1978 or Medioli et al., 1994). One main reason for these different views is the way that authors judge the intraspecific variability which especially in agglutinate polymorphic species can be large. A more theoretical problem is the question whether testate amoebae are primarily clonal organisms where a species is composed of a certain number of genetically isolated clones, or whether sexual reproduction continuously recombines the gene pool. The more important practical problem is to find the gaps in a morphological continuum where taxa can be delimited. Very helpful in this process is biometry of the tests (e.g., Schönborn et al., 1983; Lüftenegger et al., 1988; Wanner, 1988). Additional information can be gained by biometry in combination with clonal cultures where environmental factors (e.g., chemical composition of the culture medium, supply of building material) can be manipulated (Meisterfeld, 1984; Ogden & Meisterfeld, 1991; Schönborn & Peschke, 1988, 1990; Wanner & Meisterfeld, 1994; Wanner et al., 1994). Often the variability under those artificial conditions is higher than in the field. But taxonomic conclusions drawn from such experiments can be wrong, if cytological characters that allow to differentiate between taxa are ignored (e.g., Medioli et al., 1987).

Methods

Sampling of sediments

Coring of subaquatic sediments for testate amoebae analysis is done the same way as for pollen or diatoms. We refer to the appropriate chapter by Glew et al. in volume 1 of the present series for a discussion on the coring devices. In the testate amoebae literature, the use of different corers can be found in e.g., Müller, 1970; Beyens, 1984b; Patterson et al., 1985;

Medioli & Scott, 1988; Schönborn, 1990; Ellison, 1995; McCarthy et al., 1995 or Asioli et al., 1996. It is known that the choice of the coring site(s) is important for distinguishing local and regional influences on the lake system. This is illustrated by Laminger (1972, 1973a), who found different assemblages depending on the coring site in the same lake.

Storing the cores / samples

Desiccation and growth of fungi and bacteria during storage should be prevented. Action by these organisms may destroy the cement of agglutinate tests. A cool room or refrigerator with temperatures at 4 °C is preferred.

Preparation and processing of the samples

The superficial layers of the core should be discarded to obtain samples free from contamination. The sampling resolution of the sediment is a function of the aim of the study, and the distance between samples is taken accordingly. If laminations are obvious, their orientation and thickness should be taken into account.

A. **Samples sizes** of 0.5 to 2.0 g wet weight (Ellison, 1995; Ellison & Ogden, 1987; Ruzicka, 1982) and air dried samples of 0.1 to 10 g (Schönborn, 1990) or sediment volumes of 5.5 cm^3 (Charman et al., 1998) to 10 cm^3 (Medioli & Scott, 1988; McCarthy et al., 1995), are typically processed. These samples are dispersed in about 250 ml distilled water. Since the concentration of shells is unknown, one should try to process a fair amount, depending on the available quantity. In sediments rich in very fine material like clay the visibility of the tests is limited. Ellison & Ogden (1987) found that the optimal sample size is inversely related to the amount of clay sized matter present.

B. **Treatment** of the sample with a high–speed blender for 4 seconds (Ruzicka, 1982) can be considered to free all tests from the sediment. But Schönborn (1990) states that this can be harmful to the largest *Difflugia* species such as *Difflugia oblonga*. *Lycopodium* spores which are available in tablet form (T. Person, University of Lund, Sweden) can be added to the sediment. With this internal standard it is possible to estimate test concentrations (Stockmarr, 1971).

If the sample contains much organic material, boiling it in 5% KOH for a few minutes may clear up the solution (Schönborn, 1984). Ellison & Ogden (1987) judge this method to be too severe, destroying many tests. According to Hendon & Charman (1997), KOH digestion does not destroy tests but damages them, making identification more difficult, although a higher number of tests may be retrieved. KOH can also discolor tests, obscuring the difference between e.g., *Heleopera petricola* and its subspecies *amethystea*. To avoid such problems, boiling should be done in distilled water. The addition of a detergent (e.g., metaphosphates or polyphosphates 2–6%) can be helpful.

Chemical treatment as done for the preparation of diatoms (Battarbee et al., this volume) or pollen (Bennet & Willis, this volume) is not appropriate. In a number of studies, testate amoebae have been recorded in pollen slides. This does not produce a reliable picture of the testate amoebae thanatocoenoses since many species do not survive pollen preparation methods. Nevertheless, some interesting results have been obtained this way (e.g. Müller, 1970; Van Geel, 1976; Van Geel et al., 1981; Lopez-Saez et al., 1998) by recording resistant taxa such as *Amphitrema flava*. In general, due to the delicate nature of many testate amoebae sample preparation should be as gentle as possible (usually without chemical treatment).

C. Given the size range of app. 10--400 μm of tests, sieving has the advantage that smaller taxa are more easily found when large particles are removed, and that larger tests (>200 μm) which are often rare, can be concentrated. Researchers differ on the number of sieving steps and on the mesh sizes used. Ellison (1995) washed once with water through a 40 μm screen, Schönborn (1990) used 174 μm and 44 μm meshes. Although most testate amoebae are found in the fraction between 174 μm and 44 μm, all fractions must be examined. Charman et al. (1998) sieved the fraction below 63 μm through a 15 μm screen to remove fine particles and mineral material. Tolonen (1986) proposed to use 30 μm and 200 μm screens. Available standard mesh sizes differ in several countries. For counting with an inverted microscope (see below) we recommend using a large screen size of 200–250 μm first, but marker grains are sieved through, and thus not found in the fraction left on the screen. A larger size eliminates this problem, since few taxa exceed 400 μm but has the disadvantage that large rare species are not concentrated. Even more crucial is the size of the finest sieve. We recommend a 10–15 μm screen (Supplier: in Europe Union Seidengaze, Germany, in North America a 11 μm net filter is available from Millipore), which will retain almost all tests, except perhaps some *Cryptodifflugia* and *Pseudodifflugia* species. If concentration of fractions is necessary centrifugation (5 min at 500 rpm) is the mildest way, although increasing the rotation rate up to 3000 rpm for 5 minutes can be used without obvious signs of damage. To remove the fine particles Medioli & Scott (1988) propose an alternative procedure which basically consists of mixing the sample in water, 30 sec. settling and decantation of the supernatant. This is repeated until the supernatant water seems to be clear. The disadvantage of this method is that almost all small and most proteinaceous species are systematically lost (Medioli & Scott, 1988).

D. The fractions are stored in closed vials with distilled water (inverted and standard microscope) or glycerol (standard microscope) and preserved from bacterial decomposition by adding 7% phenol or 3% neutral formaldehyde.

E. Different fractions can be counted separately, but Ellison (1995) found this impractical. It is possible to count a larger volume of the largest fraction, to obtain a better representation of the usually less abundant large testate amoebae. This should then be taken in account when performing the abundancies calculations.

For standard microscopy, a drop of the suspension is put on the slide and covered with a coverglass. Nail varnish can be used for sealing the margins of the coverslip to avoid drying of the sample. At least 150 tests are counted for each sample.

For inverted microscopy a modified Utermöhl method is used (Utermöhl, 1958; Hasle, 1978). Sieving and diluting are performed in a volume specific way to allow for the estimation of concentrations. The whole size fraction is suspended in a known volume of water and while mixing on a shaker, subsamples are taken with a wide bore pipette and allowed to settle for 1 h in a chamber. The content of the chamber corresponds to a known fraction of the original sample. Flat counting chambers with a coverslip as bottom are handy. Settling chambers are commercially available or can be easily made from plastic rings that fit into holder of the microscope stage. In these chambers the shells can be turned to observe the different views like the shape of the aperture or can be picked for SEM or permanent mounts. The number of chambers that should be analyzed depends on the density of the tests. The minimum is 150 shells.

F. Concentrations of tests are expressed as numbers per gram or volume of sediment. However, this makes a direct comparison between studies difficult, as different techniques are often used. It is also difficult to relate modern and fossil concentration data, since the concentration of fossil shells can be altered by an unknown decay rate (Charman & Warner, 1997). A relative representation can thus be favored in specific studies, despite the loss of information.

G. For the construction of the diagrams and statistical methods we refer e.g. to Birks (1986). The ordering of the taxa, however, is preferentially according to their vertical distribution in the core (see Fig. 8), or according to their ecological requirements (see Fig. 11), if known. In a peat profile the moisture optimum of the taxa is a good criterion. The recommended procedure is synthesized in the following scheme:

1) Subsample the sediment by taking a known wet or dry weight (0.5 to 2.0 g.) or a known volume (5.5 cm^3). Place this subsample in a beaker with 250 ml distilled water.

2) Treat with a high-speed blender for 4 seconds. Do not spill material here.

3) If using a standard microscope, add 1 to 3 tablets of *Lycopodium* as a marker to perform a quantitative analysis. This is not necessary with an inverted microscope.

4) Boil for 10 minutes, stir frequently. This step is necessary when organic material is present in the sample, and also speeds up the dissolution of the *Lycopodium* tablets.

5) Sieve the sample. Wash first through a sieve with mesh width of maximum 400 μm, and proceed with the filtrate through a finer sieve with mesh size 10 or 15 μm.

6) If you use an inverted microscope, go to step 9. For standard microscopy, the different fractions (>400 μm, the one between 400 μm and 15 or 10 μm, and the smallest one) are now concentrated through centrifugation. Centrifuge for 5 minutes using a speed of at least 500 rpm to an upper limit of 3000 rpm.

7) Pour off the supernatant, and store the concentrates in glycerol in closed vials. Add 3% neutral formaldehyde or 7% phenol for preservation.

8) For analysis, a drop of the concentrate is put onto a microscopic slide (76 mm × 26 mm) and covered with a coverslip. Sealing with nail varnish is advised to delay desiccation. This is however not so easy done with greater coverglasses. At least 150 test are counted.

9) Each of the fractions is resuspended in 100–200 ml distilled water.

10) Mix with a shaker and meanwhile take subsamples with a pipette which is emptied in a counting chamber.

11) Count the number of test in the chamber. At least 150 test should be counted.

Peat samples

For the preparation of peat samples, see Hendon & Charman (1997, p. 204). They discuss different methods and conclude with a procedure that is recommended. The main difference with the method proposed here for sediments is the staining of the material with safranine to ease the distinction between organic particles and tests. This is performed in step 7, after pouring away the supernatant. Consequently the material is washed twice before storage.

Observation and identification

Identification is often based on characters such as the shape of the shell or the kind and pattern of body plates. Dimensions are often very useful. The variability of these characters differs between different taxa. Although not present in fossil material characters of the cell are sometimes essential for a correct identification. For more detailed observations of Arcellinida species, embedding in resin (e.g., Euparal CHROMA, refractive index 1.535) is recommended to clear the opaque shells and to inspect the internal morphology of some Centropyxidae, *Pontigulasia, Zivkovicia* or Plagiopyxidae. Most species can be identified using bright field microscopy, but for small transparent shells phase-contrast or interference-contrast is often necessary. For bright field microscopy, it may help to mount empty tests of Euglyphida in media with a high refraction index such as Naphrax or Styrax. Scanning electron microscopy (SEM) is the best means to document test morphology and to study the organic cement matrix or the apertural plates of Euglyphidae. For most agglutinate tests standard methods are sufficient. *Arcella* species are not always very well preserved and critical point drying can be helpful.

Only the more common species have thus far been studied by scanning electron microscopy, and it has not always been easy to coordinate these new findings with earlier descriptions. These problems are partly caused by the limitations of light microscopy, but more often by inadequate diagnoses. Most species found in sediments can be identified to the species level from empty shells alone. Problematic and sometimes impossible is the distinction between small *Difflugia* species (lobopodia), *Cryptodifflugia* (reticulolobopodia) and *Pseudodifflugia* (filopodia) or *Phryganella* and some *Cyclopyxis* where the type of pseudopodia must be known, or between some *Difflugia* and *Netzelia* with lobed aperture where the type of nucleus (e.g., vesicular or ovular, number of nucleoli) is important (Ogden & Meisterfeld, 1989).

Literature

No modern monographs that treat the majority of the species are available, but Leidy (1879), Penard (1902), Cash & Hopkinson (1905, 1909), Cash et al. (1915), Wailes & Hopkinson

(1919) are still useful. Keys to almost all genera can be found in Meisterfeld (2001 a, b). Many aquatic and planktonic species can be determined using Grospietsch (1972a, b) and Harnisch (1958), but the descriptions of most soil species are too recent to be included in these publications. Bonnet & Thomas (1960) describe about a hundred taxa, often soil and moss species. Corbet (1973) can be used as an introduction to the *Sphagnum* fauna. Recently, Medioli et al. (1994) have published a guide to sediment testate amoebae but their classification of *Cucurbitella, Difflugia* and *Netzelia* is not universally accepted because they often ignore cytological characters like nuclear structure, the presence of symbiotic algae or the mode of test building. Schönborn (1966a) contains information on the general biology and ecology. Ogden & Hedley (1980) illustrate 95 mainly aquatic and moss species by SEM micrographs. Decloitre (1962; 1976a, b; 1977a, b; 1978; 1979a, b; 1981; 1982; 1986) has published a series of compilations, although without any attempt to revise the numerous insufficiently described taxa. These papers, however, help to provide access to the scattered literature. The most extensive bibliography is the one compiled by Medioli et al. (1999), which is accessible in the internet journal "Palaeontologica Electronica" (http://www-odp.tamu.edu/paleo).

Applications

Introduction

The testate amoebae fauna of lakes has been considered in a number of studies, which range from faunistic lists to production and population dynamics assessments. The study of Penard (1902) on Lac Léman (Switzerland) was a landmark for the systematic probing of a limnic system, while the work of Schönborn (1962) on the oligotrophic Lake Stechlin (Germany) became a classic for its ecological approach. Numerous studies since have appeared from all over the world (Table I).

An important advantage for (paleo)ecological work is the short generation time, which makes testate amoebae, like other protists, sensitive indicators of short-lived environmental changes. The generation time depends on many factors, such as temperature, nutritional conditions, food quality and availability, and population density (Schönborn, 1992). In a humus microcosm the generation time of *Phryganella acropodia* ranged from 4 hours to 7.4 days. In the course of one year, this species underwent 19 generations in the river Saale, and *Centropyxis aculeata* no less than 60 (Schönborn, 1981).

An important topic is the preservation of shells in sediments. In soils under aerobic conditions, shells decompose within a few weeks (Lousier & Parkinson, 1981; Meisterfeld & Heisterbaum, 1986; Coûteaux, 1992). The anaerobic conditions in bogs and most lake sediments allow a much longer survival time. Often the decrease of tests with depth follows an exponential curve (e.g., Ruzicka, 1982; Schönborn, 1990). Damage of the shell makes identification difficult. Examples of partly decomposed and broken tests are given in Schönborn (1973, 1984, 1990). Decay rates in peat, are largely unknown (Charman & Warner, 1997). Finding the real reason of fluctuations in shell concentrations can be problematic, since they can be caused by differences in decay rates, or by real changes in population size. Interdisciplinary research allowing to compare the quantitative and qualitative behavior of other organisms and with lithological data in the same sequence, can help to solve this problem. Reviews on paleoecological testate amoebae analyses have

Table I. Representative studies on testate amoebae in lakes.

Lake	References
Africa	
Lakes in Congo	Chardez (1964), Stepanek (1963)
Arctic	
Canadian High Arctic	Beyens et al. (1991)
Northeast Greenland	Trappeniers et al. (1999)
Antarctica	
South Georgia	Beyens et al. (1995)
Asia	
Lake Dong Hu (China)	Shen (1980)
Lake Biwa (Japan)	Mori & Miura (1980)
Europe	
Lake Constance (Germany)	Stepanek (1968), Laminger (1972, 1973a)
Italian Lakes	Asioli, Medioli & Patterson (1996)
Lake Léman (Switzerland)	Penard (1902)
Lago Maggiore (Italy)	Grospietsch (1957)
Lake Ohrid (Macedonia)	Golemansky (1967)
Lakes Shabla and Ezerets (Bulgaria)	Todorov & Golemansky (1998)
Lake Stechlin (Germany)	Schönborn (1962)
Lakes in Swedish Lappland	Schönborn (1975)
Lake Vlasina (Yugoslavia)	Ogden (1984)
Vorderer Finstertaler See (Austria)	Laminger (1973b)
Vranov Reservoir (Czechoslovakia)	Stepanek (1967)
North America	
Flathead Lake (USA)	Laminger et al. (1979)
Lake Erie	Scott & Medioli (1983)
Lake Washington (USA)	Lena (1982)
South America	
Lake Cocococha (Peru)	Haman & Kohl (1994)
Lago San Roque (Argentina)	Vucetich (1976)
Lago Valencia (Venezuela)	Grospietsch (1975)

previously been published by Harnisch (1927), Grospietsch (1952, 1972c), Beyens (1984a), Tolonen (1986), Ellison & Ogden (1987), Medioli & Scott (1988), Warner (1988), Medioli et al. (1994).

We discuss below the information that can be derived from (sub)fossil testate amoebae assemblages by looking at their recent ecology, and illustrate how this can be used in paleolimnological studies.

Indication of pH

Although the influence of pH on the testate amoebae fauna was doubted in earlier literature, recent studies contradict this (e.g., Tolonen et al., 1992; Charman & Warner, 1997).
Waterbodies in the same area but with different pH harbour different testate amoebae
assemblages (Beyens et al., 1995). The effect of artificial acidification in a running water
system on testate amoebae communities was studied by Costan & Planas (1986). In this
short-term experiment, running for 24 hours after the first acid treatment, the pH declined
from 6.1–7.1 to 3.9–4.1. The most obvious result was a decrease of overall abundances
of testate amoebae. No marked decrease in diversity was observed, perhaps due to the
short experimental period. Ellison (1995), in his study of the lakes of the English Lake
District, gives a list of pH-indicator taxa. He distinguish taxa indicating a low-pH (<6.2)
such as *Cyclopyxis arcelloides, Difflugia elegans* and *Nebela* species. As indicators for a
high-pH (>6.2) he mentions e.g., *Centropyxis* species, *Cyclopyxis kahli*. Some taxa are
regarded as being ubiquitous with respect to pH: *Difflugia curvicaulis* and *Netzelia* species
belong to this category. By plotting the percentages of low-pH and high-pH taxa against
time, he documented paleoclimatological changes and anthropogenic changes in catchment
processes. Although these papers indicate a strong relationship between pH and testacean
community structure, it is possible that the effect of pH is mainly indirect by influencing
the composition of available food organisms. An obvious effect of pH is the lack of species
with calcareous tests in acid waters. In calcareous waters, during periods of high primary
production, calcite is precipitated and incorporated in the agglutinate tests of species as
Difflugia limnetica. Such species act thus as indicator species for calcite precipitation
(Casper & Schönborn, 1985).

Indication of trophic status

In their work on lakes in northern Germany, Schönborn et al. (1965) established a classification of these waterbodies according to the profundal testate amoebae assemblages. They
recognized three main assemblages, indicative of the C/N ratio and the trophic state. The
densities of testate amoebae are considerably smaller in eutrophic lakes than in oligotrophic
ones (Schönborn, 1962). The principal factor involved is the oxygen content of the waters
overlying the sediments (Tolonen, 1986). Table II shows the characteristic assemblages
and their differentiating taxa, but also some ubiquitous taxa found in all three lake types.
Indicator taxa are shown Figures 2, 3 & 4.

This framework is used by several researchers in the interpretation of fossil data. Ruzicka (1982), in a major work on the fossil testate amoebae from Lake Krotten (Austria),
found that the oligotrophic *Arcella hemisphaerica* group characterizes the entire 7.5 m long
core, starting no later than the beginning of the Late Glacial. Only during Younger Dryas
times were dystrophic conditions inferred by the occurrence of the *Arcella gibbosa* group.
A distinct shift in trophy was shown by Schönborn (1984) in the history of a lake in Poland.
The occurrence of *Centropyxis aerophila* from the Younger Dryas to the Boreal indicated
oligotrophic conditions. From the end of the Subboreal, *Difflugia urceolata* appeared,
indicating significant eutrophication. Pollen analysis indicated this to be caused by human
settlement and pastoral husbandry. However, indicator species for eutrophication are not
always present, as noticed by Schönborn (1973) (see Fig. 5) in a study of Lake Latnjajaure

Table II. Distribution of indicator species in lakes of different trophic conditions; compiled from Schönborn (1967) and Laminger (1973a, b) and our own data. Symbols: *** = common, ** = moderately common, * = rare, + = single records only, − = missing.

	Types of Lakes			
Species	oligotrophic	mesotrophic	dystrophic	eutrophic
Arcella hemisphaerica	***	*	*	*
Centropyxis aerophila	***	−	−	−
C. platystoma	***	**	*	*
C. plagiostoma	***	*	*	−
Centropyxis orbicularis	***	−	−	−
Cyclopyxis kahli	***	−	−	−
Difflugia lemani	***	−	−	−
D. finstertaliensis	***	−	−	−
Nebela (Argynnia) dentistoma	**	−	***	−
N. vitrea	*	*	**	−
Arcella gibbosa	−	−	***	−
Heleopera petricola	*	*	***	*
Hyalosphenia cuneata	***	*	**	−
Lesquereusia spiralis	*	*	***	+
Nebela collaris	*	−	***	−
Difflugia amphora	+	+	−	***
D. corona	*	*	−	***
D. urceolata	−	*	−	***
D. acuminata	***	***	**	***
D. elegans	***	***	***	***
Cyphoderia spp.	**	**	−	**

(Swedish Lapland). Periods of early eutrophication are marked by a decrease of *Centropyxis aerophila* and an increase of some *Difflugia* species.

Indications for changes in catchment area

Increased run-off and erosion can be related to human activities, or to natural processes (e.g., climatic change, giving way to soil mobilization). A higher input of eroded soil includes a higher allochthonous fraction of terrestrial testate amoebae (Ruzicka, 1982). Many soil testate amoebae have a typical test morphology (Fig. 6) often characterized by a plagiostom or cryptostom aperture of reduced size. The fraction of these substrate specific allochthonous lifeforms can be estimated by grouping the tests according to the aperture-types of Bonnet (1961, 1964, 1975), (see Fig. 7).

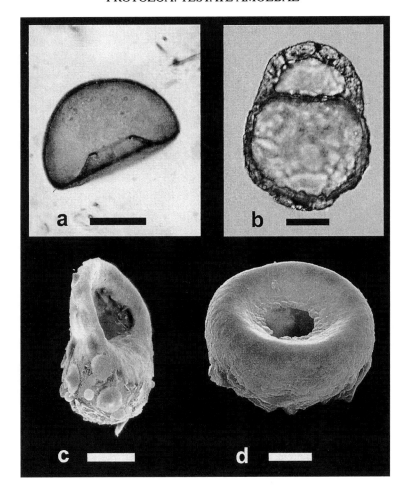

Figure 2. Indicator species for oligotrophic lakes. a) *Arcella hemisphaerica*; b) *Centropyxis aerophila*; c) *C. platystoma*; d) *Cyclopyxis kahli*. (Scale bar = 20 µm).

Scott & Medioli (1983), in a study of Lake Erie (North America), related the occurrence and numbers of *Difflugia bidens* to increased sediment input following human settlement and a change in the nutrient level of the lake. This behavior of *Difflugia bidens* is supported by Patterson et al. (1985). They found this species at a level corresponding with a prehistoric forest fire, leading to a high clastic input. In the Ullswater core (U.K.), Ellison (1995) calculated sedimentation rates from the density of testate amoebae, assuming that relative rates of sedimentation were reciprocal to test density. These rates reflected the expansion of the deciduous forest cover and an increase in runoff with increased erosion of soils in the catchment area after the *Ulmus* decline (ca 5000 ^{14}C yrs BP). Numbers of tests per gram of sediment ranged from less than 1000 to more than 8000, and averaged about 2500 throughout the core. For the assumptions and limits of this method, see Ellison (1995).

Figure 3. Characteristic species of dystrophic lakẹs: a) *Arcella gibbosa*; b) *Nebela (Argynnia) dentistoma*; c) *Heleopera petricola*; d) *Lesquereusia spiralis*. (Scale bar =30 μm).

Indication of climatic changes

Evidence for relations between climate and testate amoebae assemblage composition is not firmly established. Patterson et al. (1985) note that "Arcellacean assemblages illustrate few changes… even through climatic changes". This is confirmed by Medioli & Scott (1988), but these authors remark that testate amoebae require a minimum temperature at some time of the year to reproduce successfully. This should provide a hint to a possible filter effect by water temperatures in different biogeographic zones. McCarthy et al. (1995) compared paleotemperatures and paleo-precipitation levels derived from pollen data with

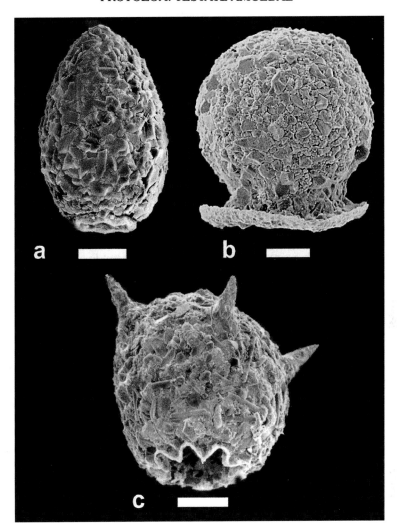

Figure 4. Characteristic species of eutrophic lakes: a) *Difflugia amphora*; b) *D. urceolata*; c) *D. corona*. (Scale bar = 50 μm).

the testate amoebae assemblages from the same sediments. This allowed them to conclude that these protists did, indeed, respond to climatic change. A succession of assemblages was established for the past 12,000 years (see Fig. 8). The authors explain the behavior of some taxa on the basis of their tolerance to harsh conditions, including slightly brackish water, cold arctic temperatures, and hyper-oligotrophic conditions. With the start of the climatic warming, sudden changes also occur in the testate amoebae assemblages. *Difflugia oblonga* increased and remained common through the Holocene, while numbers of *Centropyxis aculeata* fell, although this taxon regained some importance during the late Holocene cooling. These changes might be a response to warming or to increased deposition of organic

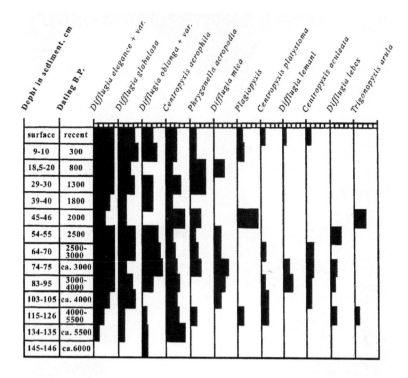

Figure 5. The diagram from Lake Latnjajaure in Swedish Lapland (Schönborn 1973, redrawn in Tolonen, 1986). *Centropyxis aerophila* indicates the oligotrophic status of the lake throughout its history. Around 2000 BP the climate deteriorated, resulting in a summer ice-cover. These arctic-alpine conditions are reflected in the testate amoebae assemblages, and the retreat of most *Difflugia* species is explained this way. This corresponds with recent observations on assemblages from arctic waterbodies (Trappeniers et al., 1999). On the same level, terrestial taxa as *Trigonopyxis* arrive in the sediment. Reproduced from Tolonen (1986) Rhizopod analysis. In Berglund (ed.) Handbook of Holocene Palaeoecology and Palaeohydrology. J. Wiley & Sons, Chichester: 645–666. (With permission of Wiley & Sons.).

matter. Overall species diversity increased sharply at the beginning of the Holocene, which corresponded well with data on diversity in modern assemblages from polar regions and other biozones (Beyens & Chardez, 1995).

In a study of Late Glacial to Subatlantic sediments from Lake Krotten (Austria), Ruzicka (1982) grouped the testate amoebae according to their habitat: species living in lakes, in mosses, and in soils. The relative importance of these groups could be correlated to the fluctuating amount of precipitation. Dry periods induce the development of mosses around the lake. Concomitantly the share of moss-dwelling taxa and those living on higher plants becomes more important. During periods of higher humidity, lake-dwelling taxa are more prominent. In some samples soil-dwelling species dominate, which is explained by a sudden and high amount of precipitation, leading to an input of taxa from different habitats. Indicator species for *Sphagnum*-dominated biotopes (Fig. 9) form a morphologically and ecologically well-defined assemblage. It is interesting to note that McCarthy et al. (1995)

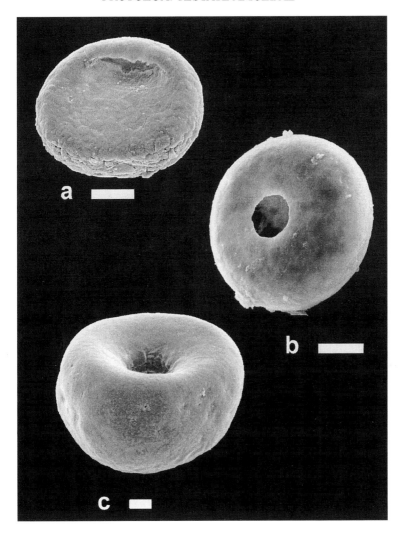

Figure 6. Indicator species for allochthonous input of soil. a) *Plagiopyxis declivis*; b) *Geopyxella sylvicola*; c) *Cyclopyxis puteus*. (Scale bar = 20 μm). Note the cryptostom aperture in *Plagiopyxis* and the tubular opening in *Cyclopyxis puteus*.

explained increases of the bog species *Heleopera sphagni* and *Nebela collaris* in terms of increased precipitation, as this resulted in paludification of the surroundings.

The relation between temperature and the occurrence of certain taxa deserves some attention. Medioli & Scott (1988) observed no correlation between assemblages and water temperature in sediments from Lake Erie (North America). Patterson et al. (1985), however, found that the division between two major assemblages in some small lakes appears to be a function of seasonal temperatures. One assemblage occurs when the summer water temperatures reach values greater than 18 °C, and the sediment surface

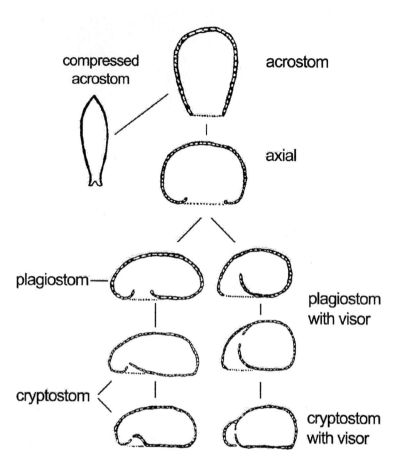

Figure 7. Pseudostome-types after Bonnet (1964). Acrostom tests are common in aquatic habitats, compressed ones dominate wet mosses, and plagiostom or cryptostom types are characteristic for drier terrestrial habitats like soils.

remains above the thermocline. The other assemblage lives below the thermocline where the temperature does not exceed 6 °C. Attention must be paid to other possible parameters that can influence assemblage structure, for instance littoral or profundal habitat linked with oxygen depletion.

Some researchers suggest a certain degree of thermophily for some taxa, and testate amoebae should, thus, reflect climatic changes after deglaciation (McCarthy et al., 1995; Collins et al., 1990). Possible thermophilous taxa according to these authors are *Difflugia bacillifera* and *Lesquereusia spiralis*. However, recent studies (Beyens & Chardez, 1995) recorded *D. bacillifera* and *L. spiralis* in the Arctic, an observation which does not support this hypothesis. Also *Difflugia oblonga*, although becoming important in the Holocene history of some lakes in Atlantic Canada as described by McCarthy et al. (1995), can hardly be considered a thermophilous taxon, since Schönborn (1966b, 1975) mentions it

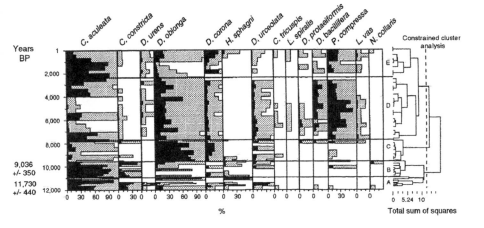

Figure 8. Testate amoebae diagram constructed by McCarthy et al. (1995). Boundaries are drawn according the cluster analysis. The succession of the taxa is readily shown. Interesting is the relation between *Centropyxis* and *Difflugia*. It seems that *Centropyxis* is the first colonizer of lakes after deglaciation, which is in agreement with the present community structure of testate amoebae from aquatic environments in the Arctic (Trappeniers et al., 1999). Stippled area is 10× exaggeration. Reproduced from McCarthy, F., E. Collins, J. McAndrews, H. Kerr, D. Scott & F. Medioli, 1995. A comparison of postglacial arcellacean ("Thecamoebian") and pollen succession in Atlantic Canada, illustrating the potential of arcellaceans for paleoclimatic reconstruction. J. Paleontol. 69: 980–993, with permission of The Paleontological Society.

as a dominant species on Spitsbergen. The clue to distinguishing colder climate regimes could be on the community level where not only single indicator species but whole assemblages with their different spectra of dominant species is taken into account. In aquatic environments, for instance, *Difflugia* is the most important genus in temperate and tropic climates, but is less dominant in an arctic context. Some taxa, such as *Centropyxis gasparella* (Chardez & Beyens, 1988) and *Centropyxis pontigulasiformis* (Beyens & Chardez, 1986), do appear to be confined to arctic regions, and can be considered as biogeographic indicator taxa.

The number of taxa occurring in the Arctic and the Antarctic is substantially lower than in other climatic regions (Smith, 1992; Smith & Wilkinson, 1986; Beyens & Chardez, 1995). One might expect that even minor climatic fluctuations such as the Little Ice Age might exert some pressure on the diversity of the testate amoebae assemblages. This is the explanation Ruzicka (1982) gives for an important decrease in the numbers of taxa and individuals in the Lake Krotten (Austria) sediments, with the onset of this cooling period. Schönborn (1973) states that during the cold phase in the North near the beginning of Christian Era, the Latnjajaure (Swedish Lapland) mountain lake was ice covered even during the summer. According to him, this is reflected in a decrease of *Difflugia* species, and an increase of *Centropyxis aerophila*. In modern arctic aquatic assemblages, the species richness of *Centropyxis* (including *Cyclopyxis*) is, indeed, higher than that of *Difflugia* (Beyens & Chardez, 1995; Beyens et al., 1986, 1991). Likewise, a transition occurs in the Ullswater (U.K.) core (Ellison, 1995) from *Centropyxis-Cyclopyxis* dominated assemblages to the

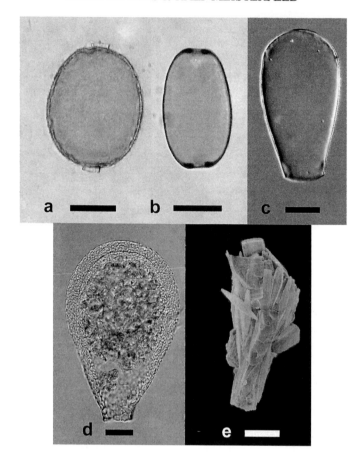

Figure 9. Indicator species for *Sphagnum* contact and paludification. a) *Amphitrema wrightianum*; b) *Amphitrema (Archerella) flavum*; c) *Hyalosphenia papilio*; d) *Nebela carinata*; e) *Difflugia bacillifera*. (Scale bar = 30). These species are also indicators for water table fluctuation in peatlands.

predominance of *Difflugia* after the start of the Holocene hypsithermal warming period (±9000 years BP).

Indications for transition between marine and freshwater environments

Sea-level changes have been assessed by means of diatoms (Denys & De Wolf, 1999) and foraminifera (Scott & Medioli, 1978). Although testate amoebae are mainly restricted to terrestrial and freshwater habitats, some species are observed in the interstitial water of marine sandy beaches (e.g., Chardez & Thomas, 1980; Golemansky, 1971, 1981, 1990, 1991). These "beach dwellers" were found to prefer homogenous fine grained sand rather than heterogeneous coarse grains, and are oligo-, meso-, or euhaline (Golemansky, 1994). Holocene marine to freshwater transitions in lake sediments from New Brunswick and

Nova Scotia (Canada) are dominated by *Centropyxis aculeata* (Patterson et al., 1985). These authors consider "arcellaceans" to mark the transition from extreme tidal conditions (high marsh) to freshwater conditions. Medioli & Scott (1988) quote higher percentages of *Centropyxis aculeata* and *Centropyxis constricta*, together with lowered total numbers of other testate amoebae, as indicators of brackish water conditions. This does not necessarily mean that they are specific indicators (both *Centropyxis* species mentioned here live quite profusely in entirely freshwater conditions), but that these ubiquitous freshwater species are the only ones found in this study that can colonize unstable environments with variable salinity. The indicator potential of a species depends on the type of environmental gradient. Indicators are not necessarily taxa that only occur in a well described environmental context. If the gradient is steep, as in the transition zone between marine and freshwater sediments where one group of organisms (Foraminifera) is replaced by another one (testate amoebae), even species with a broad ecological amplitude can be useful qualitative indicators. But if the environmental gradient is flat other indicator species or groups of species are needed with a narrow ecological amplitude that respond closely to the gradient.

The possibility of using testate amoebae in sea-level studies was addressed by Charman et al. (1998), who analysed surface-sediment samples from a salt-marsh transect in the Taf estuary in South Wales, U.K. Small testate amoebae showed a clear zonation across the marsh surface, which they related to elevation and position within the tidal frame. In the same study, a comparison was made with the zonation of foraminifera which showed that testate amoebae revealed a more detailed and informative zonation in the uppermost part of the high marsh zone, while they also occur in the supratidal zone. Testate amoebae were not found in samples dominated by polyhalobous diatoms. Yet some psammobiontic testate amoebae were discovered in interstitial waters with a salt concentration reaching 40‰ and a high number of taxa can tolerate moderate salinities. Salinity seems, thus, to play an important role in limiting testate amoebae distributions, with different species having subtly different responses to this factor (Charman et al., 1998).

Indication for water table depth and surface moisture in peatlands

Lake basins become filled with sediment and organic material and can end up as peatlands. It is in the study of peat that rhizopod analysis has its historical strongholds. Numerous studies have been devoted to the ecology of *Sphagnum*-dwelling taxa (Schönborn, 1963; Meisterfeld, 1977, 1979; Tolonen et al., 1992, 1994), exploring the relation of the moisture content of the *Sphagnum*-substrate and the autecology and synecology of the species. Earlier paleo-ecological studies (Beyens, 1985; Beyens & Chardez, 1987) used relative measures, dry versus wet, for reconstructing the surface wetness of the bog surface. During the last few years, research has concentrated on reconstructing the water table in Holocene peatlands more precisely (Warner & Charman, 1994; Charman & Warner, 1992, 1997; Butler et al., 1996; Bobrov et al., 1999; Woodland et al., 1998). Using statistical models Charman & Warner (1997) were able to calculate optimum water table for individual species. These data than can be used to reconstruct water tables from species assemblages where the water table is not known, as in fossil samples. By this approach, the evolution of the surface wetness changes could be expressed as mean annual water table depth changes and coupled to paleoclimate (Charman et al., 1999) (see Fig. 10). The information thus obtained can be used as tools in peatland management (Warner & Chmielewski, 1992). A succession from

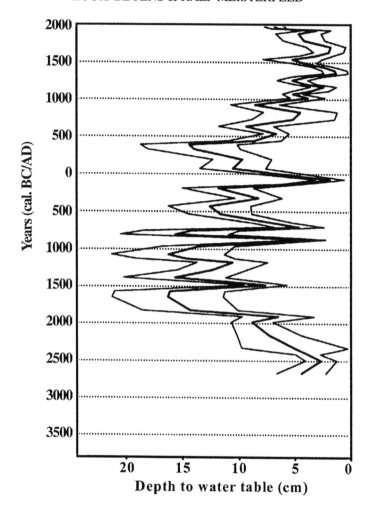

Figure 10. Diagram constructed by Charman et al. (1999), showing the evolution of the changes in annual water-table depth (thick-line), with 2σ error (thin) lines, estimated from testate amoebae assemblages. Reproduced from Charman, D., D. Hendon & S. Packman, 1999. Multiproxy surface wetness records from replicate cores on an ombrotrophic mire: implications for Holocene palaeoclimate records. J. Quarter. Sci. 14: 451–463, with permission of J. Wiley & Sons, Ltd.

the limnic stage to peat bog was reconstructed by Beyens (1984b), by combining data from pollen, diatoms and testate amoebae (Fig. 11).

Other Protozoa

Remains of other freshwater Protozoa groups other than testate amoebae are less frequently encountered in freshwater sediments. Already Lagerheim (1901) discussed the presence of Heliozoa and Tintinnids in Scandinavian lake sediments. Plates of the Heliozoan species

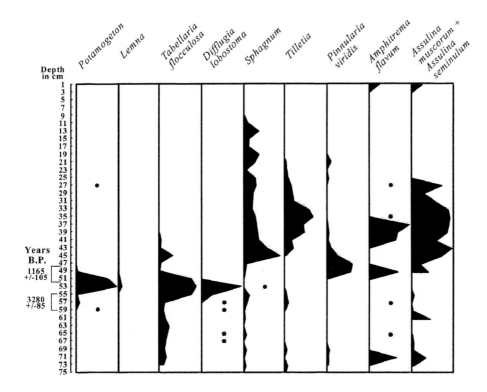

Figure 11. Diagram illustrating succession from an open water ditch to *Sphagnum* bog. The creation of waterbodies by digging activities is not a recent phenomenom. In medieval Europe, peat was extracted, which could result in
limnic conditions. The open water ditch, made by digging, was overgrown with *Potamogeton* sp., while *Difflugia
lobostoma* lived as epiphytes on the submerged stems. The diatom *Tabellaria flocculosa* flourished too in this
environment. On the watersurface *Lemna* was growing. Later *Sphagnum* mosses (mostly *Sph. cuspidatum*) began
to colonize the ditch, untill a new bog was formed. This was indicated by typical *Sphagnum* bog testate amoebae
such as *Amphitrema flavum*, and the moss dwellers *Assulina muscorum* and A. seminulum. From Beyens (1984b).

Clathrulina elegans were observed in Subboreal peat by Hesmer (1929), and Douglas
& Smol (1987) portray plates of a few species. Tintinnids are shell-building planktonic
ciliates, and some species as *Codonella cratera* (see Fig. 12) could possibly be mistaken
by the inexperienced observer for a *Difflugia*. The Tintinnina are mostly marine organisms,
with only a few living in freshwater ecosystems. These can be abundant in many lakes,
and are mentioned as (sub)fossils by Lagerheim (1901) and Frey (1964). The distribution
is probably cosmopolitan. Ecological data are given in Foissner et al. (1991). Their fossil
history goes back at least to the Ordovician. A review of these organisms in the perspective
of paleontological research is given in Pokorny (1975). Other protozoans with resistant
structures or stages like Suctoria with their permanent cysts are potentially useful candidates
for paleolimnological research, but are little studied.

Figure 12. Codonella cratera Scale bar = 20 μm. (modified from Foissner, W., 1994. Progress in Taxonomy of Planktonic Freshwater Ciliates. Marine Microbial Food Webs 8: 9–35). With permission of Elsevier.

Conclusions and future directions

In comparison to diatoms, relatively few paleolimnological studies are based on testate amoebae analysis. The use of testate amoebae as paleolimnological indicators is hampered by taxonomic problems, and the lack of a modern identification handbook to the species level. The ecology and biogeography of recent testate amoebae is insufficiently known, and the use of ecological data between different biogeographic zones is not without problems. The present concern on biodiversity should be an occasion for a greater exploration of the world of the free-living protozoa. Freshwater biotopes of the Southern Hemisphere such as the highly diverse East African or South American lakes, are poorly known and even North America is, in comparison with Europe, less completely studied (< 10% of the records). The number of testate amoebae taxa is relatively small, allowing one to obtain a working knowledge comparatively easily. In limnic environments they can yield information on chemical parameters (e.g., pH, nutrient status) and on changes in the catchment area such as paludification. They offer unique information in bog environments, especially about water table depth and surface moisture. The future should see the development of reliable transfer functions for the different water characteristics (e.g., water temperature and pH). The ratio of littoral-epiphytic taxa versus profundal ones could be developed into an index for habitat type within a lake environment. On a global scale, the relation between assemblages structures and major biogeographic zones and regional climate might provide some promising data to help interpret past environments.

Summary

Among the freshwater protozoa, only the testate amoebae are well represented in paleolimnological studies. They are a polyphyletic group, including lobose and filose amoebae. At the moment, more than 1900 species and subspecies are described but a thorough systematic revision is much needed. A short introduction to the systematic explains some of the problems. The architecture and composition of the shell are key elements in the taxonomy of all shelled amoebae, and this permits the identification of most (sub)fossils to the species level. Sampling procedures are similar to those used for diatoms or pollen, but preparation methods are different due to the vulnerability of the test to strong oxidizing or acid agents. The different methods to purify the sample from fine sediments, and to quantify the testate amoebae are discussed and explained. The use of testate amoebae in paleolimnological research is reviewed via case-studies. Testate amoebae are indicator organisms for limnological characteristics such as pH and trophic status. The proportion of allochthonous tests (soil or terrestrial moss testate amoebae) in the sediment reveal eventual changes in the catchment area caused by human action, or by natural processes such as fluctuating precipitation regimes. The role of climatic change in shaping testate amoebae assemblages is not yet entirely clear. Testate amoebae show detailed and informative zonation in the transition between marine and freshwater environments, yielding thus a suitable tool for assessing sea level changes. Finally, since lakes can evolve to bogs, the possibility of using testate amoebae in reconstructing surface wetness and water table is briefly discussed. This aspect is of special interest because the reconstruction of water table changes is methodologically the most advanced.

Acknowledgements

The manuscript has been improved by comments of Prof. Dr. W. De Smet and Dr. L. Denys and by the suggestions of the referees Prof. H. J. B. Birks, Prof. W. M. Last, Prof. R. T. Patterson and Prof. J. Smol. Mrs. D. Richers helped with the literature database and Lic. I. Van Dyck reworked some of the figures. Mr. M. Heuer kindly corrected the English text.

References

Asioli, A., F. Medioli & R. T. Patterson, 1996. Thecamoebians as a tool for reconstruction of paleoenvironments in some Italian lakes in the foothills of the southern Alps (Orta, Varese and Candia). J. Foramin. Res. 26: 248–263.

Beyens, L., 1984a. A concise survey of testate amoebae analysis. Bulletin de la Société belge de Géologie 93: 261–266.

Beyens, L., 1984b. Paleoecologische en paleoklimatologische aspecten van de holocene ontwikkeling van de Antwerpse Noorderkempen. (Palaeoecological and palaeoclimatological aspects of the Holocene evolution of the Antwerpse Noorderkempen). Academiae Analecta, Mededelingen-Wetenschappen 46: 15–56.

Beyens, L., 1985. On the subboreal climate of the Belgian Campine as deduced from diatom and testate amoebae analyses. Rev. Palaeobot. Palynol. 46: 9–31.

Beyens, L. & D. Chardez with collaboration of P. De Bock, 1986. Some new and rare testate amoebae from the Arctic. Acta Protozoologica 25: 81–89.

Beyens, L. & D. Chardez, 1987. Evidence from testate amoebae for changes in some local hydrological conditions between c. 5000 BP and c. 3800 BP on Edgeøya (Svalbard). Polar Res. 5 n.s.: 165–169.

Beyens, L. & D. Chardez, 1995. An annotated list of testate amoebae observed in the Arctic between the longitudes 27° E and 168° W. Arch. Protistenkd. 146: 219–233.

Beyens, L. & D. Chardez, 1997. New testate amoebae taxa from the polar regions. Acta Protozoologica 36: 137–142.

Beyens, L., D. Chardez & R. De Landtsheer with collaboration of D. De Baere, 1986. Testate amoebae communities from aquatic habitats in the Arctic. Polar Biol. 6: 197–205.

Beyens, L., D. Chardez & D. De Baere, 1991. Ecology of aquatic testate amoebae in coastal lowlands of Devon Island (Canadian High Arctic). Arch. Protistenkd. 140: 23–33.

Beyens, L., D. Chardez, D. De Baere & C. Verbruggen, 1995. The aquatic testate amoebae fauna of the Strømness Bay area, South Georgia. Antarctic Science 7: 3–8.

Bhattacharya, D., T. Helmchen & M. Melkonian, 1995. Molecular evolutionary analyses of nuclear-encoded small subunit ribosomal RNA identify an independent rhizopod Lineage containing the Euglyphina and the Chlorarachniophyta. J. Euk. Microbiol. 42: 65–69.

Birks, H. J. & H. H. Birks, 1980. Quaternary Palaeoecology. Edward Arnold, London, 289 pp.

Birks, H. J., 1986. Numerical zonation comparison and correlation of Quartenary pollen-stratigraphical data. In Berglund (ed.) Handbook of Holocene Palaeoecology and Palaeohydrology. J. Wiley & Sons, Chichester: 743–774.

Bobrov, A., D. Charman & B. Warner, 1999. Ecology of testate amoebae (Protozoa: Rhizopoda) on peatlands in Western Russia with special attention to niche separation in closely related taxa. Protist. 150: 125–136.

Bonnet, L., 1961. Caractéres Génèraux des populations thécamoebienns endogées. Pedobiologia 1: 6–24.

Bonnet, L., 1964. Le peuplement thécamoebien des sols. Rev. Ecol. Biol. sol 1: 123–408.

Bonnet, L., 1975. Types morphologiques, écologie et évolution de la théque chez les thécamoebiens. Protistologica 11: 363–378.

Bonnet, L. & R. Thomas, 1960. Faune terrestre et d'eau douce des Pyrénées-Orientales. Thécamoebiens du sol. Vie et Milieu 11 Suppl.: 1–103.

Bovee, E. C., 1985. Class lobosea carpenter, 1861. In Lee, J. J. et al. (eds.) An Illustrated Guide to the Protozoa: 158–211. Society of Protozoologists, Lawrence, Kansas, 629 pp.

Buttler, A., B. Warner, P. Grosvernier & Y. Matthey, 1996. Vertical patterns of testate amoebae (Protozoa: Rhizopoda) and peat-forming vegetation on cutover bogs in the Jura, Switzerland. New Phytol. 134: 371–382.

Cash, J. & J. Hopkinson, 1905. The British Freshwater Rhizopoda and Heliozoa I. Ray Soc. London, 148 pp.

Cash, J. & J. Hopkinson, 1909. The British Freshwater Rhizopoda and Heliozoa II. Ray Soc. London, 166 pp.

Cash, J., G. H. Wailes & J. Hopkinson, 1915. The British Freshwater Rhizopoda and Heliozoa III. Ray Soc. London, 156 pp.

Casper, S. J. & W. Schönborn, 1985. *Difflugia limnetica* (Levander) Penard (Protozoa: Testacea) as indicator organism of calcite precipitation in Lake Stechlin, GDR. Arch. Protistenkd. 130: 305–211.

Cavalier-Smith, T., 1997. Amoeboflagellates and mitochondrial cristae in eukaryote evolution: Megasystematics of the new protozoan subkingdoms Eozoa and Neozoa. Arch. Protistenkd. 147: 237–258.

Cavalier-Smith, T. & E. E. Chao, 1997. Sarcomonad ribosomal RNA sequences, rhizopod phylogeny, and the origin of euglyphid amoebae. Arch. Protistenkd. 147: 227–236.

Chardez, D., 1964. Thécamoebiens (Rhizopodes Testacés). In Symoens, J. J. (ed.) Exploration Hydrobiol du Bassin du Lac Bangweolo et du Luapula 10, 2: 1–77.

Chardez, D. & L. Beyens, 1988. *Centropyxis gasparella* sp.nov. and *Parmulina louisi* sp.nov., new testate amoebae from the Canadian High (Devon Island, N.W.T.). Arch. Protistenkd. 136: 337–344.

Chardez, D. & R. Thomas, 1980. Thecamoebiens du Mesopsammon des plages de Lacanau et Leporge-Ocean (Gironde, France). Acta Protozoologica 19: 277–285.

Charman, D. & B. Warner, 1992. Relationship between testate amoebae and micro-environmental parameters on a forested peatland in northeastern Ontario. Can. J. Zool. 70: 2474–2482.

Charman, D. & B. Warner, 1997. The ecology of testate amoebae (Protozoa: Rhizopoda) in oceanic peatlands in Newfoundland, Canada: Modelling hydrological relationships for palaeoenvironmental reconstruction. Ecoscience 4: 555–562.

Charman, D., H. Roe & W. R. Gehrels, 1998. The use of testate amoebae in studies of sea-level change: a case study from the Taf Estuary, south Wales, UK. The Holocene 8: 209–218.

Charman, D., D. Hendon & S. Packman, 1999. Multiproxy surface wetness records from replicate cores on an ombrotrophic mire: implications for Holocene palaeoclimate records. J. Quarter. Sci. 14: 451–463.

Collins, E., F. McCarthy, F. Medioli, D. Scott & C. Honig, 1990. Biogeographic distribution of modern thecamoebians in a transect along the Eastern North American Coast. In Hemleben et al. (eds.) Paleoecology, Biostratigraphy, Paleoceanography and Taxonomy of Agglutinated Foraminifera. Kluwer Academic Publishers: 783–792.

Corbet, S., 1973. An illustrated introduction to the testate rhizopods in Sphagnum with special reference to the area around Malham tarn, Yorkshire. Field Studies 3: 801–838.

Costan, G. & D. Planas, 1986. Effects of a short-term experimental acidification on a microinvertebrate community. Rhizopoda, Testacea. Can. J. Zool. 64: 1224–1230.

Coûteaux, M., 1992. Decomposition of cells and empty shells of testate amoebae in an organic acid soil sterilized by propylene oxide fumigation, autoclaving and y-ray irradiation. Biol. Fertil. Soils 12: 290–294.

Decloitre, L., 1962. Le genre *Euglypha* DUJARDIN. Arch. Protistenkd. 106: 51–100.

Decloitre, L., 1976a. Le genre *Arcella* EHRENBERG. Compléments à jour au 31. Décembre 1974 de la monographie du genre parue en 1928. Arch. Protistenkd. 118: 291–309.

Decloitre, L., 1976b. Le genre *Euglypha*. Compléments à jour au 31. Décembre 1974 de la monographie du genre parue en 1962. Arch. Protistenkd. 118: 18–33.

Decloitre, L., 1977a. Le genre *Cyclopyxis*. Compléments à jour au 31. Décembre 1974 de la monographie du genre parue en 1929. Arch. Protistenkd. 119: 31–53.

Decloitre, L., 1977b. Le genre *Nebela*. Compléments à jour au 31. Décembre 1974 du genre parue en 1936. Arch. Protistenkd. 119: 325–352.

Decloitre, L., 1978. Le genre *Centropyxis* I. Compléments à jour au 31.12.1974 de la monographie du genre parue en 1929. Arch. Protistenkd. 120: 63–85.

Decloitre, L., 1979a. Le genre *Centropyxis* II. Compléments à jour au 31. Décembre 1974 de la monographie du genre parue en 1929. Arch. Protistenkd. 121: 162–192.

Decloitre, L., 1979b. Mises à jour au 31.12.1978 des mises à jour au 31.12.1974 concernant les genres *Arcella; Centropyxis; Cyclopyxis; Euglypha* et *Nebela*. Arch. Protistenkd. 122: 387–397.

Decloitre, L., 1981. Le Genre *Trinema* DUJARDIN, 1941. Rèvision à jour au 31.XII.1979. Arch. Protistenkd. 124: 193–218.

Decloitre, L., 1982. Compléments aux publications précédentes. Mise à jour au 31.XII. 1981 des genres *Arcella, Centropyxis, Cyclopyxis, Euglypha, Nebela* et *Trinema*. Arch. Protistenkd. 126: 393–407.

Decloitre, L., 1986. Compléments aux publications précédentes. Mise à jour au 31. XII. 1984 des genres *Arcella, Centropyxis, Cyclopyxis, Euglypha* et *Nebela*. Arch. Protistenkd. 132: 131–136.

Decloitre, L., 1986. Statistique mondial des Thécamoebiens. Ann. Soc. Sci. Nat. Archeol. Toulon et Var 37.

Denys, L. & H. De Wolf, 1999. Diatoms as indicators of coastal paleoenvironments and relative sea-level change. In Stoermer, E. & J. Smol (eds.) The Diatoms. Applications for the Environment and Earth Sciences. Cambridge Univ. Press, Cambridge: 277–297.

Douglas, M. & J. Smol, 1987. Siliceous protozoan plates in lake sediments. Hydrobiologia 154: 13–23.

Ellison, R., 1995. Paleolimnological analysis of Ullswater using testate amoebae. J. Paleolimnol. 13: 51–63.

Ellison, R. L. & C. Ogden, 1987. A guide to the study of identification of fossil testate amoebae in Quaternary lake sediments. Int. Revue ges. Hydrobiol. 72: 639–652.

Foissner, W., 1994. Progress in taxonomy of planktonic freshwater ciliates. Marine Microbial Food Webs 8: 9–35.

Foissner, W., H. Blatterer, H. Berger & F. Kohmann, 1991. Taxonomische und ökologische Revision der Ciliaten des Saprobiensystems. Band I: Cyrtophorida, Oligotrichida, Hypotrichida, Colpodea. Informationsberichte des Bayer. Landesamtes für Wasserwirtschaft Heft 1: 1–471.

Frey, D., 1964. Remains of animals in Quaternary lake and bog sediments and their intepretation. In Elster H. & W. Ohle (eds.) Ergebnisse der Limnologie. Arch. Hydrobiol. Beiheft 2, 114 pp.

Golemansky, V., 1967. Matériaux sur la systématique et l'écologie des Thécamoebiens (Protozoa, Rhizopoda) du lac d'Ohrid. Sect. Sci. Nat. Univ. Skopje 14: 3–26.

Golemansky, V., 1971. Taxonomische und zoogeographische Notizen über die thekamöbe Fauna (Rhizopoda, Testacea) der Küstengrundgewässer der sowjetischen Fernostküste (Japanisches Meer) und der Westküste Kanadas (Stiller Ozean). Arch. Protistenkd. 113: 235–249.

Golemansky, V., 1981. Description de trois thécamoebiens (Protozoa, Rhizopoda) nouveaux des eaux souterreines littorales des mers. Acta Protozoologica 20: 115–119.

Golemansky, V., 1990. Interstitial testate amoebas (Rhizopoda: Testacea) from the mediterranean basin. Stygologia 5: 49–54.

Golemansky, V., 1991. Thécamoebiens mésopsammiques (Rhizopoda: Arcellinida, Gromida & Monothalamida) du sublittoral marin de l'Atlantique dans la région de Roscoff (France). Arch. Protistenkd. 140: 35–43.

Golemansky, V., 1994. On some ecological preferences of marine interstitial testate amoebas. Arch. Protistenkd. 144: 424–432.

Griffin, J. L., 1972. Movement, fine structure and fusion of pseudopods of an enclosed amoeba, *Difflugiella* sp. J. Cell Sci. 10: 563–583.

Grospietsch, Th., 1952. Die Rhizopodenanalyse als Hilfsmittel der Moorforschung. Die Naturwiss. 39: 318–323.

Grospietsch, Th., 1957. Beitrag zur Rhizopodenfauna des Lago Maggiore. Arch. Hydrobiol. 53: 323–331.

Grospietsch, Th., 1972a. Wechseltierchen. Kosmos Gesellschaft der Naturfreunde, Franckh'sche Verlagshandlung, Stuttgart, 87 pp.

Grospietsch, Th., 1972b. Testacea und Heliozoa. In Elster, H.-J. & W. Ohle (eds.) Das Zooplankton der Binnengewässer. I. Schweizerbarth, Stuttgart: 1–30.

Grospietsch, Th., 1972c. Neue Ergebnisse der Rhizopodenanalyse für die Moorforschung. Verh. Int. Verein. Limnol. 18: 1031–1038.

Grospietsch, Th., 1975. Beitrag zur Kenntnis der Testaceen-Fauna des Lago Valencia (Venezuela). Verh. Internat. Verein. Limnol. 19: 2778–2784.

Haman, D. & B. Kohl, 1994. A Thecamoebinid Assemblage from Lake Cocococha, Tambopata Reserve, Madre de Dios Province, Southeastern Peru. J. Foramin. Res. 24: 226–232.

Harnisch, O., 1927. Einige Daten zur rezenten und fossilen testaceen Rhizopodenfauna der Sphagnen. Arch. Hydrobiologie XVIII: 345–360.

Harnisch, O., 1958. Rhizopoda. In Brohmer, P., P. Ehrmann & G. Ulmer (eds.) Tierwelt Mitteleuropas. 1, 1b: 1–75.

Hasle, G. R., 1978. The inverted-microscope method. In Sournia, A. (ed.) Phytoplankton Manual, 88–96. Monographs on Oceanographic Methodology 6, UNESCO, Paris, France. 337 p.

Hedley, R. H., C. G. Ogden & N. J. Mordan, 1977. Biology and fine structure of *Cryptodifflugia oviformis* (Rhizopodea: Protozoa). Bull. Br. Mus. Nat. Hist. Zool. 30: 313–328.

Hendon, D. & D. Charman, 1997. The preparation of testate amoebae (Protozoa: Rhizopoda) samples from peat. The Holocene 7: 199–205.

Hesmer, H., 1929. Mikrofossilien in Torfen. Paläontologische Ztschr. 11: 245–257.

Lagerheim, G., 1901. Om lämningar af Rhizopoden, Heliozoer och Tintinnider i Sveriges och Finlands lakustrina Kvartäraflagringar. Geol. Fören. Stockh. Förhandl. 23: 469–520.

Laminger, H., 1972. Die profundale Testaceenfauna (Protozoa, Rhizopoda) älterer und jüngerer Bodensee-Sedimenten. Arch. Hydrobiol. 70: 108–129.

Laminger, H., 1973a. Quantitative Untersuchungen über die Testaceenfauna (Protozoa, Rhizopoda) in den jüngsten Bodensee-Sedimenten. Biol. Jb. Dodonaea 41: 126–146.

Laminger, H., 1973b. Untersuchungen über Abundanz und Biomasse der sedimentbewohnenden Testaceen (Protozoa, Rhizopoda) in einem Hochgebirgssee (Vorderer Finstertaler See, Kühtai, Tirol). Int. Rev. ges. Hydrobiol. 58: 543–568.

Laminger, H., R. Zisette, S. Phillips & F. Breidigam, 1979. Contribution to the Knowledge of the Protozoan Fauna of Montana (USA): I. Testate Amoebae (Rhizopods) in the Surrounding of Flathead Lake Valley. Hydrobiologia 65: 257–271.

Lee, J. J., S. H. Hutner & E. C. Bovee, (eds.), 1985. An illustrated Guide to the Protozoa. Lawrence, Kansas, 629 pp.

Leidy, J., 1879. Fresh-water rhizopods of North America. Government Printing Office, Washington, 324 pp.

Lena, H., 1982. Benthic Testacida (Rhizopoda, Protozoa) of Lake Washington, Brevard County, Florida. Biol. Sci. 45: 101–106.

Loeblich, A. R., JR. & H. Tappan, 1964. Sarcodina Chiefly "Thecamoebians" and foraminiferida. In Moore, R. C. (ed.) Treatise on Invertebrate Paleontology. Geol. Soc. Am. and Univ. Kansas Press, Lawrence, Kansas: Part C, Protista 2: 1–54.

Lopez-Saez, J., B. van Geel, S. Farbos-Texier & M. Diot, 1998. Remarques paléoécologiques à propos de quelques palynomorphes non-polliniques provenant de sédiments quaternaires en France. Revue Paléobiol. 17: 445–459.

Lousier, J. & D. Parkinson, 1981. The disappearance of the empty tests of litter- and soil testate amoebae (Testacea, Rhizopoda, Protozoa). Arch. Protistenkd. 124: 312–336.

Lüftenegger, G., W. Petz, H. Berger, W. Foissner & H. Adam, 1988. Morphologic and Biometric Characterization of Twenty-four Soil Testate Amoebae (Protozoa, Rhizopoda). Arch. Protistenkd. 136: 153–189.

McCarthy, F., E. Collins, J. McAndrews, H. Kerr, D. Scott & F. Medioli, 1995. A comparison of postglacial arcellacean ("Thecamoebian") and pollen succession in atlantic Canada, illustrating the potential of arcellaceans for paleoclimatic reconstruction. J. Paleontol. 69: 980–993.

Medioli, F., D. Scott & B. H. Abbott, 1987. A case study of protozoan intraclonal variability taxonomic implications. J. Foramin. Res. 17, 1: 28–47.

Medioli, F. & D. Scott, 1988. Lacustrine Thecamoebians (mainly Arcellaceans) as potential tools for paleolimnological interpretations. Palaeogeogr., Palaeoclim., Palaeoecol. 62: 361–386.

Medioli, F., D. Scott, E. Collins & J. Wall, 1990. Thecamoebians from the early Cretaceous deposits of Ruby Creek, Alberta (Canada). In Hemleben et al. (eds.) Paleoecology. Biostratigraphy. Paleoceanography and Taxonomy of Agglutinated Foraminifera: 793–812.

Medioli, F., A. Asioli & G. Parenti, 1994. Manuale per l'identificazione e la classificazione delle tecamebe con informazioni sul loro significato paleoecologico e stratigrafico. Palaeopelagos 4: 317–364.

Medioli, F., D. Scott, E. Collins, S. Asioli & E. Reinhardt, 1999. The thecamoebian bibliography. Palaeontologia Electron. 3, 1:161 pp.
http://www-odp.tamu.edu/paleo/1999_1/biblio/issue1_99.htm.

Meisterfeld, R., 1977. Die horizontale und vertikale Verteilung der Testaceen (Rhizopoda, Testacea) in Sphagnum. Arch. Hydrobiol. 79: 319–356.

Meisterfeld, R., 1979. Clusteranalytische Differenzierung der Testaceenzönosen (Rhizopoda, Testacea) in Sphagnum. Arch. Protistenkd. 121: 270–307.

Meisterfeld, R., 1984. Taxonomic problems in 'Difflugia' species with lobed aperture–biometry and impact of supplied building material. J. Protozool. 31 (Suppl.) 62A: 226.

Meisterfeld, R., 2001a. Order Arcellinida. In Lee, J. J. et al. (eds.) The Illustrated Guide to the Protozoa. Second Edition. Allen Press, Lawrence, 1400 pp.

Meisterfeld, R., 2001b. Testate amoebae with filopodia. In Lee, J. J. et al. (eds.) The Illustrated Guide to the Protozoa. Second Edition. Allen Press, Lawrence, 1400 pp.

Meisterfeld, R. & M. Heisterbaum, 1986. The decay of empty tests of testate amoebae (Rhizopoda, Protozoa). Symposia Biologica Hungarica 33: 285–390.

Meisterfeld, R. & L. Tan, 1998. First Records of Testate Amoebae (Protozoa, Rhizopoda) from Mount Buffalo National Park, Victoria: Preliminary Notes. Vic. Naturalist 115: 231–238.

Mignot, J. P. & I. B. Raikov, 1992. Evidence of meiosis in the Testate Amoeba Arcella. J. Protozool. 39, 2: 287–289.

Mori, S. & T. Miura, 1980. List of plant and animal species living in Lake Biwa. Mem. Fac. Sci. Kyoto Univ., Ser. Biol. 8: 1–33.

Müller, H., 1970. Ökologische Veränderungen im Otterstedter See im Laufe der Nacheiszeit. Ber. Naturhist. Ges. 114: 33–47.

Ogden, C. G., 1984. Shell structure of some testate Amoebae from Britain (Protozoa Rhizopoda). J. Nat. Hist. 18: 341–361.

Ogden, C. G., 1990. The structure of the shell wall in testate amoebae and the importance of the organic cement matrix. In Claugher, D. (ed.) Scanning Electron Microscopy in Taxonomy and Functional Morphology. Syst. Ass. Special Vol., Clarendon Press, Oxford: 41: 235–257.

Ogden, C. G. & R. L. Ellison, 1988. The value of the organic cement matrix in the identification of the shells of fossil testate amoebae. J. Micropalaeontol. 7: 233–240.

Ogden, C. G. & R. H. Hedley, 1980. An Atlas of Freshwater Testate Amoebae. Oxford University Press, Oxford, 222 pp.

Ogden, C. G. & R. Meisterfeld, 1989. The taxonomy and systematics of some species of Cucurbitella, Difflugia and Netzelia (Protozoa: Rhizopoda); with an evaluation of diagnostic characters. Europ. J. Protistol. 25: 109–128.

Ogden, C. G. & R. Meisterfeld, 1991. The biology and ultrastructure of the testate amoeba, Difflugia lucida Penard (Protozoa, Rhizopoda). Europ. J. Protistol. 26: 256–269.

Ogden, C. G. & P. Pitta, 1990. Biology and ultrastructure of the mycophagous, soil testate amoeba, *Phryganella acropodia* (Rhizopoda, Protozoa). Biol. Fertil. Soils 9: 101–109.

Page, F. C., 1987. The classification of "naked" amoebae (Phylum, Rhizopoda). Arch. Protistenkd. 133: 199–217.

Patterson, R., K. MacKinnon, D. Scott & F. Medioli, 1985. Arcellaceans in small lakes of New Brunswick and Nova Scotia: modern distribution and holocene stratigraphic changes. J. Foramin. Res. 15: 114–137.

Penard, E., 1902. Faune rhizopodique du bassin de Léman. Kündig, Genf, 714 pp.

Poinar, G. O., B. M. Waggoner & U. Bauer, 1993. Terrestrial Soft-Bodied Protists and Other Microorganisms in Triassic Amber. Science 259: 222–224.

Pokorny, V., 1975. Grundzüge der Zoologischen Mikropaläontologie. Band 1. O. Koeltz Science Publ., Königstein, 582 pp.

Ruzicka, E., 1982. Die subfossilen Testaceen des Krottensees (Salzburg, Österreich). Limnologica 14: 49–88.

Schönborn, W., 1962. Die Ökologie der Testaceen im oligotrophen See, dargestellt am Beispiel des Grossen Stechlinsees. Limnologica 1: 111–182.

Schönborn, W., 1963. Die Stratigraphie lebender Testaceen im Sphagnetum der Hochmoore. Limnologica 1: 315–321.

Schönborn, W., 1966a. Beschalte Amöben (Testacea). Die neue Brehm-Bücherei, Ziemsen Verlag, Wittenberg Lutherstadt, 112 pp.

Schönborn, W., 1966b. Beitrag zur Ökologie und Systematik der Testaceen Spitzbergens. Limnologica 4: 463–470.

Schönborn, W., 1967. Taxozönotik der beschalten Süßwasser-Rhizopoden. Eine raumstruktur-analytische Untersuchung über Lebensraumerweiterung und Evolution bei der Mikrofauna. Limnologica 5: 159–207.

Schönborn, W., 1973. Paläolimnologische Studien an Testaceen aus Bohrkernen des Latnjajaure (Abisko-Gebiet; Schwedisch-Lappland). Hydrobiologia 42: 63–75.

Schönborn, W., 1975. Studien über die Testaceenbesiedlung der Seen und Tümpel des Abiskogebietes (Schwedisch-Lappland). Hydrobiologia 46: 115–139.

Schönborn, W., 1981. Population dynamics and production of Testacea in the River Saale. Zool. Jb. Syst. 108: 301–313.

Schönborn, W., 1984. Studies on Remains of Testacea in Cores of the Great Woryty Lake (NE-Poland). Limnologica 16: 185–190.

Schönborn, W., 1989. The topophenetic analysis as a method to elucidate the phylogeny of testate amoebae (Protozoa, Testacealobosia and Testaceafilosia). Arch. Protistenkd. 127: 223–245.

Schönborn, W., 1990. Analyse subfossiler Protozoenschalen der Sedimente eines kleinen sauren Waldsees (Kleiner Barsch-See, Nördliche DDR). Limnologica 21: 137–145.

Schönborn, W., 1992. Comparative studies on the production biology of protozoan communities in freshwater and soil ecosystems. Arch. Protistenkd. 141: 187–214.

Schönborn, W., D. Flössner & G. Proft, 1965. Die limnologische Charakterisierung des Profundals einiger norddeutscher Seen mit Hilfe von Testaceen-Gemeinschaften. Limnologica 3: 371–380.

Schönborn, W., W. Foissner & R. Meisterfeld, 1983. Licht-und Rasterelektronenmikroskopische Untersuchungen zur Schalenmorphologie und Rassenbildung bodenbewohnender Testaceen (Protozoa, Rhizopoda) sowie Vorschläge zur biometrischen Charakterisierung von Testaceen-Schalen. Protistologica 19: 553–566.

Schönborn, W. & T. Peschke, 1988. Biometric studies on species, races, ecophenotypes and individual variations of soil-inhabiting testacea (Protozoa, Rhizopoda), including *Trigonopyxis minuta* n.sp. and *Corythion asperulum* n.sp. Arch. Protistenkd. 136: 345–363.

Schönborn, W. & T. Peschke, 1990. Evolutionary studies on the Assulina-Valkanovia complex (Rhizopoda, Testaceafilosia) in Sphagnum and soil. Biol. Fertil. Soils 9, 2: 95–100.

Scott, D. & F. Medioli, 1978. Vertical zonation of marsh foraminifera as accurate indicators of former sea-levels. Nature 272: 538–541.

Scott, D. & F. Medioli, 1983. Testate rhizopods in Lake Erie: modern distribution and stratigraphic implications. J. Paleontol. 57: 809–820.

Shen, Y., 1980. Ecological studies on the periphytic protozoa in Lake Dong Hu, Wuhan. Acta Hydrobiologica Sinica 7: 19–40.

Smith, H. G., 1992. Distribution and ecology of the testate rhizopod fauna of the continental antarctic zone. Polar Biol. 12: 629–634.

Smith, H. G. & D. Wilkinson, 1986. Biogeography of testate rhizopods in the southern temperate and antarctic zones. Colloque sur les écosystèmes terrestres subantarctiques, Paimpont, C.N.F.R.A. 58: 83–96.

Stepanek, M., 1963. Die Rhizopoden aus Katanga (Kongo-Afrika). Annls. Mus. R. Afr. Cent.(Zool.). Serie IN 8, 117: 8–91 with 20 plates.

Stepanek, M., 1967. Testacea des Benthos der Talsperre Vranov am Thayafluss. Hydrobiologia 29: 1–67.

Stepanek, M., 1968. Die Rhizopoden des Tiefenschlammes im Bodensee. Archiv Hydrobiologie suppl. 33: 442–450.

Stockmarr, J., 1971. Tablets with spores used in absolute pollen analysis. Pollen et Spores 13: 615–621.

Todorov, M. & V. Golemansky, 1998. Testate amoebae (Protozoa, Rhizopoda) of the coastal lakes Shabla and Ezerets (Northeastern Bulgaria), with a description of Pentagonia shablensis sp. nov. In Biodiversity of Shabla Lake System. Sofia, "Prof. Marin, Drinov" Acad. Publ. House: 69–90.

Tolonen, K., 1986. Rhizopod analysis. In Berglund (ed.) Handbook of Holocene Palaeoecology and Palaeohydrology. J. Wiley & Sons, Chichester: 645–666.

Tolonen, K., B. Warner & H. Vasander, 1992. Ecology of Testaceans in mires in Southern Finland: I. Autecology. Arch. Protistenkd. 142: 119–138.

Tolonen, K., B. Warner & H. Vasander, 1994. Ecology of Testaceans in mires in southern Finland: II. Multivariate analysis. Arch. Protistenkd. 144: 97–112.

Trappeniers, K., A. Van Kerckvoorde, D. Chardez, I. Nijs & L. Beyens, 1999. Ecology of testate amoebae communities from aquatic habitats in the Zackenberg area (Northeast Greenland). Polar Biology 22: 271–278.

Utermöhl, H., 1958. Zur Vervollkommnung der quantitativen Phytoplankton-Methodik. Intern. Ver. theoret. angew. Limnol., Mitt. 9: 1–39.

Van Geel, B., 1976. A paleoecological study of Holocene peat bog sections, based on the analysis of Pollen, Spores and Macro-and Microscopic Remains of Fungi, Alga, Cormophytes and Animals. Ph.D. thesis, Universiteit Amsterdam, 75 pp.

Van Geel, B., S. Bohncke & H. Dee, 1981. A palaeoecological study of an upper Late Glacial and Holocene sequence from "De Borchert", The Netherlands. Rev. Palaebot. Palynol. 31: 367–448.

Vucetich, M. C., 1976. Tecamebianos del Lago San Roque y de un Ambiente Lentico Artificial Vinculado al Mismo (Cordoba, Argentina). Limnobios 1: 29–34.

Wailes, G. H. & J. Hopkinson, 1919. The British Freshwater Rhizopoda and Heliozoa IV. Ray Soc. London, 130 pp.

Wanner, M., 1988. Biometrische und rasterelektronenmikroskopische Untersuchungen an Testaceen-Schalen (Protozoa: Rhizopoda). Arch. Protistenkd. 136: 97–106.

Wanner, M., S. Esser & R. Meisterfeld, 1994. Effects of light, temperature, fertilizers and pesticides on growth of the common freshwater and soil species Cyclopyxis kahli (Rhizopoda, Testacealobosia) interactions and adaptations. Limnologica 24: 239–250.

Wanner, M. & R. Meisterfeld, 1994. Effects of some environmental factors on the Shell Morphology of Testate Amoebae (Rhizopoda, Protozoa). Europ. J. Protistol. 30: 191–195.

Warner, B., 1988. Methods in quaternary ecology # 5. Testate amoebae (Protozoa). Geoscience Canada 15: 251–260.

Warner, B. & D. Charman, 1994. Holocene changes on a peatland in northwestern Ontario interpreted from testate amoebae analysis. Boreas 23: 270–279.

Warner, B. & J. Chmielewski, 1992. Testate amoebae as indicators of drainage in a forested mire, Northern Ontario, Canada. Arch. Protistenkd. 141: 179–183.

Wightman, W. G., D. B. Scott, F. S. Medioli & M. R. Gibling, 1994. Agglutinated foraminifera and thecamoebians from the late carboniferous Sydney coalfiled, Nova Scotia: paleoecology, paleoenvironments and paleogeographical implications. Palaeogeography, Palaeoclimatology, Palaeoecology 106: 187–202.

Woodland, W., D. Charman & P. Sims, 1998. Quantitative estimates of water tables and soil moisture in holocene peatlands from testate amoebae. The Holocene 8: 261–273.

Wylezich, C., R. Meisterfeld & M. Schlegel, 1999. Phylogeny of testate amoebae——a comparison of morphological and molecular characters. Zoology 102, Supplement II: 12.

8. DIATOMS

RICHARD W. BATTARBEE (r.battarbee@ucl.ac.uk)
VIVIENNE J. JONES (v.jones@geog.ucl.ac.uk)
ROGER J. FLOWER (r.flower@geog.ucl.ac.uk)
NIGEL G. CAMERON (n.cameron@ucl.ac.uk)
HELEN BENNION (h.bennion@ucl.ac.uk)
Environmental Change Research Centre
University College London
26 Bedford Way, London
WC1H 0AP, UK

LAURENCE CARVALHO (l.carvalho@ceh.ac.uk)
CEH Edinburgh
Bush Estate
Penicuik
Midlothian EH26 0QB
UK

STEPHEN JUGGINS (stephen.juggins@ncl.ac.uk)
Department of Geography
University of Newcastle
Newcastle upon Tyne
NE1 7RU, UK

Keywords: Diatoms, lakes, lake sediments, environmental reconstruction, surface water acidification, eutrophication, climate change, transfer functions

Introduction

Diatoms are classified as algae, Division Bacillariophyta. They are unicellular, eukaryotic organisms characterised by their siliceous cell walls and their yellow-brown pigmentation. Each diatom cell consists of two more or less identical thecae, one slightly larger than the other (Fig. 1). The valve face of each theca is intricately patterned allowing even most fossil taxa to be identified at the specific level. Cells are mainly solitary, but some taxa form colonies. For microscopic examination, cells are usually cleaned to remove their organic contents and allow details of the siliceous component of the cell wall (often called the frustule) to be revealed.

J. P. Smol, H. J. B. Birks & W. M. Last (eds.), 2001. *Tracking Environmental Change Using Lake Sediments.*
Volume 3: Terrestrial, Algal, and Siliceous Indicators. Kluwer Academic Publishers, Dordrecht, The Netherlands.

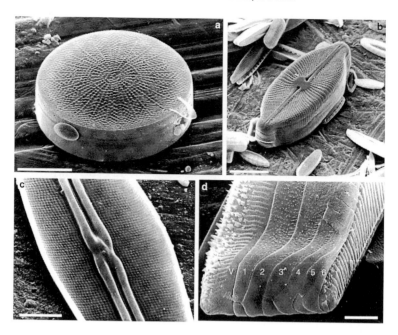

Figure 1. Shape and form in diatoms: (a) *Actinocyclus* Ehrenb., whole frustule of a large centric diatom with circular valves and a shallow girdle. Scale bar 20 μm; (b) *Cosmioneis* Mann & Stickle, whole frustule of an elongate pennate diatom with a pair of raphe fissures along the apical axis of each valve, separated by a central nodule. Scale bar 10 μm; (c) *Frustulia* Rabenhorst, inside valve view showing raphe ends, central nodule and internal occlusion of areolae. Scale bar 4 μm; (d) *Eunotia* Ehrenb. Girdle view showing the end of the frustule and details of the valve (V) end and six girdle bands. Scale bar 5 μm (modified from Mann, 1994).

Although they may have an older origin (Round, 1981a, 1981b; Round & Crawford, 1981; Mann & Marchant, 1989; Medlin et al., 1993), the first definitive evidence for diatoms in the geological record is from the early Jurassic (Rothpletz, 1896, 1900), and the oldest, well-preserved, fully-silicified flora is from the early Cretaceous (Harwood & Gersonde, 1990). The earliest known freshwater diatoms do not occur until the Eocene (e.g., Lohman & Andrews, 1968), but by the Miocene both marine and freshwater floras are diverse and many taxa have forms very similar to modern living species.

The first diatom genus was described in 1791 as *Bacillaria* Gmelin, with *Vibrio (Bacillaria) paxallifer* used as the type, a diatom first identified by O. F. Müller in 1783 and placed in *Vibrio*, and regarded by Müller as animals (see Round et al., 1990). Diatom classification progressed rapidly in the nineteenth century as microscopes and microsope lenses improved, and as collections extended to all parts of the world during the period of European colonialism and global exploration. By the late nineteenth century and early twentieth century attention was being given to aspects of diatom ecology (e.g., Cleve, 1894–95; Lauterborn, 1896), to their value as indicators of water pollution (Kolkwitz & Marsson, 1908) and to their potential as indicators of environmental change from sediment records (e.g., Cleve-Euler, 1922; Nipkow, 1920). In the following decades diatom analysis of lake sediments became increasingly common addressing in particular questions of lake ontogeny (e.g., Pennington, 1943; Nygaard, 1956; Round, 1957; Haworth,

1969; Digerfeldt, 1972) and shoreline displacement (e.g., Halden, 1929; Florin, 1944, 1946; Miller, 1964; Alhonen, 1971; Eronen, 1974; Renberg, 1976). However it was only towards the end of the twentieth century that public concern for water quality stimulated the use of diatom analysis as a primary tool for reconstructing surface water acidification (e.g., Renberg & Hellerberg, 1982; Flower & Battarbee, 1983; Charles, 1985; Birks et al., 1990a), eutrophication (Bradbury, 1975; Battarbee, 1978a; Brugam, 1979; Whitmore, 1989; Anderson & Rippey 1994; Bennion et al., 1996; Hall & Smol, 1992; Jones & Juggins, 1995; Wunsam & Schmidt, 1995) and climate change (Gasse, 1987; Fritz et al., 1991; Cumming et al., 1995; Laird et al., 1996, Smol & Cumming 2000). These and other uses of diatom analysis have recently been reviewed in detail by Stoermer & Smol (1999).

Developments in diatom analysis have also been promoted by improvements in sediment coring (e.g., Mackereth, 1969; Wright, 1980; Renberg, 1981; Glew, 1991) and dating (Pennington et al., 1973; Appleby et al., 1986; Appleby & Oldfield, 1978) and in the availability of powerful numerical techniques (ter Braak, 1986; Birks, 1995, 1998) that together enable robust quantitative reconstruction of environmental change to be made (see Battarbee, 1991, and Stoermer & Smol, 1999, for further details).

Today diatom analysis is a widely used technique. This chapter reviews the methods used in analysing and interpreting the diatom record of lake sediments. It updates an earlier paper (Battarbee, 1986a) and draws on recent major contributions to the diatom literature, principally on the biology of diatoms (Round et al., 1990) and on the use of diatom analysis for environmental reconstruction (Stoermer & Smol, 1999).

Biology

A full description of the biology of diatoms is beyond the scope of this chapter. The most accessible accounts are given in Round et al. (1990) and van den Hoek et al. (1995). However, a brief review is presented here as there are some key aspects of diatom biology that are essential for diatomists working with sediment records, especially in the context of understanding the range of morphological variation exhibited by fossil assemblages.

The diatom frustule

Much of the uniqueness of diatoms is related to silica. The cell contents are similar to other eukaryotic algae including the nucleus, choloroplasts, mitochondria, etc. But silica leads to the rigidity of the cell walls, constrains aspects of reproduction, and leads to the preservation of diatom frustules as fossils. Moreover the detailed construction of the silica wall provides the characters most commonly used in taxonomy—in both living and fossil material.

In all diatoms the cell wall is made up of several siliceous elements as illustrated in Figure 2. The thecae have two main components, the valve and the cingulum. The valves carry most of the taxonomic features used in standard floras and it is desirable in microscope preparations for the valve view to be clear to the observer. The cingulum comprises a single or sometimes multiple series of bands that are formed during the process of cell division, allowing the internal formation of daughter cells (Fig. 2). These bands together form the

Valve

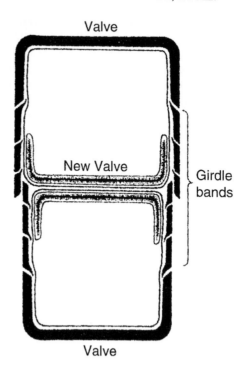

Valve

Figure 2. Diagrammatic cross-section through a recently divided diatom cell, showing parent valves, back to back position of the daughter valves and the overlapping sets of girdle bands associated with the epi- and hypovalves of the parent cell (redrawn from Mann, 1994; with permission).

girdle of the cell. They are not identical and sometimes are given names depending on their position with respect to the valve, but taken together they are usually called copulae or girdle bands. Diatoms lying in girdle view (e.g., Figs. 1d) are less easy to identify as girdle bands are much less intricately patterned than valves.

Whereas the living cell maintains all these rigid siliceous elements in position by organic membranes inside and outside the cell, most but not all dead cells become disarticulated in sediments and standard cleaning of samples (see below) can hasten the process. As long as the material is not damaged in cleaning this is advantageous for microscopy as separated valves are more likely to present their valve faces to the observer than intact cells. Girdle bands are often thinly silicified and easily broken and dissolved in sediments and usually disregarded in microscope counting. However, some are more robust and remain intact (e.g., *Tabellaria* spp.) and can be identified and counted separately from the valve (e.g., *Tetracyclus*).

Whilst most diatoms are solitary, some form colonies. These are of different kinds; in the case of *Tabellaria* they can be star-shaped, zig-zag, coiled or straight with individual cells linked together at their apices by muco-polysaccharide strands or pads. Other taxa such as most *Fragilaria* and *Aulacoseira* spp. form chains linked together by spines. In the former case colony structure rarely survives in the sediment, but in the latter case remnants of colonies of different length are commonly encountered. In some species of *Aulacoseira*

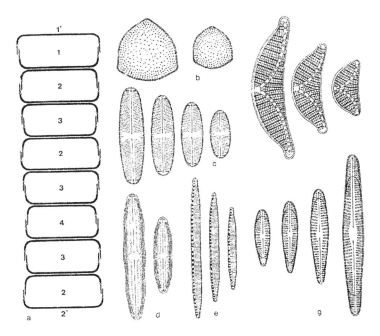

Figure 3. (a) Change in valve size following normal mitotic cell division. 1' and 2' indicate the epivalve and hypovalve of the original cell. (b)–(g) Examples of changes in valve size and shape following division: (b) *Stictodiscus*, (c) *Sellaphora*, (d) *Brachysira*, (e) *Nitzschia*, (f) *Epithemia* and (g) *Rhoicosphenia*. (from Round et al., 1990; with permission).

(e.g., *A. granulata*) morphologically distinct separation valves are produced that allow growing colonies to divide into shorter filaments (Crawford, 1979).

Life-cycles and size reduction

Diatoms are unusual in that vegetative cell division involves a successive diminution in population mean cell size as daughter cells are generated by the laying down of daughter thecae back to back within the parent cell (Fig. 2), constrained by the relative rigidity of the cingulum wall. This size reduction process was independently reported by MacDonald (1869) and by Pfitzer (1869) and has subsequently become known as the MacDonald-Pfitzer rule. Size is restored after sexual reproduction involving gamete production, auxosporulation, and the production of new silicified initial cells. Reproduction in diatoms varies considerably amongst species; reviews are given by Round et al. (1990), Mann (1993) and van den Hoek et al. (1995). Although it is now clear that the McDonald-Pfitzer rule cannot be applied mechanically to all taxa (e.g., Jewson, 1992a, b), these size reduction—size regeneration reproductive processes are responsible for controlling the size ranges of diatoms. In centric diatoms overall shape is unaltered, but in pennate diatoms reduction is greater along one axis so that shape as well as size is altered during mitosis. Consequently diatom floras usually present characteristic size ranges for individual taxa and illustrate the shapes characteristic of the different sizes (Fig. 3).

Valve ultrastructure and key taxonomic characters

Although the overall shape and structure of diatom cells are important for generic identification of diatoms (see Barber & Haworth, 1981), species and variety level taxonomy is
based on the intricate detail of the cell wall, especially the valve. Whilst most key features of
diatom valves can be recognised in the light microscope (LM), use of the scanning electron
microscopy (SEM) since the 1960s has revealed both finer features and new structures and
given rise to new terminology. These are presented by Ross & Sims (1972), Anonymous
(1975) and Ross et al. (1979) and reviewed also in Round et al. (1990). Only key characters
are presented here.

Almost all freshwater diatoms contain areolae that appear to be simple perforations
through the cell wall (Fig. 1a, b) although internally they are occluded by a finely perforated
velum (Fig. 1c). Areolae are usually arranged in rows (or striae) running at right angles to
the valve margin (Figs 1 and 3). For many taxa in the LM the individual areolae cannot be
observed, and the striae appear as continuous lines. In other cases the striae themselves are so
fine or close together, that they also cannot be resolved by LM and the valve surface appears
hyaline. SEM images (Fig. 1) show otherwise. Striae characteristics are used extensively
for identification purposes. The spacing between striae (striae density) is fairly consistent
within a species but varies between species, and is consequently a taxonomic character used
almost universally in diatom floras. It is usually expressed as the number of striae in 10 μm
along the length of the cell in pennate taxa and along the tangent and in the centre radially
in centric taxa. For pennate raphid diatoms Droop (1993) argues that the location for the
most constant striae density measurement should be made close to the raphe or sternum on
the primary side of the valve. The main reason for this is that silica deposition begins close
to the raphe on the primary side and spreads outwards to the margins and round the poles
to meet on the secondary side of the valve. Variation in density is also likely to occur along
the length of a pennate form so measurements between individuals should be at the same
place for comparison, and for larger species they should be recorded at different positions.

Some diatoms, especially centric forms, have tube-like processes that penetrate the basal
siliceous layer. They are often difficult to observe with the LM but are very prominent
in internal SEM images. There are two types, labiate processes (or rimoportulae) and
strutted processes (or fultoportulae). Labiate processes are restricted to centric and araphid
diatoms and usually occur singly on one or both valves in a marginal (centric) or polar
(araphid) position. The external expression of a labiate process is relatively indistinct but
internally it can be quite prominent as a strong slit-shaped projection. Strutted processes
only occur in centric diatoms. They are more numerous than labiate processes and occur
both on the valve face and around the margin of the valve. Each process consists of a
tube that penetrates the cell wall and has an internal projection surrounded by between 2
and 5 satellite pores. Taxonomically these features can be very important, especially for
Cyclotella and *Stephanodiscus* identification, but use of the SEM is needed for their value
to be realised fully.

One of the most important features of some pennate diatoms is the raphe, its presence,
absence, shape and type. It is used as a major character in separating sub-orders (Hustedt,
1930) or classes (Round et al., 1990) in classification systems. It is an elongated fissure,
or pair of fissures through the valve wall. There are two types, the naviculoid raphe and
the canal raphe. The naviculoid raphe (Fig. 1b) runs along the central axis of the valve, on

one valve in some genera, but normally on both valves. On each valve the raphe is divided into two, separated by the central nodule (Fig. 1b, c). The cleft of the raphe is usually in the shape of a v on its side (>) in cross-section for part or all of its length. The ends of the raphe, especially at the apices, often have characteristic terminal shapes and structures that are also of taxonomic value. The canal or fibulate raphe is a tubular passage running along the inner side of the valve, separated from the rest of the interior by siliceous elements called fibulae ("keel punctae" in many old floras). In most cases the canal raphe runs along the margin of the valve rather than along the central axis (e.g., Fig. 3e).

Diatom systematics

In the early 19th century it was debated whether diatoms should be placed in the animal kingdom or plant kingdom because of their motility. Eventually, because of the presence of chloroplasts they were placed in the plant kingdom. Today they are placed together with other unicellular eukaryotic organisms in the Kingdom Protoctista (van den Hoek et al., 1995). Although early diatom classification schemes included features of the living cell, by the turn of the nineteenth century diatomists developed classification systems almost solely based on shape, symmetry and ornamentation of the siliceous cell wall as revealed by cleaned specimens from light microscopy (e.g., Hustedt, 1930). The rise of electron microscopy revealed much finer detail of the cell wall, such as the importance of processes (Hasle, 1977) which led to further refinements (e.g., Simonsen, 1979) and enabled the use of a much larger range of ultrastructural characters to be used in identification (e.g., Round et al., 1990). More recent studies are placing greater emphasis on living material (Cox, 1996) and on molecular techniques (Medlin et al., 1993) and these are showing faults and weaknesses in the previous systems, leading to the creation of new genera and the restoration of old genera. The classification of Round et al. (1990) is gradually being adopted, and Table 1 shows an updated version of this classification for freshwater taxa (from Cox, 1996).

Although these developments are moving us closer to a natural classification, they pose problems for the diatom analyst as taxa undergo rapid nomenclatural change and floras based on different premises and traditions co-exist with different names. This is especially acute at the present time with the juxtaposition of the Krammer & Lange-Bertalot (1986–1991) floras and the Round et al. (1990) generic atlas. Up-to-date diatom checklists that provide lists of synonyms and translation between names e.g., DIATCODE (see below) are essential.

Distribution and ecology

Habitats and lifeforms

Diatoms occur throughout the world growing in almost all aquatic environments. In lakes diatoms are found in abundance in both planktonic and benthic habitats, and together form the source communities for the sediment record. By and large the species found in different habitats within a lake are characteristic of those habitats, although many species can be found in more than one habitat. The general features of the different habitats for diatoms and other algae are described by Round (1981c) .

Table I. Taxonomic position of freshwater genera within the classification system of Round et al. (1990).

BACILLARIOPHYTA

COSCINODISCOPHYCEAE

    Thalassiosirales

        Thalassiosiraceae: *Thalassiosira*

        Skeletonemaceae: *Skeletonema*

        Stephanodiscaceae: *Cyclotella, Stephanodiscus*

    MELOSIRALES

        Melosiraceae: *Melosira*

    PARALIALES

        Paraliaceae: *Ellerbeckia*

    AULACOSIRALES

        Aulacosiraceae: *Aulacoseira*

    ORTHOSEIRALES

        Orthoseiraceae: *Orthoseira*

    RHIZOSOLENIALES

        Rhizosoleniaceae: *Urosolenia*

    CHAETOCEROTALES

        Chaetocerotaceae: *Chaetoceros*

        Acanthocerataceae: *Acanthoceros*

FRAGILARIOPHYCEAE

    FRAGILARIALES

        Fragilariaceae: *Fragilaria, Centronella, Asterionella, Staurosirella, Staurosira, Pseudostaurosira, Punctastriata, Fragilariaforma, Martyana, Diatoma, Hannaea, Meridion, Synedra*

    TABELLARIALES

        Tabellariaceae: *Tabellaria, Tetracyclus*

BACILLARIOPHYCEAE

    EUNOTIALES

        Eunotiaceae: *Eunotia, Semiorbis*

        Peroniaceae: *Peronia*

    MASTOGLOIALES

        Mastogloiaceae: *Aneumastus, Mastogloia*

    CYMBELLALES

        Rhoicospheniaceae: *Rhoicosphenia*

        Anomoeoneidaceae: *Anomoeoneis*

        Cymbellaceae: *Placoneis, Cymbella, Encyonema*

        Gomphonemataceae: *Gomphonema, Didymosphenia, Reimeria*

Table I. Taxonomic position of freshwater genera within the classification system of
Round et al. (1990) (continued).

ACHNANTHALES

    Achnanthaceae: *Achnanthes*

    Cocconeidaceae: *Cocconeis*

    Achnanthidiaceae: *Achnanthidium, Eucocconeis*

NAVICULALES

    Cavinulaceae: *Cavinula*

    Cosmioneidaceae: *Cosmioneis*

    Diadesmidiaceae: *Diadesmis, Luticola*

    Amphipleuraceae: *Amphipleura, Frustulia*

    Brachysiraceae: *Brachysira*

    Neidiaceae: *Neidium*

    Sellaphoraceae: *Sellaphora, Fallacia*

    Pinnulariaceae: *Pinnularia, Caloneis*

    Diploneidaceae: *Diploneis*

    Naviculaceae: *Navicula*

    Pleurosigmataceae: *Gyrosigma*

    Stauroneidaceae: *Stauroneis, Craticula*

THALASSIOPHYSALES

    Catenulaceae: *Amphora*

BACILLARIALES

    Bacillariaceae: *Bacillaria, Denticula, Hantzschia, Tryblionella, Nitzschia*

RHOPALODIALES

    Rhopalodiaceae: *Epithemia, Rhopalodia*

SURIRELLALES

    Entomoneidaceae: *Entomoneis*

    Surirellaceae: *Stenopterobia, Surirella, Campylodiscus, Cymatopleura*

Plankton: The plankton can be categorised into those taxa that spend their whole life-
cycle suspended in the water column (euplankton or holoplankton taxa), those that have
some of their life-cycle resting on the sediment (meroplankton), and those that have their
true habitat in the benthos but can be found resuspended in the water column (tychoplankton
or pseudoplankton). Truly euplanktonic species are common in marine environments, but
in lakes many planktonic taxa are known to survive unfavourable conditions by entering a
resting phase, and are therefore mainly meroplanktonic. In some cases (e.g., *Chaetoceros*
species), morphologically distinct resting spores are formed, as in the Chrysophyceae (see
Zeeb & Smol, this volume), but more commonly there is no major alteration of the cell-wall
and physiological dormancy is achieved by rearrangement of organelles, as in *Aulacoseira*

subarctica (Lund, 1954, 1955). Tychoplanktonic taxa are extremely common in lakes. This may be fully expected in small lakes where benthic diatoms are especially abundant and easily detached and resuspended into the water column. However, these taxa are also common in large lakes. Even for Lake Baikal between 4 and 10% of the total diatom population caught in sediment traps originated from the marginal benthos (D. B. Ryves, pers. commun.). It is consequently important not to assume that all taxa collected in a plankton sample are planktonic diatoms.

Benthos: Benthic diatoms are those associated with substrates around the margins of lakes. Their extension into deeper water depends on the availability of suitable substrates and the extent of light penetration. The main habitat types are the epilithon (attached to stones), the epiphyton (attached to plants), the epipsammon (attached to sand grains), and the epipelon (associated with the mud). The epilithon and epiphyton have many species in common, but especially large fixed stones are more permanent and allow more complex assemblages to develop over time, combining mixtures of both living and dead cells. Epiphytic communities, on the other hand, depend on the availability of host aquatic plants that, in temperate lakes, die back over winter and limit the possibility of long-term community development.

The epipsammon is a very distinctive community, composed mainly of very small firmly attached adnate taxa that are capable of surviving potentially long periods in dark, anoxic environments, probably by entering a resting phase (Jewson & Lowry, 1993) until the sand grain is mixed back into the photic zone by wave activity. The epipelon is also a specialised flora, adapted to low light conditions and composed mainly of motile raphid taxa that can move through the interstitial waters of the mud surface to avoid burial from sediment deposition. In shallow lakes or lakes with transparent water, epipelic taxa may be living *in situ* amongst dead or resting cells derived from other habitats in the lake, and sampling the true epipelon in these habitats can be difficult (see below).

Physico-chemical variables

The kinds of diatoms found in lakes depend on the range and extent of habitats available for growth, as described above, and also on the combination of physical, chemical and biological conditions that prevail in the water column in general and in these habitats specifically. It is beyond the scope of this chapter to review the full range of relationships put forward in the literature. The purpose here is to document some of the key controls on diatom composition that palaeolimnologists endeavour to identify from sediment records. Depending on lake type these include mainly physical (temperature, light, turbulence, etc.) and chemical (pH, DOC, nutrients, salinity) factors. Biological controls on diatom composition such as mutualism (e.g., Cattaneo & Kalff, 1979, Jones et al., 2000), grazing (e.g., Kairesalo & Koskimies 1987, Underwood & Thomas 1990), and parasitism (e.g., Canter 1979) are also important, but changes in these factors through time are difficult to identify with any certainty from the sediment record, and are consequently not discussed here.

Physical factors

Temperature: There is a substantial older literature (e.g., Hustedt, 1956) that describes the distribution of diatoms in relation to temperature and there has been a number of recent studies using contemporary surface sediments that show temperature to be a potentially important variable in explaining differences in diatom composition between lakes (Vyvermann & Sabbe, 1995; Pienitz et al., 1995; Lotter et al., 1997; Weckström et al., 1997). In addition, culture studies often show clear differential growth rate responses of taxa to temperature (e.g., Suzuki & Takahashi, 1995) and studies of the endemic planktonic diatoms from Lake Baikal show a close coupling between the temperature responses of taxa in culture and life-cycle strategies in the lake in relation especially to water temperature and ice-cover (D. H. Jewson, pers. commun.; Richardson et al., 2000).

These studies indicate the potential of diatoms for reconstructing past temperature from sediment cores using a transfer function approach (e.g., Korhola et al., 2000). However, temperature change has a major influence on the behaviour of other physical and chemical variables in lake systems, such as ice-cover, stratification, pH, nutrient cycling, etc. (Battarbee, 2000, Anderson 2000) that co-vary with temperature individualistically from lake to lake. It is consequently difficult to separate the specific influence of temperature on diatom composition from that of other variables, but this is a key area of current research.

Light: Light is a major regulator of photosynthesis and consequently plays an important role in driving algal productivity and determining species composition in lakes (cf. Reynolds, 1984). In temperate lakes light duration and intensity usually control the timing of the spring diatom bloom (e.g., Maberly et al., 1994) and the attenuation of light with depth, as a result of suspended particulates or water colour, defines the limit of the euphotic zone in which net photosynthesis takes place. As diatom phytoplankton has differing abilities to adapt to low light conditions (e.g., Talling, 1957), changes in light climate not only have a potential impact on plankton biomass but can also influence the composition of the plankton. In addition water column transparency controls the extent of habitats capable of supporting benthic diatom communities. The role of changing dissolved organic carbon (DOC) in boreal lakes is especially significant (Schindler et al., 1996; Vinebrooke & Leavitt, 1996). Despite the importance of light, it is not an easy variable to reconstruct in any quantitative sense from lake sediment records. For boreal lakes, one way is to use changes in diatom-inferred DOC coupled to an optical model (Pienitz & Vincent, 2000). In other cases past changes in light climate might be inferred qualitatively where, for example, there is evidence for eutrophication that leads to an increase in the abundance of algal cells in the water column or where there is evidence from the sediments for soil inwash that leads to an increase in suspended clays and silts.

Turbulence: Whilst turbulent mixing may influence the composition of benthic taxa in a lake and be responsible for resuspending benthic taxa into the water column, its main influence is on the composition of the phytoplankton. Given that planktonic diatoms are non-motile and have a specific gravity greater than 1.0 it is not surprising that they are more abundant during periods of water-column mixing and that different taxa have evolved competitive physiological, morphological and life-cycle strategies to cope with buoyancy problems (Lund, 1955, 1959a; Reynolds, 1984). Indeed in lakes that stratify, the seasonal succession of diatom plankton is often closely controlled by mixing and stratification that in turn are controlled by temperature and wind during the growing period (e.g., Reynolds, 1973). The responsiveness of planktonic populations to these influences potentially allows past changes in stratification intensity to be inferred from the sediment record, an approach pioneered by Bradbury (Bradbury & Dieterich-Rurup, 1993).

Ice cover: Ice cover can influence diatom assemblages in several direct and indirect ways. Irradiation is significantly reduced by ice, although under clear or snow-free ice, light is often adequate to allow growth, and convective mixing generated by radiative heating can allow planktonic diatom crops to be supported (e.g., Catalan & Camarero 1991). In the case of Lake Baikal in Russia endemic diatoms appear to have evolved life-cycle strategies that are dependent on ice-cover (D.H. Jewson, pers. commun.). Opaque or snow-covered ice severely restricts light penetration and growth is usually prevented (Smol 1988).

The extent of ice-cover on lakes may also influence lake-water chemistry especially through changes in oxygen consumption at the mud-water interface that controls nutrient re-cycling and alkalinity generation (Psenner, 1988; Catalan & Camarero, 1993), with potential impacts on diatom productivity and species composition. In extreme environments ice may be present throughout most of the year and variations in climate from year to year may lead to significant changes in the availability of both planktonic and benthic habitats available for diatom growth (Douglas & Smol, 1999).

On this basis, longer-term changes in ice-cover associated with sustained changes in climate will cause major changes in diatom abundance and composition in lakes that should be clearly recorded in lake sediments (cf. Douglas et al., 1994).

Chemical factors

There are many specific chemical influences on diatom composition that have been reported in the literature (e.g., Cholnoky 1968, Patrick, 1977), but there are a relatively small number of over-riding controls that influence individual diatom ranges and the overall composition of diatom assemblages. These include nutrients, pH, and conductivity/salinity. Their importance is apparent from the classical literature and from the more recent studies of diatom-environment relationships using multivariate statistical techniques such as canonical correspondence analysis (CCA).

Nutrients: The key limiting nutrients in surface waters for algal growth are N and P. However, in the case of diatoms, the availability of dissolved silica (SiO_2) is also important, both in regulating the size of the phytoplankton crop, especially in spring in temperate lakes, and in influencing species composition through the differing abilities of planktonic diatoms to compete for silicon and phosphorus (Tilman et al., 1982; Kilham, 1984; Kilham et al., 1986; 1996). Although silica concentrations can change through time, depending on catchment weathering rates and internal silica re-cycling, the main variable responsible for driving diatom productivity and species change is usually P, and this is the variable that most diatomists seek to reconstruct in studies of lake eutrophication (Whitmore, 1989; Anderson & Rippey, 1994; Bennion et al., 1996; Hall & Smol, 1992, 1999). Increasing the availability of P usually causes most alteration to phytoplankton communities, but benthic communities can also be affected as a result of changing habitat availability (e.g., the type and distribution of aquatic macrophytes) and through increased shading by plankton crops. Typically this is reflected by an increase in the ratio of planktonic to non-planktonic diatoms in sediment cores (Battarbee, 1986b), even where the production of benthic diatoms has also increased (Oldfield et al., 1983). Attempts in the past to use the ratio of Araphidinate to Centric diatoms (the A:C ratio) as an indicator of nutrient enrichment (cf. Stockner & Benson 1967) are now largely discredited (Brugam 1979, Brugam & Patterson 1983) and are no longer recommended. For details see Battarbee (1986a).

pH: pH is probably the single most important controlling variable on species composition in freshwater systems, and its significance has been long recognised in the diatom literature (e.g., Hustedt, 1937–1939). The strength of this relationship is particularly illustrated by canonical correspondence analyses of surface sediment training sets from soft water lakes where pH and its correlates explain more of the variance between assemblages than any other physico-chemical variable (e.g., Battarbee et al., 1999 and references therein).

There is no detailed ecophysiological understanding of how pH influences the growth and competitive abilities of individual diatom taxa. However, pH controls many chemical and biochemical processes and reactions including the carbonate-bicarbonate balance in lakes, the availability of nutrients for algal uptake, and the solubility of metals, especially toxic ones, (e.g., aluminium), and the activity of specific enzymes such as the phosphatases.

Although rarely used now in any quantitative way this strong relationship between diatoms and pH gave rise to a pH classification of diatom taxa and the use of diatoms by palaeolimnologists to reconstruct past lake pH. This was based on various indices (e.g., Index α, Index B) derived from calculations of the ratio of taxa in different pH groups (Nygaard, 1956; Meriläinen, 1967; Renberg & Hellberg, 1982) or the use of the pH groups as explanatory variables in multiple regression analyses (e.g., Charles, 1985; Flower, 1986; Davis & Anderson, 1985). Details of these earlier methods can be found in Battarbee (1986a) and the history of this classification has been reviewed by Battarbee et al. (1986b). However, modern studies show that there is a constant diatom species turnover along the pH gradient and that forcing diatoms into pH categories as required by the classification system is unnatural and can lead to anomalies in pH reconstruction. Today pH reconstruction is used in studies of lake ontogeny (Whitehead et al., 1989; Renberg, 1990a; Jones et al., 1989), surface water acidification (Battarbee et al., 1990; Charles & Whitehead, 1986; Cumming et al., 1994) and in climate change (Psenner & Schmidt, 1992) using transfer functions based on estimates of the pH optima of individual taxa (see below).

DOC: Although there is little older literature on the relationship between diatoms and dissolved organic carbon (DOC), ordination of diatom assemblages that include samples from acidic brown-water lakes often show that a significant proportion of the variation between samples can be explained by this variable independent of pH and other variables (Birks et al., 1990b; Kingston & Birks, 1990; Pienitz & Smol, 1993; Korsman & Birks, 1996). However, despite the strong statistical relationship, the ecophysiological mechanisms responsible are not clear and DOC optima between datasets are not always repeatable (cf. Kingston & Birks, 1990; Dalton, 2000). In some cases high concentrations of DOC may offer protection from trace metal toxicity, or, especially in high latitudes, from UV radiation. Alternatively its importance may be in attenuating light and influencing the balance between planktonic and benthic production and favouring benthic taxa capable of growth in low light intensity. Despite these mechanistic uncertainties, studies of lakes in the Canadian Arctic in particular have shown the potential of diatom-DOC relationships to reconstruct vegetation and climate change (Pienitz et al., 1999) and the past optical qualities of the water column (Pienitz & Vincent, 2000).

Salinity (*athalassic*): Although pH is the most important variable in some salt lake systems, especially those dominated by sodium carbonate-bicarbonate chemistry (e.g., Gasse, 1986), diatom composition in most salt lakes is usually controlled by the ionic strength of the water measured as salinity or conductivity. There is also good evidence both from culture and field survey that this relationship is also moderated by brine type (Schmid, 1995; Fritz et al., 1993; Gasse et al., 1983). Diatoms found in salt lakes are predominantly euryhaline and in many cases are the same as those found in estuarine or coastal waters where brackish water conditions exist. However, despite their tolerance to salinity fluctuation, different taxa show clearly defined optima along the salinity gradient, a response that allows robust diatom-salinity transfer functions to be constructed (Fritz et al., 1991; Cumming et al., 1995; Gasse et al., 1995). Palaeosalinity inferred from diatom records in salt lakes has been used extensively in climate change studies (Gasse et al., 1997; Laird et al., 1996, 1998a, 1998b, Fritz et al. 1999), although in some lakes and lake districts its usefulness is constrained by poor preservation (Barker et al., 1994; Reed, 1998; Ryves 1994).

   Salinity (thalassic): There has been a long history in the use of diatoms as indicators of salinity and sea-level change in coastal and estuarine environments (Cleve, 1899, and reviewed by Juggins, 1992 and Stoermer & Smol 1999). Classification of diatoms in relation to salinity in such environments was first established by Heiden (1902, cited in Brockmann, 1954) and developed principally by Hustedt (1957), Simonsen (1962), Werff & Huls (1957–1974), and Vos & de Wolff (1988, 1993a), with the latter authors also including information on the life-form of taxa as an aide to identifying particular coastal sedimentary environments (e.g., Vos & de Wolff, 1993b). These classification schemes have been used extensively to provide qualitative reconstructions of changes in salinity, sea-level and coastal geomorphology (reviewed by Denys & de Wolf, 1999). In many cases these qualitative reconstructions of salinity are sufficient to identify major transgressive / regressive events and index points that are used to construct relative sea-level curves (Shennan et al., 1983). Where more precise reconstructions are required several authors have used collections of modern diatoms to define quantitative transfer functions for salinity (Juggins, 1992) and tidal-level (Zong & Horton, 1999) to provide high-resolution, quantitative reconstructions of relative sea-level change. Despite problems in separating allochthonous from autochthonous valves, there is clear potential for quantitative environmental reconstructions in the coastal zone using the transfer function approach, for both salinity and relative sea-level, and also for inferring past nutrient levels (N and P) in studies of coastal eutrophication (Weckström et al., unpublished; Clarke et al., unpublished).

Taphonomy and preservation

A key issue in diatom analysis is the accuracy with which diatom assemblages in sediments reflect or represent the composition of the source communities and habitats from which they were derived. However, one of the difficulties in assessing the quality of the fossil record is establishing in the first place the abundance and composition of both the living and fossil diatom assemblages needed to make the comparison. Comparisons of abundance are especially difficult because of problems in estimating annual production in benthic habitats and problems of spatial variability of diatom accumulation rates in the sediment (Battarbee, 1978b; Anderson, 1989). Comparisons of composition are easier and there have been studies showing good agreement for both plankton-dominated systems (e.g., Battarbee, 1979, 1981; Haworth, 1980) and benthic-dominated systems (DeNicola, 1986; Jones & Flower, 1986; Cameron, 1995). However the number of studies is very limited and there are sites where the comparison is less good, for example in Lake Baikal where

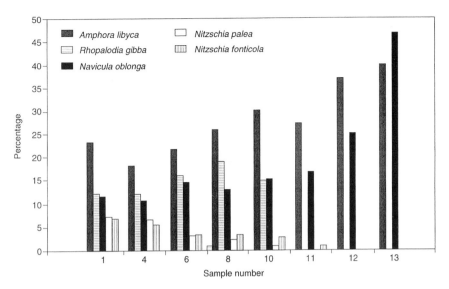

Figure 4. Experimental dissolution of a mixed sample of diatoms from lakes in the Northern Great Plains, USA showing progressive loss of taxa and changes in percentage abundance over the four week period of the experiment. The experiment was carried out at 25 °C in distilled water buffered at pH 10 (from Ryves, 1994).

comparisons between the water column and sediment record show substantial differences both in abundance and composition (Battarbee et al., unpublished). In extreme cases diatoms are completely lacking from the sediment record due to dissolution problems.

The processes responsible for the differences between living and fossil assemblages include removal via the outflow, contamination from upstream sources, resuspension and reworking of older sediments within the lake basin, bioturbation within the sediment column and poor preservation as a result of dissolution (Battarbee, 1986a). The most serious concern is usually over dissolution. Partial or complete dissolution of frustules may occur in the water column or in the sediments. In most cases the main site of dissolution is the surface sediment. Girdle bands, the least silicified parts of valve walls, and the entire valves of lightly silicified taxa are especially vulnerable. Losses lead not only to a reduction in the number of valves accumulating but, because dissolution is differential between species, it leads also to biases in the composition of the sediment assemblage towards more robust taxa (cf. Fig. 4), a bias that may be constant or may vary through time, and one that has implications for the accuracy of transfer functions that are based on compositional data (cf. Ryves 1994).

Although there are exceptions to the rule, preservation is usually best in cold, soft water lakes typical of boreal latitudes (Cameron, 1995) and poorest in warm alkaline or saline lakes in low latitudes (e.g., Barker, 1992; Ryves 1994). The factors controlling opaline silica dissolution are outlined by Hurd (1972) and Hurd & Birdwhistell (1983). Variations in preservation between lakes and at different times in lake history can be the result of many influences including the silica content of the cell wall (Jorgensen, 1955), temperature and pH (Rippey, 1977, 1983), the concentration gradient of the dissolved silica between mud and water (Tessenow, 1966), benthic macroinvertebrate activity (Tessenow, 1964), water depth (Jousé, 1966), sediment accumulation rate (Bradbury & Winter, 1976) and the availability of polyvalent cations for adsorption to the cell wall to provide a protective coating (Lewin, 1961).

Field sampling and coring

Diatomists working with lake sediment records need to sample both living and fossil diatoms to develop transfer functions and to understand the relationship between source communities and fossil assemblages. Ideally all diatom habitats should be sampled at several times during the year and water samples for chemical analysis should be taken at the same time. In this way it is possible to construct a species list of all diatoms growing in the lake and to make comparisons between the living flora and the assemblage of diatoms found in the surface sediments at the point where sediment cores are likely to be taken. Such a comparison might be used to show which habitats are the most important sources of diatoms to the sediment (e.g., Jones & Flower, 1986) and the extent to which diatoms are under-represented especially as a result of poor preservation. It can also reveal the extent to which the samples are contaminated by non-contemporary reworked material.

Plankton

Plankton is usually well mixed within a lake and estimates of the species composition of the plankton should be obtained from samples collected from the open water, preferably in a central position in the lake to maximise representation of true planktonic taxa and minimise contamination by suspended benthic taxa. In productive lakes the volume of water required is quite small (e.g., less than 250 ml) but for oligotrophic lakes a much higher volume is needed (e.g., greater than 1 litre). Because the composition of the plankton is constantly changing, samples need to be taken frequently (e.g., every two weeks if possible) during the year. If total cell numbers are needed then the sampling strategy needs to take into consideration the possibility that diatoms are not evenly distributed with depth in the water column. An integrated sample can then be obtained using a weighted plastic tube (Lund & Talling, 1957). Lugol's iodine should be added as a preservative (Wetzel & Likens, 1991:140). For compositional analysis the sample should be concentrated by settling and prepared for microscopy as outlined below. For cell number counting the diatoms in a sample of known volume are allowed to settle in a sedimentation chamber (Lund, 1959b) and then counted directly without further treatment using an inverted microscope (Lund et al., 1958).

Benthos

Epilithon: The epilithon is relatively easy to sample and suitable submerged stones occur in almost all lakes. Select one or more suitable large stones from a water depth sufficient to ensure that the stone has been permanently submerged and scrape (with a sharp knife) or brush (with a toothbrush) the epilithon from the stone surface into a tray, transfer to a plastic sample tube, and add a few drops of Lugol's iodine.

Epiphyton: Epiphyton sampling depends to some extent on the architecture of the host plant. For submerged and floating leaf plants, both leaves and stem from the upper portions of the plants should be sampled, taking care to avoid contamination from sediments that might be disturbed. If necessary the plant fragments can be cut into small pieces, then placed in a sterile container and covered by methanol. For emergent plants the upper part of the plant should be cut away at a level a few centimetres below the surface of the water and then a number of stem samples approximately 5 cm long can be cut. Each stem should then be carefully scraped with a razor blade or very sharp knife to remove diatoms that are tightly bound to the plant surface, washed into a sterile container and covered with methanol. Older or dead stems should be selected where possible. Where epiphytic material is loosely bound to the stem and may be dislodged by lifting from the water, a measuring cyclinder can be inverted over the stem in the water before cutting off the sample.

Epipsammon: Epipsammon can be collected by skimming the upper centimetre of sand surfaces in lake margins with a small sample tube. The tube should be only half filled. As the sample will contain plankton, epipelon and other diatoms not belonging to the true epipsammon, these should be removed by stoppering the sample tube, vigorously shaking the sample, allowing the sand to settle and pouring off the supernatant. This should be repeated with clean water until the supernatant is clear. As epipsammic diatoms are usually firmly attached to the substrate these should be effectively retained by this procedure. Fill the tube with methanol and return to the laboratory.

Epipelon: The epipelon is the most difficult community to sample, partly because of the range of water depths that epipelon can be found, but mainly because of the difficulty of separating true epipelic forms from dead diatoms from other habitats. The issues are fully discussed by Eaton & Moss (1966) who also describe appropriate sampling methods. For epipelic sampling in shallow water surface, sediment is drawn up using a 1 cm diameter clear plastic tube under capillary action and the sample is released into a Petri dish along with overlying water. For deeper samples a sediment grab or corer is needed and the uppermost half centimetre is removed and transferred to a Petri dish. Squares of lens tissue paper with a surface area of approximately $2 \, cm^2$ are placed on the surface of the sample and the dish is left in bright light overnight to allow the motile epipelic diatoms to migrate to the tissue paper. The lens tissues are then removed into sterile sample containers and covered with methanol.

Sediment traps

Whilst sediment traps are used for a wide variety of limnological and palaeolimnological studies, they can be of special value for diatomists especially as a means of providing an integrated sample of the present day lake flora that can be directly compared with sediment core samples on the one hand and with plankton and benthic samples on the other (e.g., Cameron, 1995). There are many trap designs including simple open traps e.g., Bloesch & Burns (1980) and more sophisticated sequencing traps. Open tube traps need to have a high aspect ratio ($> 5 : 1$) to avoid resuspension and loss of collected sediment, and should be placed in replicate (often 4) in the centre of the lake at different depths in the water column. An upper array of traps should be positioned within the photic zone and lower traps above the sediment surface in the profundal zone. Sequencing traps contain a carousel of sample containers that are rotated beneath a collecting funnel and are programmed to open and close at pre-defined intervals (usually two weeks or one month). As for open traps these are placed in deep water and ideally include arrays in both the upper and lower parts of the water column. Sequencing traps are excellent for remotely sampling plankton, tracking the seasonal succession of diatom crops and assessing the efficiency of their sedimentation from the water column. Sequencing trap arrays can also be combined with open traps and with other instrumentation (e.g., thermistors) at the same anchorage.

Sediment coring

Sediment coring for most aspects of diatom analysis follows the procedures used in other palaeolimnological studies (e.g., Wright, 1980). However, somewhat different procedures may be required when constructing modern training sets needed to develop diatom transfer functions (see below). In these cases short sediment cores are taken specifically to obtain a sample that best represents the modern flora of the lake. It is consequently essential to only use coring devices that sample the mud-water interface with the minimum of disturbance. Small light-weight gravity samplers e.g., the Glew corer (Glew, 1991; Glew et al., 2001) are the most convenient. The thickness of surface sediment to be retained for training set purposes then depends on sediment accumulation rate. The aim is for a sample that integrates the diatom flora over the time period (usually a year) that is represented by the water chemistry samples used in the transfer function. However, in most cases, the rate of sediment accumulation is not known at the time of sampling and given potential problems of bioturbation, resuspension, and compaction, it is not possible to match exactly the time ranges of the water and sediment samples. Whilst these problems deserve more attention they are invariably ignored and it is assumed that the uppermost half or one centimetre sediment sample can be taken to represent the modern flora in this way.

Laboratory procedures

Sample preparation

There are a number of ways of treating diatoms for slide preparation. The precise combination of steps must be determined experimentally in relation to the individual characteristics of the sample being analysed. Care should be taken at all stages not to lose or damage valves, and occasional checks for such events should be made by examining treated material before, during and after the various stages of preparation. It should be noted that delicate spines and processes are easily destroyed and that vigorous stirring and rapid centrifugation can break fragile diatoms. For sediment samples, fresh material or freeze-dried material is preferred as oven-drying can cause diatom breakage (Flower, 1993). At all times care should be taken to avoid contamination. It is especially important in laboratories where diatoms from different sites are being prepared simultaneously. Clean glassware is essential and if glassware is to be re-used it should be cleaned with hot 10% Na_2CO_3. Where it is necessary to prepare a large number of sediment samples, a waterbath using disposable test tubes rather than beakers may be preferable (Renberg, 1990b). It is important to pay attention to safety issues, especially the proper use of safety apparel and fume cupboards. Hydrogen peroxide is a very powerful oxidising agent and contact with the skin should be avoided. Strong acids require very careful use and some diatom mountants contain organic solvents that also require handling in a fume cupboard.

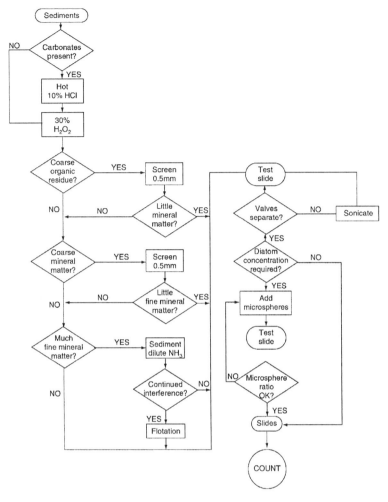

Figure 5. Flow chart for diatom sample and slide preparation (re-drawn from Battarbee, 1986a).

Different fractions of the sediment may need removing in order to obtain a satisfactory slide preparation, although in very diatom-rich material chemical pre-treatment may not be necessary at all. Treatment techniques commonly used are as follows (see Fig. 5):

Removing salts soluble in hydrochloric acid: Carbonate and many metal salts and oxides can be removed by treatment with dilute hydrochloric acid. For non-calcareous sediments this stage can be omitted. Using a fume cupboard, place a small quantity of sediment (usually about 1 g of wet sediment is sufficient) into a beaker and slowly add 10% HCl. Agitate and heat gently for about 15 minutes. Centrifuge or settle and wash in distilled water.

Removing organic matter: Organic matter can be removed by oxidation by hydrogen peroxide, by strong acids or by incineration. For the hydrogen peroxide method, use a wide heatproof beaker in a fume cupboard, and add a small amount of 30% H_2O_2. When foaming has subsided, add more hydrogen peroxide and heat gently on a hot-plate until all organic matter has been removed. Cold digests are also possible, and may be advisable where a vigorous reaction is expected. If there is a coarse organic residue, sieve using a 0.5 mm screen. For some sediments removal of the organic matter is the only step required, in which case the sample residue should be thoroughly washed (centrifuge or settle and wash with distilled water at least three times) and suspended in distilled water ready for slide preparation. Where the use of hydrogen peroxide results in excessive foaming, be ready to dilute or transfer the sample into a larger and wider beaker.

The strong acid method (van der Werff, 1955) has the advantage of not causing foaming but all calcareous compounds must be removed first or gypsum crystals will form. Add concentrated sulphuric acid until the volume is twice the original sample. Add enough potassium dichromate to make a saturated solution (a few orange crystals are formed) and then add oxalic acid until the solution becomes clear. Add distilled water and leave to settle overnight. Decant and wash as above.

The incineration method is quick and has the advantage of leaving frustules intact. It is appropriate especially for fresh material or peaty samples where mineral material is absent. A drop of sample material which has been washed and subsequently allowed to settle is dried on a clean cover slip and then heated over a spirit or gas flame until it eventually becomes grey or white. This is then mounted as usual.

Removing minerogenic matter: Where minerogenic material prevents good slide preparation, attempts can be made to reduce the mineral material by physical methods. Coarse-grained mineral matter may be removed by sieving (mesh size not less than 0.5 mm) or by gentle swilling in a beaker followed by decanting the diatoms together with the finer mineral fractions (Brander, 1936). Clays can be partially removed in a related manner, allowing diatoms and coarser mineral particles to sediment before decanting and discarding suspended clay. This technique may be improved by allowing the sedimentation to take place in dilute ammonium hydroxide solution.

Diatoms can also be separated from mineral particles by flotation in heavy liquids. As some of the older methods (Jousé, 1966; Knox, 1942) use hazardous chemicals, it is preferable to use sodium polytungstate (SPT) that is safe to use. Pre-treatment with hydrogen peroxide or strong acids as described above is required. Tapia & Harwood (pers. commun. 1999) provide a method as follows. Prepare a solution of SPT at a density of 2.2 g/cc and suspend the cleaned diatom sample in 10 ml of distilled water. Place 3.5 ml of SPT in the bottom of a 15 ml disposable centrifuge tube and add the 10 ml diatom sample carefully to the top of the SPT to prevent mixing of the two solutions. Fill the centrifuge tube with distilled water and centrifuge at low speed (500 rpm) for 3 minutes. By pipette, carefully remove the distilled water/SPT interface into another centrifuge tube. Remove the SPT from the diatoms by resuspending the working aliquot in 15 ml of distilled water and centrifuge at 1500 rpm for 3 minutes. Wash the residue up to four times to remove all traces of SPT.

Separating valves: After treatment and the final wash, the preparation should be checked to ensure adequate removal of unwanted material and to ensure that most frustules and colonies have been separated into single valves. If not, and it appears that identification and counting might be impaired by valves lying in girdle view, a subsample of the final suspension can be sonicated in an ultrasonic bath. This treatment is effective, but since it leads to fracturing of valves it should only be used to aid identification and not for formal counting. The prepared suspension of diatoms in distilled water is diluted to an appropriate concentration and thoroughly mixed ready for slide making.

Preparation for estimating diatom concentration: If diatom concentration is required, a known quantity of marker microspheres can be added at this stage (Battarbee & Kneen, 1982). The most appropriate microspheres are those with a size range of 5–10 μm and made of divinylbenzene (DVB) that are resistant to the organic solvents used in diatom mountants. However, as they have a much lower specific gravity than diatoms, they can only be added at the final stage of preparation to avoid loss during washing processes. A stock suspension should be made up with a concentration of about 5×10^6 spheres per ml and the concentration then precisely determined using an electronic particle counter. The addition of a small amount of ammonia solution helps to keep the spheres from clumping and a very small amount of mercuric chloride is added (< 5 ppm) to prevent the growth of micro-organisms. Store the stock suspension in the dark at 4 °C.

For the microsphere method initially add about 2 mls of 5×10^6 suspension for each 0.1 gram of dry sediment digested to a single sample in a batch, being careful to shake and sonicate the stock suspension before each use to disperse the spheres evenly. Make a test slide to check the diatom to microsphere ratio (ideally 1:1) and calculate the required amount of microsphere suspension needed before adding spheres to a whole batch of samples.

Diatom concentrations can also be derived using specially constructed evaporation trays (Battarbee, 1973) or similar random settling methods (e.g., Bodén, 1991; Scherer, 1994). Evaporation trays are best made from plexiglass (solid PVC and acrylic can also be used) and are used to settle out a diatom suspension directly onto coverslips which are nested into wells in the trays. As the area of the cover slips and the plate is known, and the sediment has been weighed, diatom concentrations can be calculated. If this method is used, it is important to check for each sample batch that the distribution of valves is indeed random as for some sediments the effects of surface tension during the final stages of evaporation can distort the final distribution on the coverslips.

Slide preparation

To prepare regular strewn mounts, with or without added microspheres, about 0.2 ml of diatom suspension is dropped carefully by pipette on to a clean coverslip and the diatoms are allowed to settle and the water evaporate at room temperature. Care needs to be taken not to disturb the coverslip. When dry, the coverslip is mounted using a resin with a high refractive index (i.e., 1.7) such as Naphrax or Hyrax. For Naphrax place one drop on a glass slide and invert the coverslip with the dried diatoms over the drop. Heat the slide on a hotplate at about 130 °C for 15 minutes to drive off the toluene in the Naphrax. Allow the slide to cool and then check that the coverslip does not move when pushed with a fingernail. If it does, the slide will need to be re-heated. All slides need to be checked to ensure that the concentration of diatoms on the coverslip is appropriate for counting; three or four diatoms per field of view is ideal, and it is wise to make several slides from each sample to provide replicates for slide archives. Store the prepared suspension in methanol in an airtight glass container in case material is needed for SEM or additional slides. Diatoms stored in water are likely to dissolve.

Preparing diatom samples for EM

The preparation of diatom material for the scanning electron microscope (SEM) is straight-forward. A small drop of prepared suspension can be evaporated directly on a specimen stub, although it is advisable to settle material on to a 11 mm diameter coverslip first and then fix the coverslip on the stub using resin. A small amount of silver should be dabbed onto the edge of the stub to ensure good electrical contact between the glass coverslip and the metal stub. The sample is then coated with gold in a sputter coater and ready for inspection. The transmission electron microscope (TEM) is less widely used by diatomists. It lacks many of the advantages of the SEM for diatom taxonomy, but its resolving power is much greater and it is especially appropriate for the inspection of thin-walled specimens. Samples are prepared by drying a drop of clean suspension on to formvar-coated copper grids.

Light microscopy

Most routine diatom analysis is based on light microscopy using oil immersion objectives and magnifications of 750 x and over. Brightfield, phase contrast or differential interference contrast illumination can be used; each has different advantages. An eyepiece reticule is needed for measuring; the reticule should be calibrated using a micrometer slide with a scale marked off in 0.01 mm divisions. A camera lucida attachment is useful and microphotography is essential. Increasingly diatomists use digital cameras and image-capturing software to store, manipulate and transfer images.

Identification and taxonomy

Most lake-sediment diatom assemblages are quite diverse and even experienced diatomists need continually to check out unknown and uncertain taxa. Fortunately, identification to the generic level is relatively straightforward, and it is usually possible on the basis of size, shape, raphe presence or absence and striae arrangement to narrow down possible alternative names quite quickly. However, this will require making measurements using a calibrated eyepiece reticule and a familiarity with the terminology used for valve shape and valve features (e.g., Anonymous, 1975; Barber & Haworth, 1981; Ross et al., 1979).

The next step is to use an appropriate flora. The most commonly used floras are those by Krammer & Lange-Bertalot (1986–1991). These are up-to-date, comprehensive and use micrographs rather than line drawings for illustration. However, a knowledge of German is required to read taxon descriptions and the floras are chiefly concerned with European species. Where possible it is important to use floras generated in the same geographical region although many such floras are incomplete. In addition to the Krammer & Lange-Bertalot floras other useful general floras include Hustedt (1930, 1927–66), Cleve-Euler (1951–55), Molder & Tynni (1967–80), Patrick & Reimer (1966, 1975), Schmidt's Atlas (1874–1959), Germain (1981) and Schoeman & Archibald (1980). Terminology should be in accordance with the proposals outlined in Anonymous (1975) and Ross et al. (1979).

There are also more specialised floras associated with diatoms from specific ecosystem types, such as acidic lakes (Camburn & Charles, 2000), or reviews of specific genera (e.g., Williams & Round, 1986, 1987), and electronic floras are increasingly becoming available that can be accessed either by CD-ROM (e.g., Campeau et al., 1999) or eventually by the Internet (e.g., Battarbee et al., 2000). Developments in taxonomy (i.e., name changes, new species descriptions or combinations) are continually taking place and the use of standard floras need to be supplemented by reference to key diatom or algological journals including *Diatom Research, Bibliotheca Diatomologica, Journal of Phycology, European Journal of Phycology, Nova Hedwigia*, and *Phycologia*.

Quite commonly diatom assemblages include taxa that are difficult to identify. Where difficulties cannot be resolved using literature sources, problem taxa need to be compared with type material archived in museum collections. The collections in the main herbaria have been described by Fryxell (1974).

From time to time, taxa are encountered that have not been previously described. For routine work these taxa should be described and illustrated and can simply be given temporary names. If it is decided to publish the description of a taxon and furnish it with a formal name, then strict rules specified by The International Code of Botanical Nomenclature must be followed to ensure that the taxon is validly named (cf., Williams, 1989). A name alone does not give an unambiguous indication of identity. Publication must be accompanied by a description together with an illustration or photograph, and these must be associated with a slide containing the type specimen deposited in a herbarium. In this way the limits of the newly defined taxa are set and the taxon is known under its first validly published name. All others names are synonyms. A comprehensive list of valid names and their synonyms (up to 1964) has been published in six volumes by VanLandingham (1967–1979). More recently coded computer checklists have been created listing published species names and synonyms. One of the most comprehensive lists, called DIATCODE, was created by Williams et al. (1988). This is regularly up-dated and freely available for inspection or downloading from the Environmental Change Research Centre (UCL) website (www. geog.ucl.ac.uk/ecrc).

In addition to diatoms that can be identified and given formal names, lake sediment samples often contain diatoms that cannot be identified at the specific level because of poor preservation (breakage and/or dissolution), because the specimen is obscured or because it appears in girdle view. In these situations diatoms should be allocated to the most precise taxonomic category possible such as genus (e.g., *Cyclotella* spp. or *Cyclotella* sp. 1) or order (e.g., unknown pennate spp.). Other conventions include the use of "cf." as in *Achnanthes* cf. *marginulata* for taxa that do not exactly fit the formal description. DIATCODE also includes a coding system for these informal names.

In cases where research projects involve more than one diatomist or more than one laboratory, problems of taxonomic consistency between laboratories become important. Taxonomic harmonisation is especially necessary in the setting up and use of transfer functions (see below) where it is essential to define the environmental ranges and optima for different taxa and where the taxonomy used in sediment core analysis must match that used in the training set. This issue was first tackled in the Surface Water Acidification Project (SWAP) (Battarbee et al., 1990) and the Paleoecological Investigation of Recent Lake Acidification (PIRLA) project (Charles & Whitehead, 1986; Kingston et al., 1992) where differences in conventions between laboratories were resolved following taxonomic workshops, slide exchanges, the circulation of agreed nomenclature (e.g., Williams et al., 1988), taxonomic protocols (Stevenson et al., 1991) and taxonomic revisions (e.g., Flower & Battarbee, 1985; Camburn & Kingston, 1986). The PIRLA project also produced a project iconograph (Camburn et al., 1986; Camburn & Charles, 2000). In the case of SWAP the effectiveness of this approach was tested by an inter-laboratory comparison of counts from test slides before and after the harmonisation workshops (Kreiser & Battarbee, 1988; Munro et al., 1990). Typical problems that may be encountered are differences in nomenclatural usage, splitting versus amalgamation of taxa, and differing criteria used in the identification of taxa. Full documentation of decisions made in harmonising taxonomy in this way is essential.

Percentage counting

Diatom counts are expressed as a concentration value (see below) or as a percentage of a total sum. Most often the sum comprises all taxa but sometimes it is useful to exclude certain groups, for example, in cases where diatoms are clearly derived from the re-working of older sediments or where some taxa are massively dominant and obscure more interesting variation in less abundant taxa. The total number of valves to be counted for each sample varies according to the purpose of the analysis and according to the need to produce statistically reliable results. The statistical precision of percentage counting depends on the frequency of the taxon in the sample count in relation to the size of the sample count. An illustration of the way in which percentages change as the total counted increases is shown in Figure 6. It can be seen that there are marked differences in the percentages between a count of 100 and 200 while there is little change between 400 and 500. A count of 300 to 600 may therefore be recommended for purposes of routine analysis. In some cases this total may be inappropriate. A lower count is often sufficient if, for example, it is only necessary to locate biostratigraphic boundaries or events in replicate cores that are to be used for stratigrapahic correlation (Battarbee et al., 1978c) or if large numbers of samples are to be counted in high resolution studies (e.g., Renberg, 1990b). On the other hand a higher count may be necessary if the complete floristic composition of the sample is to be recorded. While most of the taxa in a sample may be encountered after a count of about 400 valves (cf., Fig. 6) additional taxa are likely to be encountered if greater total numbers are counted. A large count is also necessary if the fluctuations of interesting or ecologically important taxa through a core are obscured by the mass occurrence of more common taxa. This can occur especially in large eutrophic lakes where the benthic groups may be swamped by the quantity of planktonic individuals present or in shallow alkaline lakes where massive dominance by *Fragilaria* species is not unusual.

When counting it is necessary to ensure that a representative proportion of the coverslip is examined. This can be achieved by counting from a number of randomly selected fields of view, or more usually by counting along continuous traverses. The traverses should include equal proportions of edge and centre to override any sorting that may have taken place on evaporation. The basic counting units should be the single valve so that complete frustules are counted as two. Where chains of frustules remain intact, each valve should be counted individually. When a long chain of an infrequent diatom is encountered this may lead to statistical bias, but in this case the sample count should be increased so that a more accurate estimate of the true relative frequency of the taxon is obtained.

Fragments of diatom valves are included in the counts as long as a system is adopted that excludes the possibility of double or multiple counting. The best ways of doing this partly depends on the taxon concerned. In most cases the most satisfactory solution is to count all those fragments that include the valve centre (e.g., central nodule in *Pinnularia*, central inflation in *Tabellaria*, central porate area in *Stephanodiscus*), or a single characteristic feature of the valve (e.g., the larger apical inflation in *Asterionella formosa*). Some taxa (e.g., some *Synedra* species) do not possess recognizable centres and an alternative approach is to count only fragments that include a valve end and divide the total by 2. Small diatom fragments or fragments not included in the percentage count are usually ignored, but these could be counted separately and divided into different fracture classes (e.g., Beyens & Denys, 1982).

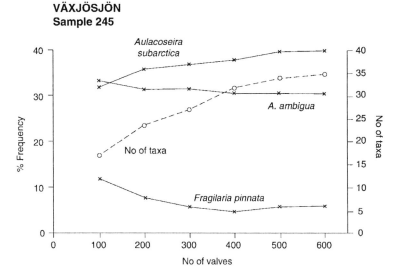

Figure 6. Graph of percentage and species number change with increasing count size (from Battarbee, 1986a).

Counting strategies also need to take into account partially dissolved valves so that a dissolution index as a measure of varying preservation can be derived. One of the simplest indices now used routinely in some laboratories is the Diatom Dissolution Index (F) developed by Flower & Likoshway (1993). This divides diatoms into two categories based on their appearance in the light microscope, non-dissolved (or "pristine") valves and partially dissolved forms according to:

$$F_i = \frac{\sum_{j}^{m} x_{ij}}{\sum_{j}^{m} X_{ij}}, \tag{1}$$

where x_{ij} is the number of pristine valves of species j (of m) counted in a sample i, compared to X_{ij}, the total of partially dissolved valves of species j. The index (F) varies between 1, perfect preservation, and 0, all valves show signs of dissolution. It is useful in comparing different assemblages within a core and between trap samples and core samples, but cannot discriminate between samples in which all valves are partly dissolved but where the degree of dissolution nevertheless varies. In these cases a somewhat more extended procedure could be used based on a classification of valves from different taxa into different dissolution stages, an approach that has been explored by Ryves (1994).

Diatom concentration and accumulation rate

It is often useful to estimate the concentration of diatoms per unit weight or unit volume of sediment. Given the sediment accumulation rate, diatom concentration values can be used to calculate the diatom accumulation rate per square cm per year. A number of ways of doing this are presented and evaluated in Battarbee (1986a) and Wolfe (1997). The simplest method recommended here is to use known quantites of plastic microspheres added as a spike to the sample. Diatom concentration is calculated according to:

$$\text{Total} = \frac{\text{microspheres introduced} \times \text{diatoms counted}}{\text{microspheres counted}}. \tag{2}$$

Diatom concentration values *per se* have limited direct palaeoecological significance but they can be very useful (see below). Their most important function is to provide the basis for calculating diatom accumulation rates (DAR). If the sediment accumulation rate is known, DAR can be calculated both for individual taxa and for all taxa combined (e.g., Battarbee, 1978a; Anderson, 1989). However, since diatoms vary considerably in size this latter statistic may not be very meaningful in palaeoproductivity terms, unless some conversion from cell number to biomass or biovolume is undertaken (Paasche, 1960). Mean diatom volumes can be computed and used to convert cell number data to cell volume data for individual taxa, and then the totals for individual taxa can be added, corrected for the rate of sediment accumulation, and expressed as diatom biovolume per cm per year. This can be plotted both against depth and, more realistically, since the depth axis does not often represent a linear timescale, against time (Battarbee, 1978a, 1978b).

Data analysis and interpretation

The aim of diatom analysis is to use the microfossil record to reconstruct changes in past diatom floras, occasionally for its own sake (e.g., in the case of rare or endemic taxa; Mackay et al., 1998) but mainly as indicators of past ecological and environmental change. This is a multi-step process involving an understanding of diatom ecology and distribution, diatom taphonomy (above), and lake sedimentology and stratigraphy. The development of sophisticated numerical procedures now allows an increasingly quantitative approach to the reconstruction of past environmental conditions, although much interpretation remains qualitative. Based on the above methods, the data generated from standard diatom analysis consist essentially of two complementary types, concentration data and compositional data.

Concentration data

Diatom concentrations vary in sediment cores as a result of a number of different factors including: the numbers of diatoms produced in the lake, the efficiency with which diatoms are transported to the sediment, the extent to which diatoms are dissolved either in the water column or in the sediment, and the rate of sediment accumulation. Consequently correctly interpreting changes in concentration needs a knowledge of all these factors. Sharp changes in diatom concentration are more likely to be the result of diatom preservation changes or changes in sediment accumulation rates rather than changes in production. In the case of changing sediment accumulation rate concentrations can be rapidly diluted by inwash of allochthonous sediment (e.g., Battarbee & Flower, 1984) and the preservation status of individual valves may not be affected. The occurrence of such events is usually clearly indicated by other changes in the core, such as in the loss-on-ignition, or mineral magnetic record. In the case of preservation changes it is very useful to have a record of the dissolution status of the diatom assemblages (e.g., the "diatom dissolution index (F or DDI)" defined above) as a down-core concordance between the DDI and the diatom concentration curve would then be expected.

In situations where changes in concentration can not be explained by dissolution or by sedimentological change, it may be useful to consider the changes as a record of diatom productivity and possibly a record of net primary productivity (e.g., Battarbee, 1973, 1978a; Bradbury & Waddington, 1973; Anderson, 1989). However, translating diatom concentration data into diatom production data is far from straightforward (e.g., Battarbee, 1978b) and requires good sediment chronology, a multi-core approach, a knowledge of diatom preservation status and an assumption that sediment focussing has been more or less constant through time. These issues have been most fully explored by Anderson (1989). The next step, from diatom production to primary production, is also problematic and requires the additional assumption that there is a linear relationship between the two, an assumption that may be valid in lakes that are not silica-limited, but certainly invalid in many eutrophic lakes where N and P are in excess of diatom uptake. Consequently, estimates of past primary productivity using diatoms are best addressed through the diatom compositional approach (e.g., using a diatom-TP transfer function) rather than through diatom concentration data.

Despite these limitations diatom concentration data do have ecological value, especially in helping to interpret diatom compositional data where changes in individual taxon abundances are otherwise not known because they are constrained by the constant 100% sum. This is important in eutrophication studies where nutrient enrichment may cause increases in the population of both benthic and planktonic taxa to occur, but where these changes, for example, might be registered by percentage counting as a relative decrease in benthos and a relative increase in plankton, since nutrient enrichment tends to cause a disproportionate increase in plankton. An example of this is given by Oldfield et al. (1983).

Compositional data—qualitative assessment

Although there has been a rapid development of quantitative reconstruction techniques over the last 20 years, some important aspects of diatom palaeolimnology remain essentially qualitative with interpretation based on an understanding of the ecology of individual taxa and groups of taxa. There are two good examples where this applies. First, diatom data may sometimes indicate major changes in habitat availability within a lake. Smol (1988) has advanced the hypothesis that the extent of winter ice and snow cover in high Arctic environments has a major impact on the balance between benthic and planktonic diatom populations in small lakes, and this has been used to explain some of the very clear species shifts observed in sediment cores from arctic (Smol, 1988; Douglas et al., 1994) and alpine lakes (Lotter et al., in press). Second, relative changes in the abundance of planktonic diatoms are not always easy to interpret as a response to a quantifiable change in a single driving variable. More often they are the result of complex interactions between variables (e.g., light, turbulence, resource limitation). This is especially well illustrated by Bradbury & Dieterich-Rurup's interpretation of data from Elk Lake, Minnesota (Bradbury & Dieterich-Rurup, 1993) who argued that the varying proportions of the main planktonic diatom taxa through the Holocene sediments of the lake were the result of the relative availability of silica and phosphorus in a manner generally predicted by resource-ratio theory (Kilham et al., 1986, 1996), and that these nutrient changes were the result of changes in water column

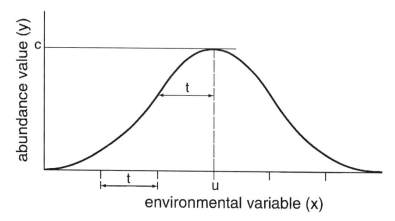

Figure 7. Gaussian response curve for the changing abundance of a species along an environmental gradient (from ter Braak, 1987b).

circulation driven by seasonal shifts in weather patterns. Rioual (2000) has also used this approach to interpret the changes in planktonic diatom dominance that are recorded in the Eemian sediments of Ribains Maar in the Massif Central, France. In both these cases powerful and robust interpretations about the direction of change can be made, but as yet there are no methods for quantifying the magnitude of change that the diatom data represent. This is an area for future research and probably requires a more process-based modelling approach than the empirical approach offered by transfer functions.

Compositional data—transfer functions

Attempts to derive quantitative estimates of environmental change have a relatively long history in diatom analysis stemming mainly from Nygaard's study of the acidification of Lake Gribsø in Denmark (Nygaard, 1956) based on the Hustedt pH classification scheme for diatoms. The underlying approach is principally the same today although much less emphasis is given to diatom classification schemes as there is great difficulty in assigning a taxon unambiguously to an individual class. As a result, classification schemes have largely been replaced by species-based approaches in which the relationship between individual diatom taxa and the environmental variable (cf., Fig. 7), or variables, of interest is encapsulated in a ecological response function, or transfer function.

Calibration training sets:

Transfer functions are most effectively derived by constructing calibration training sets that combine modern diatom and water chemistry data from selected lakes along the environ-mental gradient of interest (e.g., pH, total phosphorus, DOC, salinity) in the geographic regions that contain the study sites. Charles (1990) provides a useful guide to the compilation of training sets.

Diatom samples needed to compile training sets are usually obtained from surface sediments (often the top 0.5 to 1.0 cm level). It is often argued that these samples are the most appropriate ones to construct training sets as they are assumed to contain an integrated flora representative of the various habitats in a lake and are analogous to the sediment core assemblages used in reconstructions. However, surface samples have the disadvantage that they may also contain contaminants from older reworked sediment, and the age span of the uppermost sample is usually unknown and varies both within and between lakes. Moreover as diatoms in sediment samples may also be partially broken and dissolved, the assumption that this assemblage is fully representative of the present-day flora is not always true. An alternative is to define species optima using samples of living diatoms collected directly from benthic and planktonic habitats (e.g., Gasse et al., 1997). However, if this approach is preferred, a larger number of samples and field visits may be required to encompass the spatial and temporal variability of diatom growth at a site and to cover adequately the full range of taxa present in sediment cores.

In addition to diatom samples, training sets also require data for environmental variables that comprehensively describe the key characteristics of the catchment and the lake, especially the lake water chemistry. For very remote lakes it is often only practicable to visit the site once in which case the visit should take place in spring or autumn, before or after the summer stratification period. However, where possible, it is advisable to collect samples on at least four occasions, on a seasonal basis, in order to generate a mean or median value for each site. For nutrient training sets a monthly sampling regime is advisable.

The earliest pH training sets were relatively small, often between 30 and 40 sites (e.g., Charles, 1985; Flower, 1986). Whilst such datasets can perform well it is preferable to compile larger datasets ensuring that sites are evenly distributed along the gradient of interest and cover a long gradient to avoid artificial truncation of species' ranges. Small local datasets can be extended either by adding sites from the wider region or by combining with existing datasets. In the latter case, great care is needed to harmonise the datasets by standardising both diatom taxonomy and environmental variables as discussed above. As datasets are progressively developed, combined and published, it may be unnecessary to compile project-based training sets, but simply to construct a training set from published data available on the web, selecting sites that provide the best modern analogues for the sediment core assemblages for which reconstruction is required (Battarbee et al., 2000).

Diatom-environment training sets can be analysed statistically in very many different ways (Birks, 1995, 1998). However, a premier method that allows diatom and environmental data to be analysed together is canonical correspondence analysis (CCA) implemented by the computer program CANOCO (ter Braak, 1987a). Diagrams generated by CCA (e.g., Fig. 8) show ordination axes scores for samples and for species, and arrows for the environmental variables. The arrows indicate the direction of maximum variation of each environmental variable, and their length is directly proportional to their importance in explaining variation in the dataset. In Figure 8, axis 1 is strongly related to conductivity and anion composition, whereas axis 2 shows that pH and cation composition are also important in explaining differences between the diatom assemblages. Such ordination diagrams are also useful in summarising species-environment relationships, and allow taxa characteristic of particular chemical conditions to be identified (e.g., *Thalassiosira rudolfi* and *Navicula elkab* (hyperalkaline sodium carbonate brines), *Rhopalodia musculus* and *Navicula ammophila* (sodium chloride brines)).

For a transfer function to be developed for a particular variable, we require that the variable should explain a significant part of the total variation in the diatom data, independent

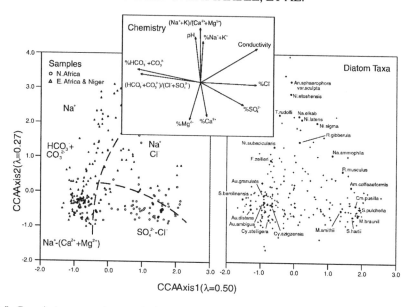

Figure 8. Canonical correspondence analysis sample- and species-environment biplot of the combined African saline lake training set (from Gasse et al., 1995).

of any additional variables. This requirement can be tested using partial canonical corre-spondence analysis, coupled with a Monte Carlo permutation test, to test the independence and relative strength of each hydrochemical variable. For example, the ten environmental variables shown in Figure 8 account for a total of 11.6% of the variance in the diatom data. Figure 9 shows how this variance is partitioned among the four hydrochemical gradients represented by these variables. Although there are confounding effects between some variables, especially conductivity and anion-type, each of the four major gradients accounts for an independent and statistically significant portion of the total variance, justifying the development of transfer functions for variables representing these four gradients (Gasse et al., 1995).

Analyses of this kind enable palaeolimnologists to identify the environmental variables that can be legitimately reconstructed on statistical grounds. Invariably the principal vari-ables that emerge are the expected ones associated with pH, conductivity and nutrient status, although DOC (Birks et al., 1990b; Kingston & Birks, 1990; Korsman & Birks, 1996), water depth (Yang & Duthie, 1995; Brugam et al., 1998) and temperature (Vyvermann & Sabbe, 1995; Pienitz et al., 1995; Lotter et al., 1997; Weckström et al., 1997) can also be important. When diatom relative abundance is plotted against these variables, a unimodal response for most taxa is expected (Fig. 10) allowing the centroid of the distribution to be calculated.

Weighted-averaging regression and calibration: There are a range of numerical tech-niques appropriate for developing diatom-based transfer functions (see review by Birks, 1995, 1998). However, the currently most widely used method, and the one that invariably performs best in empirical comparisons, is weighted averaging. The method was first used in palaeolimnology by ter Braak & Van Dam (1989) and developed by Birks et al. (1990a) during the SWAP project. It has subsequently been used for reconstruction of a wide range

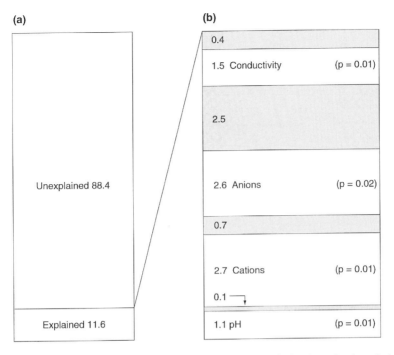

Figure 9. Results of partial CCAs of the training set shown in Figure 8, partitioning the total variance in the diatom data into (a) explained and unexplained portions, and (b) components representing the unique contributions of variables representing the conductivity, pH, cation and anion gradients (open), and correlations between gradients (shaded). Significance (p) values are based on 99 random Monte Carlo permutations.

of variables including pH (Korsman & Birks, 1996; Kingston & Birks, 1990; Dixit et al., 1991, 1993; Cameron et al., 1999), TP (Anderson et al., 1993; Bennion, 1994; Hall & Smol, 1992; Wunsam & Schmidt, 1995; Lotter et al., 1998), and salinity (Fritz et al., 1991; Gasse et al., 1995; Wilson et al., 1996). In WA the main assumption is that, at a given value of the environmental variable, taxa that have optima nearest to that value will be most abundant. The optimum (u_k) for a taxon (Fig. 7) would then be an average of all the values for that environmental variable of lakes in the training set in which the taxon occurred, weighted by its relative abundance. This is the regression step:

$$\hat{u}_k = \frac{\sum\limits_{i=1}^{n} y_{ik} x_i}{\sum\limits_{i=1}^{n} y_{ik}}, \tag{3}$$

where x_i is the value of the environmental variable at site i, and y_{ik} is the abundance of species k at site i.

Weighted averaging can also be used to estimate tolerance values (Fig. 7) for diatoms (Juggins, 1992; Birks, 1995) and these values can be used to down-weight the importance of taxa with a wide tolerance relative to those with a narrow tolerance (e.g., Birks et al., 1990a). Figure 11 shows a plot of species optima and tolerances against salinity for the northern Great Plains salinity training set (Fritz et al., 1991).

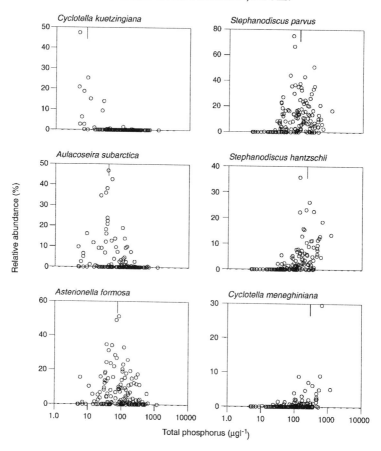

Figure 10. Relative abundance of taxa vs. total phosphorus from a training set of 146 Northwest European lakes (Bennion et al., 1996). Ticks indicate position of weighted-average optima or centroid.

The second step in the reconstruction is a calibration step where the calculated optima of the various taxa in the training set are used to calculate the value of the environmental variable at each site by taking the average of the species optima in the sample weighted by their abundance:

$$\hat{x}_i = \frac{\sum_{k=1}^{m} y_{ik} \hat{u}_k}{\sum_{k=1}^{m} y_{ik}}. \tag{4}$$

This simple approach performs well for many datasets. However, one weakness is that it ignores correlations in the species data that remain after fitting the environmental variable of interest where the species response is also influenced by secondary or "nuisance" environmental variables that are not taken into account in weighted averaging (ter Braak & Juggins, 1993; Birks, 1995). In these cases "weighted averaging—partial least squares" (WA-PLS) can be used (ter Braak & Juggins, 1993). In WA-PLS successive components are extracted from the training set to increase the predictive power of the regression model.

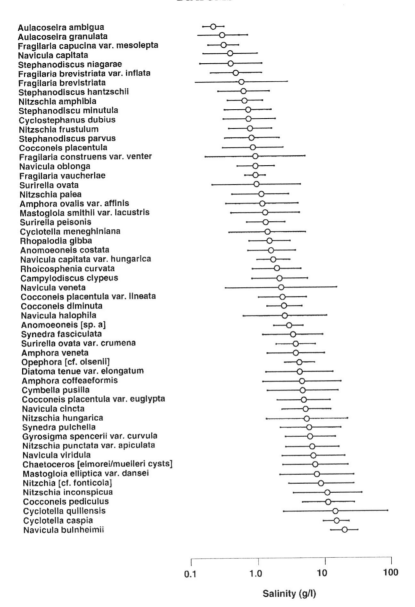

Figure 11. Taxon optima and tolerances for salinity from the Northern Great Plains training set (from Fritz et al., 1991; with permission).

The predictive ability of transfer functions can be assessed by examining the relationship between the measured and inferred values of the environmental variable of interest in the training set. However, calculating the correlation coefficient and root mean squared error (RMSE) on the basis of the training set data alone can be misleading or over-optimistic as the same data are being used to generate and evaluate the model (Birks, 1995). Birks (1995)

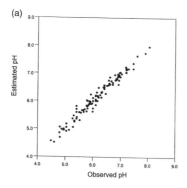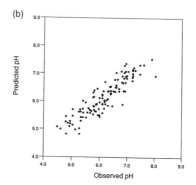

Figure 12. (a) Plot of estimated pH values based on a two component weighted averaging partial least squares (WA-PLS) regression for the 167 lakes within the SWAP calibration set against observed pH values. (b) as for (a) but plotting predicted values, using leave-one out jack-knifing, against observed pH (redrawn from Cameron et al., 1999; with permission).

referes to these as "apparent" errors. A more realistic error estimate is the root mean squared error of prediction (RMSEP) that can be calculated using intensive computer resampling, such as jackknifing and bootstrapping (Birks et al., 1990a; ter Braak & Juggins, 1993). Jackknifing is sometimes preferred as it is less computer intensive than bootstrapping. In this approach one sample is excluded from the original training set, and the inferred value of the environmental variable for that sample is based on the optima and tolerance of the taxa in the remaining samples in the training set. Jackknife and bootstrap error estimates can be made using the computer programs CALIBRATE (Juggins & ter Braak, 1999) and WACALIB (Line et al., 1994) respectively. Figure 12 provides an example of the difference between "apparent" and "prediction" errors for the AL:PE training set for pH (Cameron et al., 1999).

Once a transfer function has been established for a specific variable, past values for that variable can be reconstructed from sediment core analysis on the assumption that the diatom taxa in the core samples are well represented in the training set and diatom analysis of the core uses the same taxonomic system as that used in constructing the training set. Where core samples contain taxa that are not well represented in the training sets, reconstructed values should be treated with caution. A measure of the similarity between fossil and training set samples can be obtained using modern analogue techniques, and fossil samples with no modern analogues can similarly be treated with caution (Birks, 1995, 1998).

Modern analogue techniques: The modern analogue technique (MAT) is based on the simple idea that similar diatom assemblages reflect similar environmental conditions. Thus, if a suitable modern analogue can be found for a fossil sample, the past environment for that sample can be inferred from the environmental conditions under which the modern (analogous) sample was deposited (Birks, 1995; Birks et al., volume 5 of this series). To define an appropriate analogue, the method requires the numerical definition of similarity (or dissimilarity) between the modern and fossil samples. For diatom assemblages squared chi-squared distance is popular (e.g., Flower et al., 1997), as this has the desirable property of maximising the signal-to-noise ratio of the differences between diatom assemblages (Overpeck et al., 1985):

$$d_{ij}^2 = \sum_{k=1}^{m} \left[\frac{(y_{ik} - y_{jk})^2}{(y_{ik} + y_{jk})} \right] \tag{5}$$

where y_{ik} is the proportion of diatom taxon k in sample i, d_{ij} is the Chi-squared distance between samples i and j. The squared Chi-squared distance has the desirable property of weighting the diatom proportions so as to emphasise the signal component of the differences between assemblages at the expense of the noise component. The values of squared Chi-squared distance can vary between zero and 2, with lower values indicating more similar assemblages (Overpeck et al., 1985; Flower et al., 1997).

Where several modern analogues are found for a fossil sample, the environmental reconstruction can be based on the mean of the closely matching analogues (Birks, 1995). MAT has been used successfully in the fields of palynology and palaeoceanography to provide environmental reconstructions but has so far had limited use in palaeolimnology. One of the main reasons for this is that the methods requires an extensive training set that covers the complete range of diatom assemblages and environmental conditions found in the fossil material. Most current diatom training sets do not fulfil these criteria, and in empirical tests MAT appears to offer no advantages over WA-based approaches as a reconstruction technique (e.g., Juggins et al., 1994). However, there is clearly need for further evaluation as larger training sets are developed and merged.

Although its use as a reconstruction technique is limited at present, MAT has proved useful in assessing the reliability of WA-based reconstructions. In this case MAT is used to identify fossil samples that lack a "close" or "good" analogue in the modern training set, and reconstructions for such samples should be treated with caution. Birks (1995) discusses a number of ways in which cut-off dissimilarity values can be calculated to define "close" and "good" analogues, and Birks et al. (1990a), Jones & Juggins (1995), and Hall & Smol (1993) show how this approach can be used to evaluate core-based reconstructions.

Understanding of the past chemical and biological status is an essential pre-requisite to effective lake restoration and management. Where direct measurements of past status are unavailable, MAT can also help by identifying modern reference lakes that can be used as analogues for the pre-impacted levels of a polluted site using the principle of space for time substitution. For example, Flower et al. (1997) used this approach to identify pristine or near pristine lakes that can be used as reference sites for the pre-impacted levels of the acidified Round Loch of Glenhead in Scotland. Although there are methodological problems associated with the selection of potential analogue sites (see Flower et al., 1997 for a discussion), the method has the potential to identify whole-lake ecosystems that can serve as biological targets for currently disturbed sites.

Summary

The basic requirements for diatom analysis have changed little over the last few decades in terms of sampling, slide preparation, microscopy and taxonomy but, on the other hand, there have been major improvements in our knowledge of diatom distribution and ecology and a revolution in our ability to analyse diatom data. These changes have been driven by the increasing recognition of the practical uses of diatoms as indicators of environmental change and by the development of novel numerical and computing techniques that allow diatom-environment relationships to be quantified. However, in the future and despite the application of new techniques (e.g., Vasko et al., 2000), it is unlikely that there will be significant improvements in transfer function statistics or in the range of environmental variables for which diatoms can be confidently used. Nevertheless, there is real scope for making transfer functions much more widely applicable around the world principally through web-based information systems such as EDDI (Battarbee et al., 2000), and in using the databases generated through merged training sets to explore unresolved and vastly under-researched questions of diatom biogeography. In addition, as multi-proxy approaches in palaeolimnology become common, diatomists should be able to focus more on questions of ecological response to environmental change rather than on reconstructing environmental change *per se*. Such a move would be especially welcome as it would herald a change from purely empirical mechanistic approaches inherent in the transfer function method to approaches that require a deeper understanding of diatom habitats, life-cycles and competitive strategies and a wider consideration of the role of diatoms in the overall functioning of aquatic ecosystems

Acknowledgements

We should like to thank Tim Allott, John Birks, Catherine Dalton, Simon Dobinson, Annette Kreiser, Anson Mackay, Don Montieth, Patrick Rioual, Dave Ryves, Catherine Stickley, Pedro Tapia and especially students on the ECRC courses over the last 10 years for their help, knowingly or otherwise, in the development of the methods and ideas presented here. We are also grateful to John Smol, Bill Last and John Birks for helpful comments on the text, David Mann for permission to reproduce Figures 1 and 2 and for help with the glossary, Eileen Cox for Table 1 and Cath Pyke for producing the diagrams.

References

Alhonen, P., 1971. The Stages of the Baltic Sea as indicated by the diatom stratigraphy. Acta Bot. Fenn. 92: 1–18.

Anderson, N. J., 1989. A whole-basin diatom accumulation rate for a small eutrophic lake in Northern Ireland and its palaeoecological implications. J. Ecol. 75: 926–946.

Anderson, N. J., 2000. Diatoms, temperature and climate change. Eur. J. Phycol. 35(4): 307–314.

Anderson, N. J. & B. Rippey, 1994. Monitoring lake recovery from point-source eutrophication: the use of diatom-inferred epilimnetic total phosphorus and sediment chemistry. Freshwat. Biol. 32: 625–639.

Anderson, N. J., B. Rippey & C. E. Gibson, 1993. A comparison of sedimentary and diatom-inferred phosphorus profiles: implications for defining pre-disturbance nutrient conditions. Hydrobiologia 253: 357–366.

Anonymous, 1975. Proposals for a standardization of diatom terminology and diagnoses. Nova Hedwigia, Beih. 53: 323–354.

Appleby P. G., P. J. Nolan, D. W. Gifford, M. J. Godfrey, F. Oldfield, N. J. Anderson & R. W. Battarbee, 1986. ^{210}Pb dating by low background gamma counting. Hydrobiologia 143: 21–27.

Appleby, P. G. & F. Oldfield, 1988. Radioisotope studies of recent lake and reservoir sedimentation. In Crickmore, M. J. et al. (eds.) The Use of Nuclear Techniques in Sediment Transport and Sedimentation Problems, UNESCO.

Barber, H. G. & E. Y. Haworth, 1981. A guide to the morphology of the diatom frustule, with a key to the British Freshwater genera. Freshwater Biological Association Scientific Publication No. 44., 112 pp.

Barker, P., 1992. Differential diatom dissolution in late Quaternary sediments from Lake Manyara, Tanzania: an experimental approach. J. Paleolim. 7: 235–251.

Barker, P., J. C. Fontes, F. Gasse & J. C. Druart, 1994. Experimental dissolution of diatom silica in concentrated salt solutions and implications for paleoenvironmental reconstruction. Limnol. Oceanogr. 39: 99–110.

Battarbee, R. W., 1973. A new method for estimating absolute microfossil numbers, with special reference to diatoms. Limnol. Oceanogr. 18: 647–653.

Battarbee, R. W., 1978a. Observations on the recent history of Lough Neagh and its drainage basin. Phil. Trans. r. Soc., Lond. 281: 303–345.

Battarbee, R. W., 1978b. Relative composition, concentration and calculated influx of diatoms from a sediment core from Lough Erne, Northern Ireland. Pol. Arch. Hydrobiol. 25: 9–16.

Battarbee, R. W., 1978c. Biostratigraphical evidence for variations in the recent pattern of sediment accumulation in Lough Neagh, N. Ireland. Verh. int. Ver. Limnol. 20: 624–629.

Battarbee, R. W., 1979. Early algological records—help or hindrance to palaeolimnology? Nova Hedwigia 4: 379–394.

Battarbee, R. W., 1981. Changes in the diatom microflora of a eutrophic lake since 1900 from a comparison of old algal samples and the sedimentary record. Holarct. Ecol. 4: 73–81.

Battarbee, R. W., 1986a. Diatom analysis. In Berglund, B. E. (ed.) Handbook of Holocene Palaeoecology and Palaeohydrology. Wiley, Chichester: 527–570.

Battarbee, R. W., 1986b. The eutrophication of Lough Erne inferred from changes in the diatom assemblages of ^{210}Pb and ^{137}Cs-dated sediment cores. Proc. R. Ir. Acad. B 86: 141–168.

Battarbee, R. W., 1991. Recent palaeolimnology and diatom-based environmental reconstruction. In Shane L. C. K. & E. J. Cushing (eds.) Quaternary Landscapes. University of Minnesota Press, Minneapolis: 129–174.

Battarbee, R. W., 2000. Palaeolimnological approaches to climate change, with special regard to the biological record. Quat. Sci. Rev. 19: 197–124

Battarbee, R. W. & R. J. Flower, 1984. The inwash of catchment diatoms as a source of error in the sediment-based reconstruction of pH in an acid lake. Limnol. Oceanogr. 29: 1325–1329.

Battarbee, R. W. & M. Kneen, 1982. The use of electronically counted microspheres in absolute diatom analysis. Limnol. Oceanogr. 27: 184–188.

Battarbee, R. W., J. P. Smol & J. Meriläinen, 1986. Diatoms as indicators of pH: a historical review. In Smol, J. P., R. W. Battarbee, R. B. Davis & J. Meriläinen (eds.) Diatoms and Lake Acidity: the Use of Siliceous Algal Microfossils in Reconstructing pH. Junk, The Hague: 5–14.

Battarbee, R. W., D. F. Charles, S. Dixit & I. Renberg, 1999. Diatoms as indicators of surface water acidity. In: The Diatoms: Applications for the Environmental and Earth sciences. In Stoermer, E. F. & J. P. Smol (eds.) Cambridge University Press, Cambridge: 85–127.

Battarbee, R. W., J. Mason, I. Renberg & J. F. Talling (eds.) 1990. Palaeolimnology and Lake Acidification. The Royal Society, London, 219 pp.

Battarbee, R. W., S. Juggins, F. Gasse, N. J. Anderson, H. Bennion & N. G. Cameron, 2000. European Diatom Database (EDDI): an information system for palaeoenvironmental reconstruction. European Climate Science Conference, Vienna City Hall, Vienna, Austria, 19–23 October 1998, pp. 1–10.

Bennion, H., 1994. A diatom-phosphorus transfer function for shallow, eutrophic ponds in southeast England. Hydrobiologia 275/276: 391–410.

Bennion, H., S. Juggins & N. J. Anderson, 1996. Predicting epilimnetic phosphorus concentrations using an improved diatom-based transfer function, and its application to lake eutrophication management. Environ. Sci. Tech. 30: 2004–2007.

Berglund, B. E., 1986. Handbook of Holocene palaeoecology and palaeohydrology. John Wiley, Chichester, 896 pp.

Beyens, L. & L. Denys, 1982. Problems in diatom analysis of deposits: allochthonous valves and fragmentation. Geologie Mijnb. 61: 159–162.

Birks, H. J. B., 1995. Quantitative palaeoenvironmental reconstructions. In Maddy, D. & J. S. Brew (eds.) Statistical Modelling of Quaternary Science Data Quaternary Research Association Technical Guide 5, Cambridge: 161–254.

Birks, H. J. B., 1998. Numerical tools in palaeolimnology—progress, potentialities, and problems. J. Palaeolim. 20: 307–332.

Birks, H. J. B., J. M. Line, S. Juggins, A. C. Stevenson & C. J. F. ter Braak, 1990a. Diatoms and pH reconstruction. Phil. Trans. r. Soc., London B 327: 263–278.

Birks, H. J. B., S. Juggins & J. M. Line, 1990b. Lake surface-water chemistry reconstructions from palaeolimnological data. In Mason, B. J. (ed.) The Surface Waters Acidification Programme. Cambridge University Press, Cambridge: 301–313.

Bloesch, J. & N. M. Burns, 1980. A critical review of sedimentation trap techniques. Schweiz. Z. Hydrobiol. 42: 15–54.

Bodén, P., 1991. Reproducibility in the random settling method for quantitative diatom analysis. Micropaleontology 37: 313–319.

Bradbury, J. P., 1975. Diatom stratigraphy and human settlement in Minnesota. Geological Society of America, Special Paper 171, 74 pp.

Bradbury, J. P. & K. V. Dieterich-Rurup, 1993. Holocene diatom palaeolimnology of Elk Lake, Minnesota. In Bradbury, J. P. & W. E. Dean (eds.) Elk Lake, Minnesota: Evidence for Rapid Climate Change in the North-Central United States. Geological Society of America Special Paper, 276. Boulder, Colorado: 215–237.

Bradbury, J. P. & J. C. B. Waddington, 1973. The impact of European settlement on Shagawa Lake, Northeastern Minnesota, U. S. A. In H. Birks, J. B. & R. G. West (eds.) Quaternary Plant Ecology. Blackwell Scientific Publications, Oxford: 289–307.

Bradbury, J. P. & T. C. Winter, 1976. Areal distribution and stratigraphy of diatoms in the sediments of Lake Sallie, Minnesota. Ecology 57: 1005–1014.

Brander, G., 1936. Über das Einsammeln von Erdproben und ihre Praparation für die qualitative und quantitative Diatomeenanalyse. Bull. Comm. géol. Finl. 115: 131–144.

Brockmann, C., 1954. Die Diatomeen in den Ablagerungen der östpreussischen Haffe. Meyniana 3: 1–95

Brugam, R. B., 1979. A re-evolution of the Araphidineae/Centrales index as an indicator of lake trophic status. Freshwat. Biol. 9: 451–460.

Brugam, R. B. & C. Patterson, 1983. The A/C (Araphidineae/Centrales) ratio in high and low alkalinity lakes in eastern Minnesota. Freshw. Biol. 13: 47–55.

Brugam, R. B., K. McKeever & L. Kolesa, 1998. A diatom-inferred water depth reconstruction for an Upper Peninsula, Michigan, lake. J. Paleolim. 20: 267–276.

Camburn, K. E. & D. F. Charles, 2000. Diatoms of low-alkalinity lakes in the Northeastern United States. The Academy of Natural Sciences of Philadelphia Special Publication 18, 152 pp.

Camburn, K. E. & J. C. Kingston, 1986. The genus *Melosira* from soft-water lakes with special reference to northern Michigan, Wisconsin and Minnesota. In Smol, J. P., R. W. Battarbee, R. B. Davis & J. Meriläinen (Eds.) Diatoms and Lake Acidity. Dr. W. Junk, Dordrecht, The Netherlands: 17–34.

Camburn, K. E., J. C. Kingston D. F. & Charles, 1986. PIRLA Diatom Iconograph. PIRLA Unpublished Report Number 3. Indiana University, Bloomington.

Cameron, N. G., 1995. The representation of diatom communities by fossil assemblages in a small acid lake. J. Paleolim. 14: 185–223.

Cameron, N. G., H. J. B. Birks, V. J. Jones, F. Berge, J. Catalan, R. J. Flower, J. Garcia, B. Kawecka, K. A. Koinig, A. Marchetto, P. Sánchez-Castillo, R. Schmidt, M. Šiško, N. Solovieva, E. Štefková & M. Toro., 1999. Surface-sediment and epilithic diatom pH calibration sets for remote European mountain lakes (AL:PE Project) and their comparison with the Surface Waters Acidification Programme (SWAP) calibration set. J. Paleolim. 22: 291–317.

Campeau, S., R. Pienitz & A. Héquette, 1999. Diatoms from the Beaufort Sea coast, southern Arctic Ocean (Canada). Bibliotheca Diatomologica, Band 42. J. Cramer, Stuttgart, 244 pp.

Canter, H. M., 1979. Fungal and protozoan parasites and their importance in the ecology of the phytoplankton. Rep. Freshwat. biol. Ass. 47: 43–50.

Catalan, J. & L. Camarero, 1991. Ergoclines and biological processes in high-mountain lakes: similarities between the summer stratification and the ice-forming period in lake Redó (Pyrenees). Verh. int. Ver. Limnol. 24: 1011–1015.

Catalan, J. & L. Camarero, 1993. Seasonal changes in pH and alkalinity in two Pyrenean high-mountain lakes. Verh. int. Ver. Limnol. 25: 749–753.

Cattaneo, A. & J. Kalff, 1979. Primary production of algae growing on natural and artificial aquatic plants: a study of interactions between epiphytes and their substrate. Limnol. Oceanogr. 24: 1031–1037.

Charles, D. F., 1985. Relationships between surface sediment diatom assemblages and lakewater characteristics in Adirondack lakes. Ecology 66: 994–1011.

Charles, D. F., 1990. A checklist for describing and documenting diatom and chrysophyte calibration data sets and equations for inferring water chemistry. J. Paleolim. 3: 175–178.

Charles, D. F. & D. R. Whitehead, 1986. The PIRLA project: Paleoecological Investigation of Recent Lake Acidification. Hydrobiologia 143: 13–20.

Cholnoky, B. J., 1968. Die Ökologie der Diatomeen in Binnengewässern. Cramer, 699 pp.

Cleve, P. T. 1894–95. Synopsis of the naviculoid diatoms. Kgl. Sven. Vet. Akad. Handl., 26: 1–194, 27: 1–219.

Cleve, P. T., 1899. Postglaciala bildninggarnas klassifikation pågrund av deras fossila diatomacéer. Sver. geol. Unders. 180: 59–61.

Cleve-Euler, A., 1922. Om diatomacevegetationen och dess förandringar I Sabysjön, Uppland, samt några dämda sjöar i Salatrakten. Sver. geol. Unders. C309: 1–76.

Cleve-Euler, A., 1951–1955. Die Diatomeen von Schweden und Finnland. I-V. Kongl. svenska Vetenskapsakad. Handl. Ser. 4. 2(1): 1–163 (1951); Ser. 4, 3(3): 1–153 (1952); Ser. 4. 4(2): 1–158 (1953); Ser. 4. 4(5): 1–255 (1953); Ser. 4, 5(4): 1–232 (1955).

Cox, E. J., 1996. Identification of Freshwater Diatoms from Live Material, Chapman & Hall, New York, 158 pp.

Crawford, R. M., 1979. Filament Formation in the Diatom Genera Melosira C. A. Agardh and Paralia Hyberg. Nova Hegwigia 64: 121–133.

Cumming, B. F., K. A. Davey, J. P. Smol & H. J. B. Birks, 1994. When did acid-sensitive Adirondack lakes (New York, USA) begin to acidify and are they still acidifying? Can. J. Fish. aquat. Sci. S. 51: 1550–1568.

Cumming, B. F., S. E. Wilson, R. I. Hall & J. P. Smol, 1995. Diatoms from British Columbia (Canada) lakes and their relationship to salinity, nutrients and other limnological variables. Bibliotheca Diatomologica 31. J. Cramer, Stuttgart, 207 pp.

Dalton, C. P., 2000. Impact of catchment afforestation on lakes in the west of Ireland. Unpublished PhD Thesis, University College London, 293 pp.

Davis, R. B. & D. S. Anderson, 1985. Methods of pH calibration of sedimentary diatom remains for reconstructing history of pH in lakes. Hydrobiologia 120: 69–87.

DeNicola, D. M., 1986. The representation of living diatom communities in deep-water sedimentary diatom assemblages in two Maine (U.S.A.) lakes. In Smol, J. P., R. W. Battarbee, R. B. Davis & J. Meriläinen (eds.) Diatoms and Lake Acidity. Dr. W. Junk, Dordrecht, The Netherlands: 73–85.

Denys, L. & H. de Wolf, 1999. Diatoms as indicators of coastal paleoenvironments and relative sea-level change. In Stoermer, E. F. & J. P. Smol (eds.) The Diatoms: Applications for the Environmental and Earth Sciences. Cambridge University Press, Cambridge: 277–297.

Digerfeldt, G., 1972. The post-glacial development of Lake Trummen: regional vegetation history, water level changes and palaeolimnology. Folia limnol. scand. 16: 1–104.

Dixit, S. S., A. S. Dixit & J. P. Smol, 1991. Multivariate environmental inferences based on diatom assemblages from Sudbury (Canada) lakes. Freshwat. Biol. 26: 251–266.

Dixit, S. S., B. F. Cumming, H. J. B. Birks, J. P. Smol, J. C. Kingston, A. J. Uutala, D. F. Charles & K. E. Camburn, 1993. Diatom assemblages from Adirondack lakes (New York, USA) and the development of inference models for retrospective environmental assessment. J. Paleolim. 8: 27–47.

Douglas, M. S. V. & J. P. Smol, 1999. Freshwater diatoms as indicators of environmental change in the High Arctic. In Stoermer, E. F. & J. P. Smol (eds.) The Diatoms: Applications for the Environmental and Earth Sciences. Cambridge University Press, Cambridge: 227–244.

Douglas, M. S. V., J. P. Smol & W. Blake, 1994. Marked post-18th century environmental change in high Arctic ecosystems. Science 266: 416–419.

Droop, S. J. M., 1993. The "striatometer"—a new device for measuring striation densities. Diatom Research 8: 195–198.

Eaton, J. W. & B. Moss, 1966. The estimation of numbers and pigment content in epipelic populations. Limnol. Oceanogr. 11: 584–595.

Eronen, M., 1974. The history of the Litorina Sea and associated Holocene events. Commentat. Physico-math. 44: 79–195.

Florin, M-B., 1944. En sensubarktisk transgression i trakten av Södra Kilsbergen enligt diatomacé-succession i ormrådets högre belägna fornsjölagerföljder. Geol. Fören. Förh. 66: 417–488.

Florin, M-B., 1946. Clypeusfloran i postglaciala fornsjölagerföljder i östra Mellansverige, Geol. Fören. Förh. 68: 429–458.

Flower, R. J., 1986. The relationship between surface sediment diatom assemblages and pH in 33 Galloway lakes: some regression models for reconstructing pH and their application to sediment cores. Hydrobiologia 143: 93–103.

Flower, R. J., 1993. Diatom preservation: experiments and observations on dissolution and breakage in modern and fossil material. Hydrobiologia 269/270: 473–484.

Flower, R. J. & R. W. Battarbee, 1983. Diatom evidence for recent acidification of two Scottish lochs. Nature 20:130–133.

Flower, R. J. & R. W. Battarbee, 1985. The morphology and biostratigraphy of *Tabellaria quadriseptata Knudson* (Bacillariophyta) in acid waters and lake sediments in Galloway, south-west Scotland. Br. Phycol. J. 20: 69–79.

Flower, R. J. & Y. Likhoshway, 1993. An investigation of diatom preservation in Lake Baikal. Fifth workshop on diatom algae, March 16–20, Irkutsk, Russia, pp. 77–78.

Flower, R., S. Juggins & R. W. Battarbee, 1997. Matching diatom assemblages in lake sediment cores and modern surface sediment samples: the implications for lake conservation and restoration with special reference to acidified systems. Hydrobiologia 344: 27–40.

Fritz, S. C., S. Juggins, R. W. Battarbee & D. R. Engstrom, 1991. Reconstruction of past changes in salinity and climate using a diatom-based transfer function. Nature 352: 706–708.

Fritz, S. C., S. Juggins & R. W. Battarbee, 1993. Diatom assemblages and ionic characterization of freshwater and saline lakes of the Northern Great Plains, North America: a tool for reconstructing past salinity and climate fluctuations. Can. J. Fish. aquat. Sci. S 50: 1844–1856.

Fritz, S. C., B. F. Cumming, F. Gasse & K. Laird, 1999. Diatoms as indicators of hydrologic and climatic change in saline lakes. In: The Diatoms: Applications for the Environmental and Earth sciences. In Stoermer, E. F. & J. P. Smol (eds.) Cambridge University Press, Cambridge: 41–72.

Fryxell, G. A., 1974. Diatom collections. Nova Hedwigia 53: 355–365.

Gasse, F., 1986. East African diatoms: taxonomy, ecological distribution. Cramer, Stuttgart, 201 pp.

Gasse, F., 1987. Diatoms for reconstructing palaeoenvironments and palaeohydrology in tropical semi-arid zones. Example of some lakes from Niger since 12,000 BP. Hydrobiologia 154: 127–163.

Gasse, F., S. Juggins & L. Ben Khelifa, 1995. Diatom-based transfer functions for inferring past hydrochemical characteristics of African lakes. Palaeogeogr. Palaeoclim. Palaeoecol. 117: 31–54.

Gasse, F., J. F. Talling & P. Kilham, 1983. Diatom assemblages in East Africa: classification, distribution and ecology. Revue Hydrobiol. trop. 16: 3–34.

Gasse, F., P. Barker, P. A. Gell, S. C. Fritz & F. Chalie, 1997. Diatom-inferred salinity in palaeolakes: An indirect tracer of climate change. Quat. Sci. Rev. 16: 547–563.

Germain, H., 1981. Flore des Diatomées Paris: Soc. Nouv. Edit. Boubée, 444 pp.

Glew, J. R., 1991. Miniature gravity corer for recovering short sediment cores. J. Paleolim. 5: 285–287.

Glew, J. R., J. P. Smol & W. M. Last, 2001. Sediment core collection and extrusion. In Last, W. M. & J. P. Smol (eds.) Tracking Environmental Change in Lake Sediments. Vol 1, Basin Analysis, Coring and Chronological Techniques. Kluwer Academic Publishers, Dordrecht: in press.

Halden, B., 1929. Kvartärgeologiska diatomacéestudier belysande den postglaciala transgression a Svenska Västkusten, Geol. Fören. Förh. 51: 311–366.

Hall, R. I. & J. P. Smol, 1992. A weighted-averaging regression and calibration model for inferring total phosphorus concentration from diatoms in Biritsh Columbia (Canada) lakes. Freshwat. Biol. 27: 417–434.

Hall, R. I. & J. P. Smol, 1993. The influence of catchment size on lake trophic status during the hemlock decline and recovery (4800 to 3500 BP) in southern Ontario lakes. Hydrobiologia 269/270: 371–390.

Hall, R. I. & J. P. Smol, 1999. Diatoms as indicators of lake eutrophication. In: The Diatoms: Applications for the Environmental and Earth Sciences. In Stoermer, E. F. & J. P. Smol (eds.) Cambridge University Press, Cambridge: 128–168.

Harwood, D. M. & R. Gersonde, 1990. Lower Cretaceous diatoms from ODP leg 113 site 693 (Weddell Sea). Part 2: resting spores, chrysophycean cysts, an endoskeletal dinoflagellate, and notes on the origin of diatoms. In Barker, P. F., J. P. Kennett et al. Proc. ODP Sci. Results, 113: College Station, TX (Ocean Drilling Program): 403–425.

Hasle, G. R., 1977. The use of electron microscopy in morphological and taxonomical diatom studies. In Werner, D. (ed.) The Biology of Diatoms. Blackwell, Oxford: 18–23.

Haworth, E. Y., 1969. The diatoms of a sediment core from Blea Tarn, Langdale. J. Ecol. 57: 429–441.

Haworth, E. Y., 1980. Comparison of continuous phytoplankton records with the diatom stratigraphy in the recent sediments of Blelham Tarn. Limnol. Oceanogr. 25: 1093–1103.

Hurd, D. C., 1972. Factors affecting the solution rate of biogenic opal in seawater. Earth Planet. Sci. Letters 15: 411–417.

Hurd, D. C. & S. Birdwhistell, 1983. On producing a more general model for biogenic silica dissolution. Am. J. Sci. 283: 1–28.

Hustedt, F., 1927–66. Die Kieselalgen Deutschlands, Österreichs und der Schweiz. In Dr L. Rabenhorst's Kryptogamen-Flora von Deutschland, Österreich und der Schweiz. 7. Leipzig: Akademische Verlagsgesellschaft.

Hustedt, F., 1930. Bacillariophyta (Diatomaceae). In Pascher, A. Die Süsswasser-flora Mitteleuropas Heft 10. Jena: Gustav Fischer Verlag. 466 pp.

Hustedt, F., 1937–39. Systematische und ökologische Untersuchungen über den Diatomeen-Flora von Java, Bali, Sumatra. Arch. Hydrobiol. 15 & 16.

Hustedt, F., 1956. Kieselalgen (Diatomeen). Kosmos-Verlag Franckh, Stuttgart.

Hustedt, F., 1957. Die Diatomeenflora des Fluss-systems der Weser im Gebiet der Hansestadt Bremen. Abh. Naturw. Ver. Bremen. 34: 181–440.

Jewson, D. H., 1992a. Size reduction, reproductive strategy and the life-cycle of a centric diatom Phil. Trans. r. Soc., Lon. B 336: 191–213.

Jewson, D. H., 1992b. Life cycle of *Stephanodiscus* sp. (Bacillariophyta). J. Phycol. 28: 856–866.

Jewson, D. H. & S. Lowry, 1993. *Cymbellonitzschia diluviana* Hustedt (Bacillariophyceae): habitat and auxosporulation. Hydrobiologia 269–70: 87–96.

Jones, J. I., B. Moss, J. W. Eaton & J. O. Young, 2000. Do submerged aquatic plants influence periphyton community composition for the benefit of invertebrate mutualists. Freshwat. Biol. 43: 591–604.

Jones, V. J. & R. J. Flower, 1986. Spatial and temporal variability in periphytic diatom communities: palaeoecological significance in an acidified lake. In Smol, J. P., R. W. Battarbee, R. B. Davis & J. Meriläinen (eds.) Diatoms and Lake Acidity. Dr. W. Junk, Dordrecht, The Netherlands: 87–94.

Jones, V. & S. Juggins, 1995. The construction of a diatom-based chlorophyll a transfer function and its application at three lakes on Signy Island (maritime Antarctic) subject to differing degrees of nutrient enrichment. Freshwat. Biol. 34: 433–445.

Jones, V. J., A. C. Stevenson & R. W. Battarbee, 1989. Acidification of lakes in Galloway, south west Scotland: a diatom and pollen study of the post-glacial history of the Round Loch of Glenhead. J. Ecol. 77: 1–23.

Jousé, A., 1966. Diatomeen in Seesedimenten. Arch. Hydrobiol. 4: 1–32.

Juggins, S., 1992. Diatoms in the Thames Estuary, England: Ecology, Palaeoecology, and Salinity Transfer Function. Bibliotheca Diatomologica, Volume 25, 216 pp.

Juggins, S. & C. J. F. ter Braak, 1999. CALIBRATE Version 1.0—a program for species-environment calibration by [weighted averaging] partial least squares regression. Unpublished computer program, Department of Geography, University of Newcastle, 25 pp.

Juggins, S., R. W. Battarbee & S. C. Fritz, 1994. Diatom/salinity transfer functions and climate change: an assessment of methods and application to two Holocene sequences from the northern Great Plains. In Funnell, B. M. & R. L. F. Kay (eds.) Palaeoclimate of the Last Glacial/Interglacial Cycle. NERC Earth Sciences Directorate, Swindon: 37–41.

Jørgensen, E. G., 1955. Solubility of the silica in diatoms. Physiol. Pl. 8: 846–851.

Kairesalo, T. & I. Koskimies, 1987. Grazing by oligochaetes and snails on epiphytes. Freshwat. Biol. 17: 317–324.

Kilham, P., S. S. Kilham & R. E. Hecky, 1986. Hypothesized resource relationships among African planktonic diatoms. Limnol. Oceanogr. 31: 1169–1181.

Kilham, S. S., 1984. Silicon and phosphorus growth kinetics and competitive interactions between *Stephanodiscus minutus* and *Synedra* sp. Verh. int. Ver. Limnol. 22: 435–439.

Kilham, S. S., E. C. Theriot & S. C. Fritz, 1996. Linking planktonic diatoms and climate change in the large lakes of the Yellowstone ecosystem using resource theory. Limnol. Oceanogr. 41: 1052–1062.

Kingston, J. C. & H. J. B. Birks, 1990. Dissolved organic carbon reconstructions from diatom assemblages in PIRLA project lakes, North America. Phil. Trans. r. Soc., London, B 327: 279–288.

Kingston, J. C., B. F. Cumming, A. J. Uutala, J. P. Smol, K. E. Camburn, D. F. Charles, S. S. Dixit, & R. G. Kreis, 1992. Biological quality control and quality assurance: a case study in paleolimnological biomonitoring. In McKenzie, D. H., D. E. Hyatt & V. J. McDonald (eds.) Ecological Indicators. Elsevier Applied Science, London & New York: 1542–1543.

Knox, A. S., 1942. The use of bromoform in the separation of non-calcareous microfossils. Science 95: 307.

Kolkwitz, R. & M. Marsson, 1908. Ökologie der pflanzlichen Saprobien. Ber. dt. bot. Ges. 26a: 505–519.

Korhola, A., J. Weckström, L. Holmström, & P. Erästö, 2000. A quantitative climatic record from diatoms in Northern Fennoscandia. Quat. Res. 54, 284–294.

Korsman, T. & H. J. B. Birks, 1996. Diatom-based water chemistry reconstructions from northern Sweden: a comparison of reconstruction techniques. J. Paleolim. 15: 65–77.

Krammer, K. & H. Lange-Bertalot, 1986. Bacillariophyceae. I. Teil. Naviculaceae. In Süsswasserflora von Mitteleuropa, Band 2/1. 876 pp.

Krammer, K. & H. Lange-Bertalot, 1988. Bacillariophyceae. 2. Teil. Bacillariaceae, Epithemiaceae, Surirellaceae. In Süsswasserflora von Mitteleuropa, Band 2/2

Krammer, K. & H. Lange-Bertalot, 1991. Bacillariophyceae. 3. Teil. Zentrische Diatomeen, Diatoma, Meridion, Asterionella, Tabellaria, Fragilaria, Eunotia und Verwandte, Peronia und Actinella. In Süsswasserflora von Mitteleuropa, Band 2/4 230 pp.

Krammer, K. & H. Lange-Bertalot, 1991. Bacillariophyceae. 4. Teil. Achnanthes, Navicula, Gomphonema, Kritische Nachtraege, Literatur. In Süsswasserflora von Mitteleuropa.

Kreiser, A. M. & R. W. Battarbee, 1988. Analytical Quality Control (AQC) in diatom analysis. Proceedings of Nordic Diatomist Meeting, University of Stockholm, Department of Quaternary Geology Research Report 12, pp 41–44.

Laird, K. R., S. C. Fritz, K. A. Maasch & B. F. Cumming, B. F., 1996. Greater drought intensity and frequency before AD 1200 in the Northern Great Plains. Nature 384: 552–554.

Laird, K. R., S. C. Fritz & B. F. Cumming, 1998a. A diatom-based reconstruction of drought intensity, duration, and frequency from Moon Lake, North Dakota: a sub-decadal record of the last 2300 years. J. Paleolim. 19: 161–179.

Laird, K. R., S. C. Fritz, B. F. Cumming & E. C. Grimm, 1998b. Early-Holocene limnology and climatic variability in the Northern Great Plains. The Holocene 8: 275–285.

Lauterborn, R. 1896. Untersuchungen über Bau, Kernteilung und Bewegung der Diatomeen. Leipzig: W. Engelmann, 165 pp.

Lewin, J., 1961. The dissolution of silica from diatom walls. Geoch. Cosmoch. Acta. 21: 182–198.

Line, J. M., C. J. F. ter Braak & H. J. B. Birks, 1994. WACALIB version 3.3—a computer program to reconstruct environmental variables from fossil assemblages by weighted averaging and to derive sample-specific errors of prediction. J. Paleolim. 10: 147–152.

Lohmann, K. E. & G. W. Andrews, 1968. Late Eocene non-marine diatoms from the Beaver Divide area, Fremont County, Wyoming. U.S. Geological Survey Professional Paper, 593–E, 26 pp.

Lotter, A. F., H. J. B. Birks, W. Hofmann & A. Marchetto, 1997. Modern diatom, cladocera, chironomid and chrysophyte cyst assemblages as quantitative indicators for the reconstruction of past environmental conditions in the Alps. I. Climate. J. Paleolim. 18: 395–420.

Lotter, A. F., H. J. B. Birks, W. Hofmann & A. Marchetto, 1998. Modern diatom, cladocera, chironomid and chrysophyte cyst assemblages as quantitative indicators for the reconstruction of past environmental conditions in the Alps. II. Nutrients. J. Paleolim. 19: 443–463.

Lotter, A. F., P. G. Appleby, J. A. Dearing, J-A. Grytnes, W. Hofmann, C. Kamenik, A. Lami, D. M. Livingstone, C. Ohlendorf, N. L. Rose, M. Sturm & R. Thompson, in press. The record of the last 200 years in the sediments of Hagelseewli (2339 m asl), a high-elevation lake in the Swiss Alps. J. Paleolim.

Lund, J. W. G., 1954. The seasonal cycle of the plankton diatom Melosira italica (Ehr.) Kütz. subarctica O. Müll. J. Ecol. 42: 151–179.

Lund, J. W. G., 1955. Further observations on the seasonal cycle of Melosira italica (Ehr.) Kütz. subarctica O. Müll. J. Ecol. 43: 90–102.

Lund, J. W. G., 1959a. Buoyancy in relation to the ecology of the freshwater phytoplankton. Br. phycol. Bull. 7: 1–17.

Lund, J. W. G., 1959b. A simple counting chamber for nannoplankton. Limnol. Oceanogr. 4: 57–65.

Lund, J. W. G. & J. F. Talling, 1957. Botanical limnological methods with special reference to the algae. Bot. Rev. 23: 489–583.

Lund, J. W. G., C. Kipling & E. D. Le Cren, 1958. The inverted microscope method of estimating algal numbers and the statistical basis of estimations by counting. Hydrobiologia 11: 143–170.

Maberly, SC., M. A. Hurley, C. Butterwick, J. E. Corry, S. I. Heaney, A. E. Irish, G. H. M. Jaworski, J. W. G. Lund, C. S. Reynolds & J. V. Roscoe, 1994. The rise and fall of Asterionella formosa in the South Basin of Windermere: analysis of a 45-year series of data. Freshwat. Biol. 31: 19–34.

MacDonald, J. D., 1869. On the structure of the diatomaceous frustule, and its genetic cycle. Ann. Mag. nat. Hist. series 4, 3: 1–8.

Mackay, A. W., R. J. Flower, A. E. Kuzmina, L. Z. Granina, N. L. Rose, P. G. Appleby, J. F. Boyle & R. W. Battarbee, 1998. Diatom succession trends in recent sediments from Lake Baikal and relationship to atmospheric pollution and to climate change. Phil. Trans. r. Soc., London B 353: 1011–1055.

Mackereth, F. J. H., 1969. A short core sampler for subaqeous deposits. Limnol. Oceanogr. 14: 145–151.

Mann, D. G., 1993. Patterns of sexual reproduction in diatoms. Hydrobiologia 269/270: 11–20

Mann, D. G., 1994. The origins of shape and form in diatoms: the interplay between morphogenetic studies and systematics. In: The Linnean Society, Shape and Form in Plants and Fungi of London, pp. 17–38.

Mann, D. G. & H. J. Marchant, 1989. The origins of the diatom and its life cycle. In Leadbeater, B. S. C. & J. C. Green (eds.) The Chromophyte Algae: Problems and Perspectives. Oxford University Press, Oxford:

Medlin L. K., D. M. Williams & P. A. Sims, 1993. The evolution of the diatoms (Bacillariophyta). I. Origin of the group and assessment of the monophyly of its major divisions. European J. Phycol. 28: 261–275.

Meriläinen, J., 1967. The diatom flora and the hydrogen ion concentration of the water. Ann. bot. fenn. 4: 51–58.

Molder, K. & R. Tynni, 1967–80. Uber Finnlands rezente und subfossile Diatomeen I–XI, Comptes Rendus de la Societé Géologique de Finlande.

Miller, U., 1964. Diatom floras in the Quaternary of the Göta River Valley. Sver. geol. Unders. 44: 1–67.

Munro, M. A. R., A. M. Kreiser, R. W. Battarbee, S. Juggins, A. C. Stevenson, D. S. Anderson, N. J. Anderson, F. Berge, H. J. B. Birks, R. B. Davis, R. J. Flower, S. C. Fritz, E. Y. Haworth, V. J. Jones, J. C. Kingston & I. Renberg, 1990. Diatom quality control and data handling. Phil. Trans. r. Soc., London B 327: 257–261.

Nipkow, F., 1920. Vorlaufige Mitteilung über Untersuchungen des Sclammabsatzes im Zurichsee. Schweiz. Z. Hydrobiol. 1: 100–122.

Nygaard, G., 1956. Ancient and recent flora of diatoms and chrysophyceae in Lake Gribsø. Studies on the humic acid lake Gribsø. Folia limnol. scand. 8: 32–94.

Oldfield, F., R. W. Battarbee & J. A. Dearing, 1983. New approaches to recent environmental change. Geog. J. 149: 167–181.

Overpeck, J. T., T. Webb, & I. C. Prentice, 1985. Quantitative interpretation of fossil pollen spectra: dissimilarity coefficients and the method of modern analogs. Quat. Res. 23: 87–108.

Paasche, E., 1960. On the relationship between primary production and standing stock of phytoplankton. J. Cons. Int. Explor. Mer. 26.

Patrick, R., 1977. Ecology of freshwater diatoms—diatom communities. In Werner, D. (ed.) The Biology of Diatoms. Blackwell, Oxford: 284–332.

Patrick, R. & C. W. Reimer, 1966. The diatoms of the United States I. Acad. Nat. Sci. Philad., Monogr. 13, 688 pp.

Patrick, R. & C. W. Reimer, 1975. The diatoms of the United States II, part 1. Acad. Nat. Sci. Philad., Monogr. 13, 213 pp.

Pennington, W., 1943. Lake sediments: the bottom deposits of the N. Basin of Windermere with special reference to the diatom succession, New Phytol. 43: 1–27.

Pennington, W., R. S. Cambray & E. M. Fisher, 1973. Observations on lake sediments using fallout ^{137}Cs as a tracer. Nature 242: 324–326.

Pfitzer, E., 1869. Über den Bau und die Zellteilung der Diatomeen. Bot. Ztg. 27: 774–776.

Pienitz, R. & J. P. Smol, 1993. Diatom assemblages and their relationship to environmental variables in lakes from the boreal forest-tundra ecotone near Yellowknife, Northwest Territories, Canada. Hydrobiologia 269/270: 391–404.

Pienitz, R. & W. F. Vincent, 2000. Effect of climate change relative to ozone depletion on UV exposure in subarctic lakes. Nature 404: 484–487.

Pienitz, R., J. P. Smol & H. J. B. Birks, 1995. Assessment of freshwater diatoms as quantitative indicators of past climate change in the Yukon and Northwest Territories, Canada. J. Paleolim. 13: 21–49.

Pienitz, R., J. P. Smol & G. M. MacDonald, 1999. Paleolimnological reconstruction of Holocene climatic trends from two boreal treeline lakes, Northwest Territories, Canada. Arct. Alp. Res. 31: 82–93.

Psenner, R., 1988. Alkalinity generation in a soft-water lake: watershed and in-lake processes. Limnol. Oceanogr. 33: 1463–1475.

Psenner, R. & R. Schmidt, 1992. Climate-driven pH control of remote alpine lakes and effects of acid deposition. Nature 356: 781–783.

Reed, J. M., 1998. Diatom preservation in the recent sediment record of Spanish lakes: implications for palaeoclimate study. J. Paleolim. 19: 129–137.

Renberg, I., 1976. Palaeolimnological investigations in Lake Prästsjön. Early Norrland 9: 113–160.

Renberg, I., 1981. Improved methods for sampling, photographing and varve-counting of varved lake sediments. Boreas 10: 255–258.

Renberg, I., 1990a. A 12,600 year perspective of the acidification of Lilla Öresjön, southwest Sweden. Phil. Trans. r. Soc., London B, 327: 357–361.

Renberg, I., 1990b. A procedure for preparing large sets of diatom slides from sediment cores. J. Paleolim. 4: 87–90.

Renberg, I. & T. Hellberg, 1982. The pH history of lakes in southwestern Sweden, as calculated from the subfossil diatom flora of the sediments. Ambio 11: 30–33.

Reynolds, C. S., 1973. The seasonal periodicity of planktonic diatoms in a shallow eutrophic lake. Freshwat. Biol. 3: 89–110.

Reynolds, C. S., 1984. The Ecology of Freshwater Phytoplankton. Cambridge University Press, Cambridge, 384 pp.

Richardson, T. L., C. E. Gibson & S. I. Heaney, 2000. Temperature, growth and seasonal succession of phytoplankton in Lake Baikal, Siberia. Freshwat. Biol. 44: 431–440.

Rioual, P., 2000. Diatom assemblages and water chemistry of lakes in the French Massif Central: A methodology for reconstruction of past limnological and climate fluctuations during the Eemian period. Unpublished PhD Thesis, University College London. 509 pp.

Rippey, B., 1977. The behaviour of phosphorus and silicon in undisturbed cores of Lough Neagh sediments. In Golterman, H. L. (ed.) Interactions between Sediments and Freshwater. Dr. W. Junk, Dordrecht, The Netherlands: 348–353.

Rippey, B., 1983. A laboratory study of the silicon release process from a lake sediment (Lough Neagh, Northern Ireland). Arch. Hydrobiol. 96: 417–433.

Ross, R. & P. A. Sims, 1972. The fine structure of the frustule in centric diatoms: a suggested terminology. Br. Phycol. J. 7: 139–163.

Ross, R., E. J. Cox, N. I. Karayeva, R. Simonsen & P. A. Sims, 1979. An amended terminology for the siliceous components of the diatom cell. Nova Hedwigia, Beih. 64: 513–533.

Rothpletz, A., 1896. Über die Flywsch-Fucoiden und einige andere fossile Algen, sowie über Liasische, Diatomeen führende Hornschwämme Z. Deutsch. Geol. Ges., 48: 854–914.

Rothpletz, A., 1900. Über einen neuen jurassichen Hornschwamm und die darin eingeschlossenen Diatomeen. Z. Deutsch. Geol. Ges. 52: 154–160.

Round, F. E., 1957. The late-glacial and post-glacial diatom succession in the Kentmere Valley deposit. I Introduction, methods and flora. New Phytol. 56: 98–126.

Round, F. E., 1981a. Morphology and phyletic relationships of the silicified algae and the archetypal diatom—monophyly or polyphyly? In Simpson, T.L. & B.E. Volcani (eds.) Silicon and Siliceous Structures in Biological Systems. Springer-Verlag, New York: 97–128.

Round, F. E., 1981b. Some aspects of the origin of diatoms and their subsequent evolution. Biosystems 14: 483–486.

Round, F. E., 1981c. The Ecology of Algae. Cambridge University Press, Cambridge, 653 pp.

Round, F. E. & R. M. Crawford, 1981. The lines of evolution of the Bacillariophyhta I. Origin. Proc. r. Soc., London. B 211: 237–260.

Round, F. E., R. M. Crawford & D. G. Mann, 1990. The diatoms. Biology and morphology of the genera. Cambridge University Press, Cambridge, 747 pp.

Ryves, D. B., 1994. Diatom dissolution in saline lake sediments: an experimental study in the Great Plains of North America. Unpublished PhD Thesis, University College London. 306 pp.

Scherer, R. P., 1994. A new method for the determination of absolute abundance of diatoms and other silt-sized sedimentary particles. J. Paleolim. 12: 171–179.

Schindler, D. W., S. E. Bayley, B. R. Parker, K. G. Beaty, D. R. Cruikshank, E. J. Fee, E. U. Schindler & M. P. Stainton, 1996. The effects of climate warming on the properties of boreal lakes and streams at the Experimental Lakes Area, northwestern Ontario. Limnol. Oceanogr. 41: 1004–1017.

Schmid, A-M. M., 1995. Salt-tolerance of diatoms of the Neusiedlersee (Austria): a model study for palaeolimnological interpretations. In Robertsson, A-M., S. Hicks, A. Åkerlund, J. Risberg & T. Hackens (eds.) Landscapes and Life. PACT 50, Council of Europe: 463–470.

Schmidt, A., 1874–1959. Atlas der Diatomaceenkunde 472 plates. Leipzig: R. Reisland, Ascherleben.

Schoeman, F. R. & R. E. M. Archibald, 1980. The diatom flora of Southern Africa 6, 1–35 C. S.I.R. Special Rep. Wat. 50.

Shennan, I., M. J. Tooley, M. J. Davis & B. A. Haggart, 1983. Analysis and interpretation of Holocene sea-level data. Nature 302: 404–406.

Simonsen, R., 1962. Untersuchungen zur Systematik und Ökologie der Bodendiatomeen der westlichen Ostsee. Int. Rev. ges. Hydrobiol. Syst. Beih. 1, 144 pp.

Simonsen, R., 1979. The Diatom System: Ideas on Phylogeny. Bacillaria 2: 9–71.

Smol, J. P., 1988. Paleoclimate proxy data from freshwater arctic diatoms. Verh. int. Ver. Limnol. 23: 837–844.

Smol, J. P. & B. F. Cumming, 2000. Tracking long-term changes in climate using algal indicators in lake sediments. J. Phycol. 36: 986–1011.

Stevenson, A. C., S. Juggins, H. J. B. Birks, D. S. Anderson, N. J. Anderson, R. W. Battarbee, F. Berge, R. B. Davis, R. J. Flower, E. Y. Haworth, V. J. Jones, J. C. Kingston, A. M. Kreiser, J. M. Line, M. A. R. Munro & I. Renberg, 1991. The Surface Waters Acidification Project Palaeolimnology Programme: Modern Diatom/Lake-Water Chemistry Data-Set. London: Ensis Ltd, 86 pp.

Stockner, J. G. & W. W. Benson, 1967. The succession of diatom assemblages in the recent sediments of Lake Washington. Limnol. Oceanogr. 12: 513–532.

Stoermer, E. F. & J. P. Smol (eds.) 1999. The Diatoms: Applications for the Environmental and Earth sciences. Cambridge University Press, Cambridge, 469 pp.

Suzuki, Y. & M. Takahashi, 1995. Growth responses of several diatom species isolated form various environments to temperature. J. Phycol. 31: 880–888.

Talling, J. F., 1957. Photosynthetic characteristics of some freshwater plankton diatoms in relation to underwater radiation. New Phytol. 56: 29–50.

ter Braak, C. J. F., 1986. Canonical correspondance analysis: A new eigenvector method for multivariate direct gradient analysis. Ecology 67: 1167–1179.

ter Braak, C. J. F., 1987a. CANOCO—a FORTRAN program for canonical community ordination by [partial] [detrended] [canonical] correspondence analysis, principal components analysis and redundancy analysis (version 2.1) ITI-TNO, Wageningen, 95 pp.

ter Braak, C. J. F., 1987b. Unimodal models to relate species to environment. Unpublished PhD thesis, University of Wageningen.

ter Braak, C. J. F. & S. Juggins, 1993. Weighted averaging partial least squares regression (WA-PLS): an improved method for reconstructing environmental variables from species assemblages. Hydrobiologia 269/270: 485–502.

ter Braak, C. J. F. & H. van Dam, 1989. Inferring pH from diatoms: a comparison of old and new calibration methods. Hydrobiologia 178: 209–223.

Tessenow, U., 1964. Experimentaluntersuchungen zur Kieselsaureruckfuhrung aus dem Schlamm der Seen durch Chironomidenlarven (Plumosus-Gruppe). Arch. Hydrobiol. 60: 497–504.

Tessenow, U., 1966. Untersuchungen über den Kieselsaureaushalt der Binnengewasser. Arch. Hydrobiol. Suppl. 32: 1–136.

Tilman, D., S. S. Kilham & P. Kilham, 1982. Phytoplankton community ecology: the role of limiting nutrients. Annu. Rev. Ecol. Syst. 13: 349–372.

Underwood, G. J. C. & J. D. Thomas, 1990. Grazing interaction between pulmonate snails and epiphytic algae and bacteria. Freshwat. Biol. 23: 505–521.

Van den Hoek, C., D. G. Mann & H. M. Jahns, 1995. Algae: an Introduction to Phycology. Cambridge University Press, Cambridge, 627 pp.

van der Werff, A., 1955. A new method of concentrating and cleaning diatoms and other organisms. Verh. int. Ver. Limnol. 12: 276–277.

van der Werff, A. & H. Huls, 1957–74. Diatomeënflora van Netherland. Reprint 1976. Otto Koeltz Science Publishers, Koenigstein.

van Landingham, S. L., 1969–1979. Catalogue of the fossil and recent genera and species of diatoms and their synonyms Volumes 1–8, J. Cramer, Vaduz, 4654 pp.

Vasko, K., H. T. T. Toivonen & A. Korhola, 2000. A Bayesian multinomial Gaussian response model for organism-based environmental reconstruction. J. Paleolim. 24: 243–250.

Vinebrooke, R. D. & P. R. Leavitt, 1996. Effects of ultraviolet radiation in an alpine lake. Limnol. Oceanogr. 41: 1035–1040.

Vos, P. C. & H. de Wolf, 1988. Methodological aspects of paleo-ecological diatom research in coastal areas of the Netherlands. Geologie Mijnb. 67: 31–40.

Vos, P. C. & H. de Wolf, 1993a. Diatoms as a tool for reconstructing sedimentary environments in coastal wetlands; methodological aspects. Hydrobiologia 269/270: 285–296.

Vos, P. C. & H. de Wolf, 1993b. Reconstruction of sedimentary environments in Holocene coastal deposits of the southwest Netherlands; the Poortvliet boring, a case study of palaeoenvironmental diatom research. Hydrobiologia 269/270: 297–306.

Vyverman, W. & K. Sabbe, 1995. Diatom-temperature transfer functions based on the altitudinal zonation of diatom assemblages in Papua New Guinea: a possible tool in the reconstruction of regional palaeoclimatic changes. J. Paleolim. 13: 65–77.

Weckström, J. A., A. Korhola & T. Blom, 1997. The relationship between diatoms and water temperature in 30 subarctic Fennoscandian lakes. Arc. Alp. Res. 29: 75–92.

Wetzel, R. G. & G. E. Likens, 1991. Limnological analyses. Springer-Verlag, New York, 391 pp.

Whitehead, D. R., D. F. Charles, S. T. Jackson, J. P. Smol & D. R. Engstrom, 1989. The developmental history of Adirondack (N.Y.) lakes. J. Paleolim. 2: 185–206.

Whitmore, T. J., 1989. Florida diatom assemblages as indicators of trophic status and pH. Limnol. Oceanogr. 34: 882–895.

Williams, D. M., 1989. Publication of new and revised taxa: a guide to the International Code of Botanical Nomenclature for diatomists. J. Paleolim. 2: 55–60.

Williams, D. M. & F. E. Round, 1986. Revision of the genus Synedra Ehrenb. Diatom Research, 1: 313–339.

Williams, D. M. & F. E. Round, 1987. Revision of the genus Fragiliaria. Diatom Research, 2: 267–288.

Williams, D. M., B. Hartley, R. Ross, M. A. R. Munro, S. Juggins & R. W. Battarbee, 1988. A coded checklist of British diatoms. Ensis Ltd Publishing, London.

Wilson, S. E., B. F. Cumming & J. P. Smol, 1996. Assessing the reliability of salinity inference models from diatom assemblages: an examination of a 219 lake data set from Western North America. Can. J. Fish. aquat. Sci. S 53: 1580–1594.

Wolfe, A. P., 1997. On diatom concentrations in lake sediments: results from an inter-laboratory comparison and other tests performed on a uniform sample. J. Paleolim 18: 261–268.

Wright, H. E., 1980. Cores of soft lake sediment. Boreas 9: 107–114.

Wunsam, S. & R. Schmidt, 1995. A diatom-phosphorus transfer function for alpine and pre-alpine lakes. Memoire Istit. ital. Idrobiol. 53: 85–99.

Yang, J-R. & H.C. Duthie, 1995. Regression and weighted-averaging models relating surficial sediment diatom assemblages to water depth in Lake Ontario. J. Great Lakes Res. 21: 84–94.

Zong, Y. & B. P. Horton, 1999. Diatom-based tidal-level transfer functions as an aid in reconstructing Quaternary history of sea-level movements in the UK. J. Quat. Sci. 14: 153–167.

9. CHRYSOPHYTE SCALES AND CYSTS

BARBARA A. ZEEB (zeeb-b@rmc.ca)
JOHN P. SMOL (smolj@biology.queensu.ca)
Paleoecological Environmental Assessment
and Research Lab (PEARL)
Department Biology
Queen's University
Kingston, Ontario
K7L 3N6 Canada

Keywords: chrysophytes, scales, cysts, stomatocysts, statospores, bristles, Chrysophyceae, Synurophyceae

Introduction

Chrysophytes belong to the classes Chrysophyceae and Synurophyceae. They are a diverse group, with over 1000 described species in about 120 genera; however, it is generally believed that this is an underestimate of the true diversity of this group (Kristiansen, 1990). They are sometimes referred to as 'golden brown algae' due to a predominance of carotenoid pigments which may represent up to 75% of the total pigment content (Bold & Wynne, 1978). Most chrysophytes exist as single, flagellated cells, or as motile, spherical colonies of flagellated cells. However, some taxa consist of single amoeba-like cells, mucilaginous colonies of non-motile cells, or filamentous colonies (Preisig, 1995). Several edited volumes have been devoted to chrysophytes (e.g., Kristiansen & Andersen, 1986; Sandgren et al., 1995; Kristiansen & Cronberg, 1996) summarizing what is known about their taxonomy, biology, and paleoecological potential. Several reviews have focused specifically on their paleolimnological potential (Cronberg, 1986a,b; Smol, 1988, 1990, 1995; Sandgren, 1991; Zeeb et al., 1996a).

Unicellular and colonial forms are typically found in open water, and many taxa are typically euplanktonic (Sandgren, 1988). In fact, until recently, many reviews referred to chrysophytes as being typically euplanktonic (Kristiansen 1986; Sandgren 1988). However, benthic and epiphytic taxa are also known (Hilliard & Asmund, 1963; Dop, 1980; Cambra, 1989), and recent research has recorded surprisingly diverse periphytic assemblages in arctic peats and mosses (Douglas & Smol, 1995; Wilkinson et al., 1996; Gilbert et al., 1997).

Most known chrysophytes live in fresh water (Kristiansen, 1990; Sandgren, 1988). Exclusively marine taxa are considered rare (Gayral & Billard, 1986), but new nano- and picoplanktonic forms continue to be described (Peters & Andersen, 1993). Recent work has confirmed that a surprisingly large number of taxa can tolerate the high salinity levels

J. P. Smol, H. J. B. Birks & W. M. Last (eds.), 2001. *Tracking Environmental Change Using Lake Sediments.*
Volume 3: Terrestrial, Algal, and Siliceous Indicators. Kluwer Academic Publishers, Dordrecht, The Netherlands.

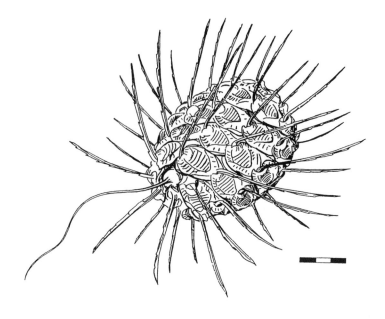

Figure 1. Three-dimensional reconstruction of a whole cell of *Mallomonas striata* depicting the distribution of scales and bristles in spiral rows (from Siver, 1991a; used with permission).

found in inland saline (athalassic) lakes (Cumming et al., 1993; Zeeb & Smol, 1995), as well as coastal estuarine and brackish waters (Dop, 1980). Chrysophytes are often efficient competitors in harsh conditions such as oligotrophy, low temperatures, and/or unpredictable climates, and many species have been reported from, for example, both Arctic and maritime Antarctic lakes and ponds (Wilkinson et al., 1996; Van de Vijver & Beyens, 1997) and alpine lakes (Lotter et al., 1997). This may partly be related to the diversity of nutritional strategies that many chrysophytes exhibit. For example, although autotrophy is clearly the dominant strategy, heterotrophy and even phagotrophy are common (Sandgren, 1988).

Chrysophyte taxa are found in many different types of limnological environments, but are often most common in somewhat acidic or nutrient-poor lakes, and are typically less abundant in very alkaline or eutrophic waters, although exceptions occur (Duff et al., 1995; Sandgren, 1988; Siver, 1995). What is most important from an ecological and paleolimnological perspective is that many taxa have fairly well defined environmental optima and tolerances, and can therefore be used in paleoenvironmental reconstructions.

Most Chrysophyceae are simply bounded by a single membrane (Pienaar, 1980). In the Synurophyceae, however, the plasma membrane has a siliceous cell covering consisting of individual plate-like scales arranged in an overlapping pattern that spiral around the cell (Fig. 1). The scales, in turn, may have associated siliceous spines or bristles (Asmund & Kristiansen, 1986; Siver, 1991a). Depending on the species, each cell may possess approximately 30 to >200 scales (Siver, 1991b) that will become incorporated into the sediment when the cell dies.

In contrast to scales, which are only produced by some taxa, all chrysophytes endogenously form a siliceous resting stage as part of their life cycle. This resting stage has often been referred to as a statospore in the older literature, although a more correct term would be a stomatocyst or simply a cyst (as it is not technically a spore). These stomatocysts may be produced asexually or sexually, with both modes of reproduction producing the identical cyst morphology for a given taxon (Sandgren, 1991). When a cyst forms, it will sink through the water column, and may germinate (chrysophyte excystment) when conditions become favourable. Regardless of germination success, the siliceous cyst will eventually become incorporated into the sediment. Sandgren (1991) provides a primer for paleolimnologists on chrysophyte reproduction and cysts.

The production of scales and stomatocysts occurs endogenously in silica deposition vesicles (SDV), which are formed from the fusion of vesicles originating from the Golgi apparatus. In the case of scales, each SDV formed in the anterior of the cell becomes compressed and is molded into the shape of a mature scale as it moves along the chloroplast endoplamic reticulum to the posterior of the cell (reviewed in Siver, 1991a). In stomatocyst formation, a SDV is formed in the anterior cytoplasm and grows to encircle the central cytoplasm with a bilayer of membrane (Sandgren, 1983). A pore forms in the SDV that eventually becomes the site of the pore of the mature cyst. Stomatocyst wall formation occurs in phases with the primary cyst wall usually being thin and unoramented. Subsequent deposition of layers of silica may give rise to secondary, tertiary, and even quaternary walls, a collar complex, and/or various types of surface ornamentation (Sandgren, 1980).

In summary, chysophytes are represented in the sedimentary records of lakes, ponds, and rivers by two forms of siliceous remains: 1) scales; and 2) stomatocysts, sometimes referred to as cysts, statospores, or statocysts. In this chapter, we briefly review the main elements of scale and cyst taxonomy, and describe how these indicators can be used in a paleolimnological study.

Taxonomy and Nomenclature

Scales

Taxonomy: The most commonly occurring scales in sediments belong to the genera *Mallomonas* and *Synura* (both of the Class Synurophyceae), but other genera such as *Chrysosphaerella, Spiniferomonas*, and *Paraphysomonas* may also be common. The current taxonomic system is based on the ultrastructure of scales and bristles. While many scales can be readily distinguished with light microscopy (LM) alone, the advent of electron microscopy (EM) allowed for more detailed observations and, in some cases, splitting of taxa. There has been some confusion with older literature that pre-dates the use of EM, as many older descriptions can not be properly linked to EM descriptions and, hence, some taxa described with LM have likely been re-described with different names using EM (Siver, 1991a). The best strategy would appear to be to seek EM confirmation of all LM identifications; however, in most cases, scales from many taxa can be identified and counted in fossil material simply using the LM, provided that the identifications have been confirmed with EM (Figs. 2 & 3). Smol (1986) provides a series of photographs and drawings noting some of the main features one can distinguish in LM preparations.

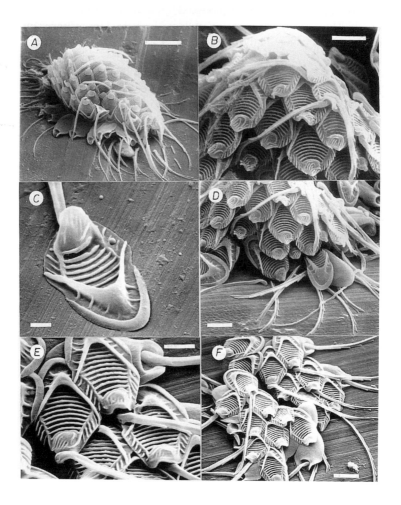

Figure 2. Scanning electron micrographs of *Mallomonas striata* var. *serrata*. A. whole cell covered with short, slightly curved bristles (scale bar = 5 μm). B. Close-up of an intact cell illustrating the pattern of overlap of the scales (scale bar = 2 μm). C. Body scale (scale bar = 1 μm). D. Scales and bristles. Bristles are curved, ribbed and unilaterally serrated (scale bar = 2 μm). E. Close-up of scales and bristles (scale bar = 1 μm). F. Scales and bristles (scale bar = 2 μm) (from Siver 1991a; used with permission).

Nomenclature: The terminology provided here is designed to familiarize researchers with the most common aspects of scale and bristle nomenclature. Details with excellent illustrations and photographs are available in, for example, Asmund & Kristiansen (1986), Siver (1991a), Takahashi (1978), and Wee (1982), as well as a large number of journal papers. We will focus on the two dominant genera found in freshwater deposits: *Mallomonas* and *Synura*. However, a considerable body of literature exists for other genera as well (e.g., Kristiansen & Andersen, 1986; Sandgren et al., 1995; Kristiansen & Cronberg, 1996).

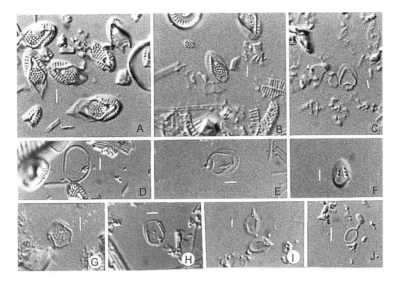

Figure 3. Light micrographs. A. *Mallomonas pseudocoronata.* B. *M. duerrschmidtiae* (this scale can be confused with the one produced by *M. crassissquama* and sometimes with *M. pseudocoronata* (for distinguishing features, see Siver et al., 1990). C. *M. acaroides.* D. *M. caudata.* E. *M. elongata.* F. *M. allorgei* (this scale is difficult to differentiate from *M. lychenensis*). G. *M. punctifera.* H. *M. transsylvanica.* I. *Synura curtispina.* J. *S. sphagnicola.* All scale bars = 2 μm. (Modified from Smol, 1986).

Table I. Comparison of features of *Mallomonas* vs. *Synura* scales (from Duff & Zeeb, 1995).

Feature	*Mallomonas*	*Synura*
Shape	Roughly circular or elliptical	Roughly circular to oval
Length	2 to 12 μm	<6 μm
Width	>1 to 7 μm	1 to 4 μm
# scales/cell	~30 to >150	~70 to >200

Mallomonas scales are generally larger, more heavily silicified, and have a more complex morphology than *Synura* scales (Table I). In both genera, there is intraspecific variation in scale size, shape, and number of scales per cell. Although the ornamentation patterns are always species-specific, an understanding of the variation on a typical cell in both ornamentation and morphology is important for correct and consistent identifications.

Mallomonas: *Mallomonas* scales are composed of a base plate that is commonly perforated with numerous tiny pores (Fig. 4a). In most species, there is a prominent V-shaped rib pointing towards the posterior (proximal) rim. When the area between the arms of the V-rib, just anterior to the point, is devoid of ornamentation, then this area is referred to as the window. Generally, there are two anterior submarginal ribs continuous with, or originating near, the distal ends of the V-rib arms and running parallel to the margins of the

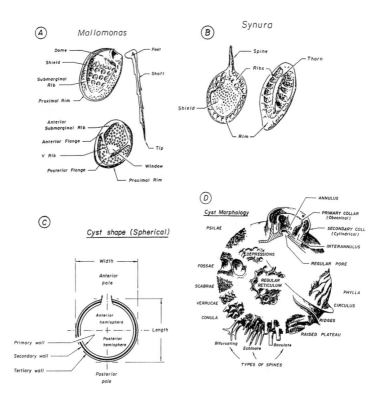

Figure 4. Line drawings of chrysophycean siliceous microfossils. A. *Mallomonas* scales and bristle. B. *Synura* scales. C. Stomatocyst body terminology. D. Stomatocyst ornamentation. (Modified from Duff & Zeeb, 1995).

scale towards the anterior region. Collectively, the V-rib and anterior submarginal ribs are referred to as the submarginal rib. The region within the submarginal rib is called the shield, whereas the area outside the rib is termed the flange. The flange is divided into a single posterior and two anterior flanges. The shield may be smooth or ornamented with papillae, ribs, etc. Shield and flange morphology are important for species identification. Moreover, the morphologies of individual scales may be slightly different, depending on which part of the organism the scale is from (e.g., apical, collar, caudal scales; see Siver, 1991a).

Not all scales have bristles. In those that do, the base plate is typically raised at the distal end of the scale to form a structure known as a dome. The foot end of the bristle is articulated in a ball and socket fashion at the dome. Bristles are divided into the foot (typically flat and bent at an angle relative to the shaft), the shaft (gently curved and can be smooth, ribbed, and/or serrated), and the tip (distal end either pointed, bifurcated, or swollen). Bristle morphology is taxonomically important for distinguishing between species with similar scales.

Synura: *Synura* scales (Fig. 4b) generally consist of an oval base plate with an upturned rim that extends around the entire posterior half of the scale (Kristiansen, 1979). The inner surface of the base plate is flat and perforated evenly with tiny pores, whereas the outer surface may be smooth or variously ornamented with thin ridges or reticula. Spines vary from tubular, terminating in minute teeth or a pointed tip, to short, thick, and conical.

The scales of *S. petersenii* Korshikov and *S. macracantha* (Petersen and Hansen) Asmund are structurally different from the general *Synura* morphology described above. These scales are elliptical or oval and consist of an evenly perforated base plate with an upturned rim and a conical thorn. The thorn is a central chamber with bilateral ribs. The inner scale surface is flat with a single large distal hole and many tiny holes, while the outer surface may be ornamented with ridges.

Stomatocysts

Taxonomy: Unlike chrysophycean scales, stomatoycsts have not yet, for the most part, been linked to the taxa that produce them. In fact, we currently estimate that only about 10–15% of cyst morphotypes have been conclusively linked to known species, and so clearly this area of research should be pursued. However, it is a slow process, as linkages between a vegetative cell and its resting stage can only be made if the vegetative cell is observed during the encystment process. This is a 'hit and miss' endeavour in field studies, and has proven problematic in laboratory cultures. Moreover, inducing cyst development in the laboratory is not, at the present time at least, an active area of research. Consequently, stomatocysts are currently primarily identified using artificial and temporary fossil names [such as the older systems of Deflandre (1932, 1936) and Nygaard (1956)] or numbers (Duff et al., 1995) rather than species. From a paleoecological perspective, this system is completely acceptable, provided that the criteria used to differentiate cysts are consistent, and the environmental optima and tolerances of different morphotypes can be discerned (Smol, 1995). However, from the perspective of linking paleolimnological studies to neo-limnological research, it clearly would be important to link as many cysts to the taxa that produce them.

Deflandre (1932, 1936) and Nygaard (1956) were two pioneering researchers who devised artificial genera (based on their form and surface ornamentation) for the stomatocysts they observed. Both of their systems were used extensively by later researchers up until the mid-1980's (e.g., Deflandre: Srivastata & Binda, 1984; Rull, 1986; Nygaard: Carney & Sandgren, 1983; Rybak 1986). However, with the continued development and accessibility of scanning electron microscopes (SEM), more detailed description of cysts could be made. Ultimately, artificial genera (which were really just temporary names) proved to be less than ideal compared to more simple numbering schemes, especially as the number and quality of cyst descriptions expanded and became more detailed.

Cronberg & Sandgren (1986) published a standardized system, developed by the International Statospore Working Group (ISWG), for describing new stomatocyst morphotypes. The system requires researchers to assign numbers to cysts and to publish descriptions that include the following information: 1) a clear identification of the original specimen on which the description is based, including electron micrographs, as well as light micrographs and line drawings where appropriate; 2) a morphological analysis of the stomatocyst that conforms to an accepted nomenclature; and 3) a discussion of pertinent ecological or stratigraphic information concerning the morphotype, and the chrysophyte species producing the stomatocyst (when this is known). For many years, the nomenclature referred to

Table II. Stomatocyst shapes.

cyst shape	Length:width ratio	position of greatest width
spherical	0.9–1.1	at equator
oval	≥1.2	at equator
oblate	≤0.8	at equator
obovate	0.9–1.1	above equator
ovate	≥1.2	below equator
pyramidal	≤0.8	varies

in '2' was loosely-defined, but generally followed accepted terminology for pollen grain descriptions. Duff et al. (1995) published the first volume of the *Atlas of Chrysophycean Cysts*. The *Atlas* provided the first detailed and illustrated terminology for describing stomatocysts, along with revised ISWG guidelines, and hundreds of scanning electron and light micrographs, to describe over 240 cyst morphotypes. A second *Atlas* with 173 new cyst descriptions, as well as some refinements to terminology, has just been completed (Wilkinson et al., 2001).

Nomenclature: Our summary of terminology provided here is meant to familiarize researchers with the most common aspects of cyst nomenclature. It is a synopsis of what is presented in Duff et al. (1995) and Wilkinson et al. (2001), and so the two *Atlas* volumes should be referred to for further details and illustrations.

Body Morphology: Typical cyst diameters range from approximately 2 to >30 μm. The cyst body is divided into the anterior and posterior hemisphere. The pore is always located in the anterior hemisphere, but not necessarily at the pole. The distance between the poles is defined as the length, whereas the width is the maximum dimension perpendicular to the length. The most common cyst shape is spherical, however various forms exist, which are defined according to their length:width ratio, and the position of the width with respect to the equator (Table II).

Pore and Collar Morphology: All stomatocysts have a single pore that is circular in shape (Table III). In some cysts, a thin shelf of silica, referred to as a pseudoannulus, projects from the inner pore margin. This pseudoannulus may be planar (flat) or swollen. Occasionally, a siliceous plug may obscure the pore (Fig. 4c).

In contrast to pores, not all cysts possess a collar. When a collar is present, it consists of a siliceous thickening that completely surrounds the pore (Table III). The basic collar shape may be modified in that the apex is acute, rounded, or ornamented; the base may have striations or struts; and a shelf connecting the inner collar margin with the outer pore margin (annulus) may be present. In some cases, more than one collar may surround the pore. The collar nearest the pore is referred to as the primary collar, and outer collars are termed secondary and tertiary (collectively they form a complex collar). The section of cyst wall between collars is termed an interannulus. A false complex collar is formed when a single collar has an apical groove giving the appearance of two separate collars. Similar

Table III. Stomatocyst pore and collar morphology.

pore morphology	
Regular	- Outer and inner pore margins are of equal diameter
	- margins may be sharply delineated or rounded
Conical	- outer pore margin has greater diameter than inner pore margin
	- pore sides are straight
Concave	- outer pore margin has greater diameter than inner pore margin
	- pore sides are curved
collar morphology	
Cylindrical	- basal diameter = apical diameter
Conical	- basal diameter > apical diameter
Obconical	- basal diameter < apical diameter

to the siliceous plug found obscuring some pores, a siliceous cap is sometimes present covering the collar or pore within the collar (Fig. 4d).

Ornamentation: The cyst body may be smooth or ornamented with a variety of siliceous projections and/or indentations (Table IV; Figs. 4d, 5 & 6). When more than one type of projection or indentation is present on the cyst body, it is referred to as compound ornamentation. Compound ornamentation is quite common and may take the form of spines combined with ridges, reticulum, and/or indentations, etc.

Methods

Because chrysophyte scales and cysts are siliceous, they can be prepared using the identical digestion and mounting techniques used for diatoms for LM, SEM, and TEM preparations (see Battarbee et al., this volume). In general, chrysophytes scales and cysts can often be more easily distinguished using LM if either phase or interference optics of some kind (e.g., Nomarski) are used; however, to some extent, this is personal preference.

In most studies, cysts and scales are counted alongside diatoms. Almost all studies simply report the relative frequency of species of scales relative to the total number of scales counted (e.g., the percentage of *Mallomonas crassisquama* scales relative to all the chrysophyte scales counted) or the percentages of a certain cyst morphotype relative to all cysts counted in a slide. In fact, all currently available surface-sediment calibration sets and the resulting transfer functions are based on percentage data. However, in some downcore studies, concentrations and accumulation rates have also been calculated, and some ratios have also been developed (discussed below).

As noted in Table I, a wide range in the number of scales per cell exists between chrysophyte taxa. Although overall species trends would be evident even if simple percentages were calculated in paleolimnological studies, scale counts could be corrected to more closely represent vegetative populations if the scale counts were "corrected" relative

Table IV. Stomatocyst ornamentation.

---

projecting elements with a roughly circular base

scabrae	- basal diameter $\leq 0.2\ \mu$m, observable only with high resolution LM
verrucae	- basal diameter $> 0.2\ \mu$m and $\geq$ its height
	- apically rounded
	- sometimes referred to as nodules
conula	- basal diameter $> 0.2\ \mu$m and $\geq$ its height
	- pointed at apex
spines	- length is $\geq$ basal diameter
	- may be baculate (cylindrical) or echinate (pointed)
	- may have primary, secondary, and even tertiary bifurcations (branches)
	- may be rope-like in appearance and/or have basal striations
raised plateaus	- basal diameter approximately 3–9 μm
	- smooth, flat-topped, spherical to oval projections (height 0.2–0.3 μm) of varying diameters
	- extremely rare

projecting elements with elongated bases

ridges	- may be straight, curved, and/or branched; sometimes lunate (semi-circular) in lateral view
	- short ridges defined as < one-half the cyst circumference
	- may have vertical striations or struts
circuli	- a ridge that has formed a closed ring
	- orientation may be latitudinal, longitudinal, or tangential
reticulum	- a network of ridges
	- spaces between the ridges are referred to as lacunae (width must be $\geq$ reticular ridge width), and may be circular or polygonal (e.g., triangular, hexagonal, etc.)
	- may be regular (lacunae all exactly the same size and shape), variable (lacunae approximately same size and shape), or irregular (lacunae vary substantially in size and shape)
phylla	- very thin (thickness <0.1 μm) sheets of silica that drape across the cyst surface
	- visible with SEM only
	- extremely rare

indentations with a roughly circular base

psilae	- basal diameter $\leq 0.2\ \mu$m, observable only with high resolution LM
	- sometimes referred to as punctae and described as microtextured
	- this ornamentation is often associated with degradation or developmental immaturity of the cyst
depressions	- basal diameter $> 0.2\ \mu$m

indentations with elongated bases

fossae	- may be straight, curved, and/or branched
	- rare

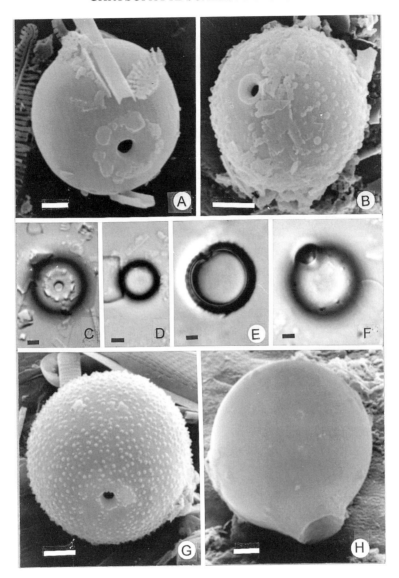

Figure 5. Chrysophycean stomatocysts. A. Cyst 205 (SEM). B. Cyst 206 (SEM). C. Cyst 205 (LM). D. Cyst 206 (LM). E. Cyst 207 (LM). F. Cyst 208 (LM). G. Cyst 207 (SEM). H. Cyst 208 (SEM). All scale bars = 2μm. (Modified from Duff & Smol, 1994).

to the number of average scales produced by each species. Siver (1991b) produced a table of correction factors for *Mallomonas* and *Synura* taxa to address this potential problem. Cumming & Smol (1993) explored the different statistical relationships one might get from transforming scale numbers to estimates of whole cells, and found that there was no statistical advantage in their data set. However, if one wishes to compare paleo-chrysophyte

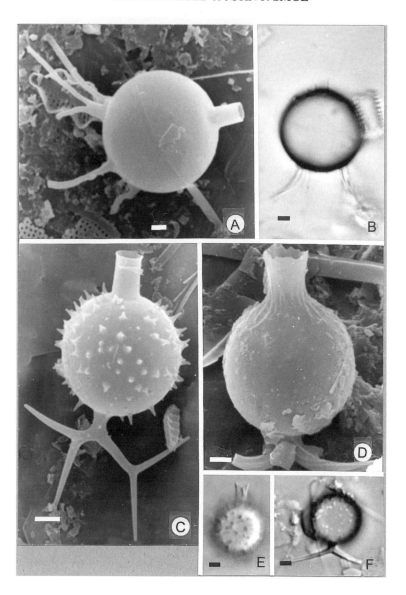

Figure 6. Chrysophycean stomatocysts. A. Cyst 128 *Forma* B (SEM). B. Cyst 128 *Forma* B (LM). C and D. Cyst 130 (SEM). E and F. Cyst 130 (LM). All scale bars = 2 μm. (Modified from Duff & Smol, 1994).

populations to actual long-term limnological chrysophyte collections, then a scale-to-cell transformation should be done.

Based on correspondence we receive, researcher have sometimes confused protozoan plates and heliozoan scales with chryophyte scales. Readers are referred to the chapter by Douglas & Smol (this volume) for a description of these similar siliceous microfossils.

Paleolimnological Applications

Chrysophyte cysts have been noted in sedimentary deposits since early in the 20th century. However, Nygaard (1956) was really the first to recognize their true potential as indicators of environmental conditions. Nygaard's classical account of the Lake Gribsø (Denmark) chrysophyte cyst flora included excellent line drawings, and demonstrated that changes in cyst distributions could be related to changes in pH, climate, and trophic status. Since his pioneering work, other studies have shown that stomatocysts can also be used as biomonitors of lake-level changes, habitat availability, metal concentrations, and salinity.

The use of scaled chrysophytes in paleolimnological studies began in earnest only over the last 20 years or so, with the almost simultaneous publication of three papers describing primarily *Mallomonas* scales in lake sediments from Finland, U.S.A., and Canada (Battarbee et al., 1990; Munch, 1980; Smol, 1980). These early studies focussed primarily on eutrophication and land-use effects. However, shortly afterwards, scaled chrysophytes were recognized as excellent indicators of trends in lakewater pH (Smol et al., 1984a,b), and soon they were incorporated into a number of paleolimnological assessments of acidification (e.g., Marsicano & Siver, 1983; Smol, 1986; Steinberg et al., 1988; Cronberg, 1990; Cumming et al., 1992, 1994; Dixit et al., 1992a,b), and many other applications (reviewed in Smol, 1995).

In contrast to scales, stomatocysts have often been clumped into one group in some paleolimnological studies, and expressed relative to the total number of diatoms counted on the same microscope slides (e.g., Grönlund et al., 1986; Smol, 1983; Stoermer et al., 1985; Harwood, 1986). For example, Smol (1985) suggested that ratio of total cysts to diatoms (i.e., the cyst to frustule ratio, expressed as a percentage) provided a quick and easy measure of the relative importances of these two algal groups in paleolimnological studies. In some temperate zone lakes, a lower percentage of chrysophyte cysts often indicates more eutrophic conditions, as these algae are often less competitive in high nutrient waters (Smol, 1985). In high polar regions, the ratio has been proposed as an estimate of lake ice cover, and hence climate (Smol 1983, 1988). More recently, Cumming et al. (1993) used the ratio of cyst to diatoms to illustrate the potential of cysts as indicators of past lakewater salinity, and Zeeb & Smol (1993) used the ratio to explore past climatic trends throughout the Holocene history of Elk Lake, Minnesota. Douglas & Smol (1995) compared the relative percentage of cysts to diatoms from epiphytic (moss), epilithic, and surface sediments from 35 high arctic ponds. They showed that the highest percentage of cysts was found in semi-aquatic mosses. Layers of cysts in laminated sediments have also been used to determine the seasonality of varve formation (e.g., Battarbee, 1981; Peglar et al., 1984; Grönlund et al., 1986). The methods used in the studies cited above are valuable in that they are generally not difficult, and can be carried out quickly to give some useful information regarding the ecological condition of the site. However, whenever an entire assemblage of diverse organisms such as chrysophytes is considered as a group, much information is inevitably lost. Considerably more and different information can be gleaned when cyst morphotypes are identified and considered individually.

As with all paleoindicators, most early studies of chrysophyte assemblages focussed on using relative abundance profiles of cysts and scales to make qualitative interpretations of past conditions or trends based on known or inferred ecological conditions. The general nature and direction of significant environmental changes can usually be obtained using this technique (e.g., trophic status: Carney & Sandgren, 1983; Zeeb et al., 1990; copper contamination: Elner & Happey-Wood, 1978; habitat availability: Hajós & Radocz, 1969; Hajós, 1973, 1974; Rybak et al., 1987).

More recently, quantitative statistical methods using surface-sediment calibration sets have been used to analyze fossil chrysophyte assemblage distributions. Multivariate analyses can be applied to identify environmental variables that appear to exert the greatest influences on cyst and scale distributions. The relationships explored most often are those related to lakewater pH, nutrients, and salinity/conductivity (Smol, 1995; Duff et al., 1995; Zeeb et al., 1996a).

Both scales and cysts appear to be particularly good indicators of lakewater pH. Scales, in particular, have been included in several multidisciplinary paleolimnological projects designed to assess the extent of anthropogenic acidification in Canada (e.g., Dixit et al., 1992a & b), the United States (e.g., Cumming et al., 1992, 1994; Siver et al., 1999), and Europe (e.g., Steinberg et al., 1988; Cronberg, 1990). Smol (1995) provides the most recent review of this work. While less has been published on cysts, recent studies using calibration sets have shown strong relationships between cysts and lakewater acidity (e.g., Carney et al., 1992; Duff et al., 1995, 1997; Facher & Schmidt, 1996).

The contribution of chrysophytes to total algal biomass tends to drop with increasing lake productivity (Sandgren, 1988; Siver, 1995). In particular, scales in the sediments of eutrophic lakes are rare and species richness is limited. While stomatocysts are generally more abundant, the flora often consists largely of unornamented morphotypes (e.g., Zeeb et al., 1990; Carney & Sandgren, 1983) making paleolimnological studies challenging. Statistical models thus far developed for total phosphorus (TP) using cysts tend to be less robust than for other variables (Duff et al., 1997; Lotter et al., 1998). Nonetheless, both scales and cysts can provide important inferences on past lake trophic status (e.g., Zeeb et al., 1994).

Chrysophytes appear to be also strongly influenced by lakewater conductivity and salinity. For example, Siver (1993) developed a model to infer lakewater specific conductivity from chrysophyte scales. Unfortunately, scales tend to be poorly preserved in some saline sites. Cysts are more robust, and Zeeb & Smol (1995) published a calibration model for inferring lakewater salinity using cysts from lakes in western Canada. Interestingly, many of the cysts in this study were heavily silicified morphotypes. Meanwhile, Duff et al. (1997) were able to develop a strong inference model for conductivity using cysts from four independent data sets in Canada, the U.S.A., and Siberia. In a surface-sediment calibration set from lakes near Arctic treeline in Canada, Taylor (1997) found that calcium, total unfiltered phosphorus, and dissolved organic carbon (DOC) explained significant proportions of the variance in cyst assemblages. Similarly, Brown et al. (1997) showed that variables such as lakewater chloride concentrations, dissolved inorganic carbon, and surface-water temperature influenced the distribution of cyst morphotypes. Although the relationships were less strong than those inferred for diatoms, these indicators could be used to provide supplemental paleoclimatic inferences. Meanwhile, Pla (1999) identified 210 cyst morphotypes from the surface sediments of 105 lakes in the central and eastern Pyrenees of Spain, and correlated these assemblages to measured environmental variables. He found that altitude and temperature (measured as mean annual air temperature) could be used to explain cyst assemblage composition. He also studied cyst distributions in sediment traps, and found that some morphotypes were typically produced in spring, summer, fall, or with a bimodal (spring and fall) distribution. Finally, Pla identified and enumerated the cyst morphotypes in Lake Redó, Spain, and related assemblage changes to temperature shifts occurring since 1781. He found that his transfer functions appeared to reliably infer past climatic changes, and also that cyst morphotypes characteristic of summer increased in frequency, over the warming trend noted since about 1940. Collectively, these data suggest that chrysophyte cysts have considerable potential as paleoclimatic indicators in alpine regions.

Aside from the more traditional paleolimnological studies discussed above, new chrysophyte applications have recently been explored, especially in the case of stomatocysts. These include the distribution of tropical (Zeeb et al., 1996c), brackish mangrove environments (Rull & Vegas-Vilarrubia, 2000), pre-Holocene (Zeeb et al., 1996b), groundwater spring outflow (Reavie et al., 2001), and terrestrial (Zeeb et al., unpublished) assemblages. In addition, stomatocyst assemblage response to polychlorinated biphenyls (PCBs) has been investigated (Betts-Piper, 2000). One promising new avenue of research is in the study of periphytic chrysophytes. Recent studies have shown that cysts are more common in the moss periphyton and epilithon of high arctic ponds than in surface sediments (Douglas & Smol, 1995; Wilkinson et al., 1996). Peat cores have likewise been found to support a rich and unique flora of cysts—many of which are highly ornamented and have not been identified elsewhere (Brown et al., 1994; Gilbert et al., 1997).

Future research directions

The number of detailed paleolimnological studies using chrysophytes increased greatly throughout the 1980's and 1990's, and there are many interesting research opportunities for the new millennium. Unfortunately, there are relatively few dedicated chrysophyte taxonomists world-wide who are currently involved in paleolimnological research. Nonetheless, with the publication of taxonomic texts such as the two volumes of the *Atlas of Chrysophycean Cysts* (Duff et al., 1995; Wilkinson et al., 2001), and the generally well understood taxonomy and ecology of most scaled chrysophytes, these indicators can now be more easily incorporated into many ongoing paleoecological studies. The fact that the same microscope preparation can be used to study all siliceous microfossils simultaneously (e.g., diatoms, sponge spicules, protozoan plates, phytoliths), makes the inclusion of chrysophyte indicators relatively easy. Such multi-indicator approaches can only help strengthen interpretations of past environments.

In many respects, though, chrysophyte-based approaches are still early in development, and can still be improved. Much progress has been made on the taxonomy of chrysophyte cyst morphotypes, but much more work is needed. Certainly any attempts to link more morphotypes with the taxa that produced them will provide an important link to phycological-based research. Although considerable progress has been made over recent years, further work on defining the ecological optima and tolerances of chrysophytes is still needed. As paleolimnologists extend their research horizons to new problems, chrysophyte scales and cysts will undoubtedly find many new applications in the years to come.

Summary

Chrysophyte algae are a diverse group of algae that may dominate the plankton of many freshwater systems, and may be especially abundant in oligotrophic and softwater lakes.

Diverse periphytic assemblages exist in some habitats. Chrysophyte microfossils include two major groups of siliceous, species-specific indicators: 1) the endogenously formed cysts (stomatocysts) that characterize this group as a whole; and 2) the often sculptured and ornamented scales that characterize important genera such as *Mallomonas* and *Synura*. Bristles and spines from scaled chrysophytes are also sometimes used. The taxonomy of scaled chrysophytes, based on scale morphologies, is fairly well established. Considerably more taxonomic work, however, is still needed on cyst morphotypes, although much progress has been made with the establishment of standard guidelines.

Paleolimnologists have only recently begun to use chrysophytes, but progress has been swift, and it is now clear that many taxa closely track important limnological variables. For example, using surface-sediment training sets, fairly robust quantitative inference functions have been constructed for variables associated with, for example, acidification, eutrophication, salinification, climatic change, and other environmental changes. Because scales and cysts are siliceous, they can be studied using the identical preparation techniques developed for diatoms, and so can be easily incorporated into ongoing paleolimnological studies. Moreover, in some environments where diatoms may be absent or rare (e.g., planktonic diatoms in very acidic lakes), chrysophytes may represent the dominant indicator group. In some cases, chrysophytes appear to be more sensitive than other indicators to subtle environmental changes. Combined studies, using multiple indicator groups such as chrysophytes, can only help strengthen paleolimnological reconstructions.

Acknowledgements

Most of the work in our lab is funded by the Natural Sciences and Engineering Research Council. We thank the members of our lab, as well as Peter Siver, for helpful comments on this chapter. Special thanks go to J. R. Glew for drafting Figure 4.

References

Asmund, B. & J. Kristiansen, 1986. The genus *Mallomonas* (Chrysophyceae). Opera Botanica 85: 1–128.

Battarbee, R. W., 1981. Diatom and Chrysophyceae microstratigraphy of the annually laminated sediments of a small meromictic lake. Striae 14: 105–109.

Battarbee, R. W., J. Mason, I. Renberg & J. F. Talling, 1990. Palaeolimnology and lake acidification. The Royal Society, London, 219 pp.

Betts-Piper, A., 2000. Chrysophyte stomatocyst-based paleolimnological investigations of environmental changes in arctic and alpine environments. M.Sc. Thesis, Queens Univ., Dept. Biology, 196 pp.

Bold, H. C. & M. J. Wynne, 1978. Introduction to the algae. Prentice-Hall, Englewood Cliffs, New Jersey, 720 pp.

Brown, K. M., M. S. V. Douglas & J. P. Smol, 1994. Siliceous microfossils in a Holocene, High Arctic peat deposit (Nordvestø, northwestern Greenland). Can. J. Bot. 72: 208–216.

Brown, K., Zeeb, B., Smol, J. P. & R. Pienitz, 1997. Taxonomy and ecological characterization of chrysophyte stomatocysts from northwestern Canada. Can. J. Bot. 75: 842–863.

Cambra, J., 1989. *Sphaeridiothrix compressa* and *Phaeothamnion articulatum*, two new records for Spanish chrysophyte flora. Beiheft zur Nova Hedwigia 95: 259–267.

Carney, H. J. & C. D. Sandgren, 1983. Chrysophycean cysts: indicators of eutrophication in the recent sediments of Frains Lake, Michigan, U.S.A. Hydrobiologia 101: 195–202.

Carney, H. J., M. C. Whiting, K. E. Duff & D. R. Whitehead, 1992. Chrysophycean cysts in Sierra Nevada (California) lake sediments: paleoecological potential. J. Paleolim. 7: 73–94.

Cronberg, G., 1986a. Chrysophycean cysts and scales in lake sediments: a review. In Kristiansen, J. & R. A. Andersen (eds.) Chrysophytes: Aspects and Problems. Cambridge University Press, Cambridge: 281–315.

Cronberg, G., 1986b. Blue-green algae, green algae and Chrysophyceae in sediments. In Berglund, B. (ed.) Handbook of Holocene Palaeoecology and Palaeohydrology. John Wiley and Sons, Chichester (UK): 507–526.

Cronberg, G., 1990. Recent acidification and changes in the subfossil chrysophyte flora of lakes in Sweden, Norway and Scotland. Phil. Trans. r. Soc. Lond. B. 327: 289–293.

Cronberg, G. & C. D. Sandgren, 1986. A proposal for the development of standardized nomenclature and termiology for chrysophycean statospores. In Kristiansen, J. & R. A. Andersen (eds.) Chrysophytes: Aspects and Problems. Cambridge University Press, Cambridge: 317–328.

Cumming, B. F., J. P. Smol, J. C. Kingston, D. F. Charles, H. J. B. Birks, K. E. Camburn, S. S. Dixit, A. J. Uutala & A. R. Selle, 1992. How much acidification has occurred in Adirondack region lakes (New York, U.S.A.) since preindustrial times? Can. J. Fish aquat. Sci. 49: 128–141.

Cumming, B. F. & J. P. Smol, 1993. Scaled chrysophytes and pH interference models: the effects of converting scale counts to cell counts and other species data transformations. J. Paleolim. 9: 147–153.

Cumming, B. F., S. E. Wilson & J. P. Smol, 1993. Paleolimnological potential of chrysophyte cysts and scales and of sponge spicules as indicators of lake salinity. Int. J. Salt Lake Res. 2: 87–92.

Cumming, B. F., K. A. Davey, J. P. Smol & H. J. B. Birks, 1994. When did acid-sensitive Adirondack lakes (New York, U.S.A.) begin to acidify and are they still acidifying? Can. J. Fish aquat. Sci. 51: 1550–1568.

Deflandre, G., 1932. Archaeomonadaceae, une famille nouvelle de Protistes fossiles marins á loge siliceuse. C.R. Acad. Sci., Paris 194: 1859–1861.

Deflandre, G., 1936. Les Flagellés fossiles. Apercu biologique et paleontologique. Role geologique. Actual. Sc. And Indust. Expos. Geol., Paris 355: 8–97.

Dixit, S. S., A. S. Dixit & J. P. Smol, 1992a. Assessment of pre-industrial changes in lakewater chemistry in Sudbury area lakes. Can. J. Fish. aquat. Sci. 49 (Supplement 1): 8–16.

Dixit, A. S., Dixit, S. S. & J. P. Smol, 1992b. Long-term trends in lake water pH and metal concentrations inferred from diatoms and chrysophytes in three lakes near Sudbury, Ontario. Can. J. Fish. aquat. Sci. 49 (Supplement. 1): 17–24.

Dop, A. J., 1980. Benthic Chrysophyceae from The Netherlands. Unpublished Ph.D. dissertation, Vrije University, Amsterdam, 141 pp.

Douglas, M. S. V. & J. P. Smol, 1995. Paleolimnological significance of observed distribution patterns of chrysophyte cysts in Arctic pond environments. J. Paleolim. 13: 1–5.

Duff, K. E. & J. P. Smol, 1994. Chrysophycean cyst flora from British Columbia (Canada) lakes. Nova Hedwigia 58: 353–389.

Duff, K. E. & B. A. Zeeb, 1995. Siliceous chrysophycean microfossils: recent advances and applications to paleoenvironmental investigations. In Babcock, L. & W. Ausich (eds.) Siliceous Microfossils. Short Courses in Paleontology, No. 8, The Paleontological Society, Knoxville, Tennessee: 139–158.

Duff, K. E., B. A. Zeeb & J. P. Smol, 1995. Atlas of Chrysophycean Cysts. Kluwer Academic Press, Dordrecht, 189 pp.

Duff, K. E., B. A. Zeeb & J. P. Smol, 1997. Chrysophyte cyst biogeographical and ecological distributions: a synthesis. J. Biogeogr. 24: 791–812.

Elner, J. K. & C. M. Happey-Wood, 1978. Diatom and chrysophycean cyst profiles in sediment cores from two linked but contrasting Welsh lakes. Br. Phycol. J. 13: 341–360.

Facher, E. & R. Schmidt, 1996. A siliceous chrysophycean cyst-based pH transfer function for Central European lakes. J. Paleolim. 16: 275–321.

Gayral, P. & C. Billard, 1986. A survey of the marine Chrysophyceae with special reference to the Sarcinochrysidales. In Kristiansen, J. & R. A. Andersen (eds.) Chrysophytes: Aspects and Problems. Cambridge University Press, Cambridge: 37–48.

Gilbert, S., B. A. Zeeb & J. P. Smol, 1997. Chrysophyte stomatocyst flora from a forest peat core in the Lena River Region, northeastern Siberia. Nova Hedwigia 64: 311–352.

Grönlund, E., H. Simola & P. Huttunen, 1986. Paleolimnological reflections of fiber-plant retting in the sediments of a small clearwater lake. Hydrobiologia 143: 425–531.

Hajós, M., 1973. Diatomées du Pannonien Inférieur provenant du bassin Néogene de Csákvár. Ile partie. Acta Botanica Academiae Scientiarum Hungaricae 18: 95–118.

Hajós, M., 1974. A pulai Put-3. Sz. fúrás felsöpannóniai képzödményeinek Diatoma flórája. Magyar Állami Földtani Intézet Évi Jelentése: 263–285.

Hajós, M. & Radócz., 1969. Diatomás rétegek a bükkalji alsópannonból. Magyar Állami Földtani Intézet Évi Jelentése: 271–297.

Harwood, D. M., 1986. Do diatoms beneath the Greenland Ice Sheet indicate interglacials warmer than present? Arctic 39: 304–308.

Hilliard, D. K. & B. Asmund, 1963. Studies on Chrysophyceae from some ponds and lakes in Alaska. II. Notes on the genera *Dinobryon, Hylobryon* and *Epipyxis* with descriptions of new species. Hydrobiologia 22: 331–397.

Kristiansen, J., 1990. Phylum Chrysophyta. In Margulis, L. et al. (eds.) Handbook of Protoctista. Jones and Bartlett Publishers, Boston: 438–453.

Kristiansen, J. 1986. Silica-scale bearing chrysophytes as environmental indicators. Br. Phycol. J. 21: 425–436.

Kristiansen, J., 1979. Problems in classification and identification of Synuraceae (Chrysophyceae). Schweiz. Z. Hydrobiol. 40: 310–319.

Kristiansen, J. & R. Andersen (eds.) 1986. Chrysophytes: Aspects and Problems. Cambridge University Press, Cambridge, 337 pp.

Kristiansen, J. & G. Cronberg (eds.) 1996. Chrysophytes: Progress and Horizons. Nova Hedwigia 114: 1–266.

Lotter, A., H. J. B. Birks, W. Hofmann & A. Marchetto, 1997. Modern diatom, cladocera, chironomid, and chrysophyte cyst assemblages as quantitative indicators for the reconstruction of past environmental conditions in the Alps. I. Climate. J. Paleolim. 18: 395–420.

Lotter, A., H. J. B. Birks, W. Hofmann & A. Marchetto, 1998. Modern diatom, cladocera, chironomid, and chrysophyte cyst assemblages as quantitative indicators for the reconstruction of past environmental conditions in the Alps. II. Nutrients. J. Paleolim. 19: 443–463.

Marsicano, L. J. & P. A. Siver, 1993. A paleolimnological assessment of lake acidification in five Connecticult lakes. J. Paleolim. 9: 209–221.

Munch, C. S., 1980. Fossil diatoms and scales of Chrysophyceae in the recent history of Hall Lake, Washington. Freshwat. Biol. 10: 61–66.

Nygaard, G., 1956. Ancient and recent flora of diatoms and Chrysophyceae in Lake Gribsø. Folia Limnol. Scand. 8: 32–262.

Peglar, S. M., S. C. Fritz, T. Alapieti, M. Saarnisto & H. J. B. Birks, 1984. Compostition and formation of laminated sediments in Diss Mere, Norfolk, England. Boreas 13: 13–28.

Peters, M. C. & R. A. Andersen, 1993. The fine structure and scale formation of *Chrysolepidomonas dendrolepidota* gen. et sp. nov. (Chrysolepidomonadaceae fam. nov., Chrsyophyceae). J. Phycol. 29: 469–475.

Pienaar, R. N., 1980. Chrysophytes. In Cox, E. R. (ed.) Phytoflagellates. Elsevier, New York: 213–242

Pla, S., 1999. The chrysophycean cysts from the Pyrenees and their applicability as paleoenvironmental indicators. Ph.D. thesis, Univ. Barcelona, Dep. D'Ecologia, 277 pp.

Preisig, H. R., 1995. A modern concept of chrysophyte classification. In Sandgren, C. D. et al. (eds.) Chrysophyte Algae: Ecology, Phylogeny and Development. Cambridge University Press, Cambridge: 46–74

Reavie, E. D., M. S. V. Douglas & N. E. Williams, 2001. Paleoecology of a groundwater outflow using siliceous microfossils. Ecoscience 8: 239–246.

Rull, V., 1986. Diatomeas y crisofíceas en los sedimentos acuáticos de una depresión cárstica del Pirineo catalán. Oecologia aquatica 8: 11–24.

Rull, V. & T. Vegas-Vilarrubia, 2000. Chrysophycean stomatocysts in a Caribbean mangrove. Hydrobiologia 428: 145–150.

Rybak, M., 1986. The chrysophycean paleocyst flora of the bottom sediments of Kortowskie Lake (Poland) and its ecological significance. Hydrobiologia 140: 67–84.

Rybak, M., I. Rybak & M. Dickman, 1987. Fossil chrysophycean cyst flora in a small meromictic lake in southern Ontario, and its paleoecological interpretation. Can. J. Bot. 65: 2425–2440.

Sandgren, C. D., 1980. Resting cyst formation in selected chrysophyte flagellates: an ultrastructural survey including a proposal for the phylogenetic significance of interspecific variations in the encystment process. Protistologica 16: 289–303.

Sandgren, C. D., 1983. Morphological variability in populations of chrysophycean resting cysts. I. Genetic (interclonal) and encystment temperature effects on morphology. J. Phycol. 19: 64–70.

Sandgren, C. D., 1988. The ecology of chrysophyte flagellates: their growth and perennation strategies as freshwater phytoplankton. In Sandgren, C. D. (ed.) Growth and Reproductive Strategies of Freshwater Phytoplankton. Cambridge University Press, Cambridge: 9–104.

Sandgren, C. D., 1991. Chrysophyte reproduction and resting cysts: a paleolimnologist's primer. J. Paleolim. 5: 1–9.

Sandgren, C. D., J. P. Smol & J. Kristiansen, 1995. Chrysophyte Algae: Ecology, phylogeny and development. Cambridge University Press, Cambridge, 399 pp.

Siver, P. A., 1991a. The Biology of *Mallomonas*: Morphology, Taxonomy and Ecology. Kluwer Academic Publishers, Dordrecht, The Netherlands, 230 pp.

Siver, P. A., 1991b. Implications for improving paleolimnological inference models utilizing scale-bearing siliceous algae: transforming scale counts to cell counts. J. Paleolim. 5: 219–225.

Siver, P. A., 1993. Inferring specific conductivity of lake water using scaled chrysophytes. Limnol. Oceanogr. 30: 1480–1492.

Siver, P. A., 1995. The distribution of chrysophytes along environmental gradients: their use as biological indicators. In Sandgren, C. D. et al. (eds.) Chrysophyte Algae: Ecology, Phylogeny and Development. Cambridge University Press, Cambridge: 232–268.

Siver, P. A., J. S. Hamer & H. Kling, 1990. The separation of *Mallomonas duerrschmidtiae* sp. nov. from *M. crassisquama* and *M. pseudocoronata*: implications for paleolimnological research. J. Phycology 26: 728–740.

Siver, P. A., A. M. Lott, E. Cash, J. Moss & L. J. Mariscano, 1999. Century changes in Connecticut, U.S.A., lakes as inferred from siliceous algal remains and their relationships to land-use changes. Limnol. Oceanogr. 44: 1928–1935.

Smol, J. P., 1980. Fossil synuracean (Chrysophyceae) scales in lake sediments: a new group of paleoindicators. Can. J. Bot. 58: 458–465.

Smol, J. P., 1983. Paleophycology of a high arctic lake near Cape Herschel, Ellesmere Island. Can. J. Bot. 61: 2195–2204.

Smol, J. P., 1985. The ratio of diatom frustules to chrysophycean statospores: a useful paleolimnological index. Hydrobiologia 123: 199–208.

Smol, J. P., 1986. Chrysophycean microfossils as indicators of lakewater pH. In Smol, J. P. et al. (eds.) Diatoms and Lake Acidity. Dr. W. Junk Publ., Dordrecht: 275–287.

Smol, J. P., 1988. Chrysophycean microfossils in paleolimnological studies. Palaeogeog. Palaeoclim. Palaeoecol. 62: 287–297.

Smol, J. P., 1990. Diatoms and chrysophytes—a useful combination in paleolimnological studies. In Simola, H. (ed.) Proceedings of the 10th International Diatom Symposium. Koeltz Scientific Books, Koenigstein: 585–592.

Smol, J. P., 1995. Application of chrysophytes to problems in paleoecology. In Sandgren, C. et al. (eds.) Chrysophyte Algae: Ecology, Phylogeny and Development, Cambridge University Press, Cambridge: 303–329.

Smol, J. P., D. F. Charles & D. R. Whitehead, 1984a. Mallomonadacean microfossils provide evidence of recent lake acidification. Nature 307: 628–630.

Smol, J. P., D. F. Charles & D. R. Whitehead, 1984b. Mallomonadacean (Chrysophyceae) assemblages and their relationships with limnological characteristics in 38 Adirondack (N.Y.) lakes. Can. J. Bot. 62: 911–923.

Srivastava, S. K. & P. L. Binda, 1984. Siliceous and silicified microfossils from the Maastrichtian Battle Formation of southern Alberta, Canada. Paleobiologie Continentale 14: 1–24.

Steinberg C., H. Hartmann & D. Krause-De llin, D., 1988. Paleoindicators of acidification in Kleiner Abersee (Federal Republic of Germany, Bavarian Forest) by chydorids, chrysophytes, and diatoms. J. Paleolim. 6: 123–140.

Stoermer, E. F., J. A. Wolin, C. L. Schelske & D. J. Conley, 1985. An assessment of ecological changes during the recent history of Lake Ontario based on siliceous algal microfossils preserved in the sediments. J. Phycol. 21: 257–276.

Takahashi, E., 1978. Electron Microscopical Studies of the Synuraceae (Chrysophyceae) in Japan. Taxonomy and Ecology. Tokai University Press, Tokyo, 194 pp.

Taylor, S. J., 1997. Taxonomy and ecological characterization of chrysophycean stomatocysts from lakes extending from the boreal forest to the arctic tundra, Canada. M.Sc. Thesis, Queen's University, Dept, Biology, 181 pp.

Van de Vijver, B. & L. Beyens, 1997. The chrysophyte stomatocyst flora of the moss vegetation from Strømness Bay Area, South Georgia. Arch. Protistenkd. 148: 505–520.

Wee, J. L., 1982. Studies on the Synuraceae (Chrysophyceae) of Iowa. Bibliotheca Phycologia 62: 1–183.

Wilkinson, A. N., B. A. Zeeb, J. P. Smol & M. S. V. Douglas, 1996. Chrysophyte stomatocyst assemblages associated with periphytic, high arctic pond environments. Nord. J. Bot. 16: 95–112.

Wilkinson, A. N., B. A. Zeeb & J. P. Smol, 2001. Atlas of chrysophycean cysts. Vol 2. Kluwer Academic Publishers, Dordrecht, 169 pp.

Zeeb, B. A., K. E. Duff & J. P. Smol, 1990. Morphological descriptions and stratigraphic profiles of chrysophycean stomatocysts from the recent sediments of Little Round Lake, Ontario. Nova Hedwigia 51: 361–380.

Zeeb, B. A. & J. P. Smol, 1993. Postglacial chrysophycean cyst record from Elk Lake, Minnesota. Geol. Soc. Amer. Special Paper 276: 239–249.

Zeeb, B. A. & J. P. Smol, 1995. A weighted-averaging regression and calibration model for inferring lakewater salinity using chrysophycean stomatocysts from lakes in western Canada. Int. J. Salt Lake Res. 4: 1–23.

Zeeb, B. A., C. E. Christie, J. P. Smol, D. L. Findlay, H. Kling & H. J. B. Birks, 1994. Responses of diatom and chrysophyte assemblages in Lake 227 to experimental eutrophication. Can. J. Fish. aquat. Sci. 51: 2300–2311.

Zeeb, B. A., K. E. Duff & J. P. Smol, 1996a. Recent advances in the use of chrysophyte stomatocysts in paleoecological studies. Beiheft zur Nova Hedwigia 114: 247–252.

Zeeb, B. A., J. P. Smol & S. L. Vanlandingham, 1996b. Pliocene chrysophycean stomatocysts from the Sonoma Volcanics, Napa County, California. Micropaleontology 42: 79–91.

Zeeb, B. A., J. P. Smol & S. P. Horn, 1996c. Chrysophycean stomatocysts from Costa Rican tropical lake sediments. Nova Hedwigia 63: 279–299.

10. EBRIDIANS

ATTE KORHOLA (atte.korhola@helsinki.fi)
Department of Ecology and Systematics
Division of Hydrobiology
University of Helsinki
P.O. Box 17 (Arkadiankatu 7)
FIN-00014 Helsinki, Finland

JOHN P. SMOL (SmolJ@BIOLOGY.QueensU.Ca)
Paleoecological Environmental Assessment
and Research Lab (PEARL)
Department of Biology
116 Barrie St., Queen's University
Kingston
Ontario K7L 3N6 Canada

Keywords: Ebridians, *Ebria tripartita*, sediment, Baltic Sea, palaeoecology, eutrophication, dissolution

Introduction

Ebridians are a group of microscopic, heterotrophic, primarily marine plankton, character-ized by a siliceous internal skeleton that is frequently preserved in sedimentary deposits. They are cosmopolitan, mainly neritic, but can also be found in estuaries, shallow em-bayments, and semi-enclosed seas such as the Baltic Sea and the Black Sea (Ernisse & McCartney, 1993; Korhola & Grönlund, 1999). Ebridians have a modern diversity of only a few species, and are not particularly significant components of modern marine plankton. However, in contrast to the diatoms, where perhaps 90% of the described species are living, most of the described ebridian taxa are fossil forms. Indeed, ebridian research is essentially palaeontologic, and they are widely used in stratigraphy in palaeoceanography and in palaeobiologic studies.

As such, the connection of ebridians to a volume on palaeolimnological methods and indicators may be a bit tenuous, but they are occasionally recorded in freshwater (likely from allochthonous sources) and brackish profiles, and so this short chapter attempts to provide a brief overview of this group which may be of use to the palaeolimnologist. Our primary aim is to attract attention to these organisms, whose remains are found regularly in the sediment record, but are often overlooked. In addition, we explore the potential usefulness and indicator value of the group in future palaeoecological studies.

J. P. Smol, H. J. B. Birks & W. M. Last (eds.), 2001. *Tracking Environmental Change Using Lake Sediments.*
Volume 3: Terrestrial, Algal, and Siliceous Indicators. Kluwer Academic Publishers, Dordrecht, The Netherlands.

We focus in particular on one species, *Ebria tripartita* (Schumann) Lemmermann, and one environment, namely the brackish-water Baltic Sea, where this species has been studied quite intensively during recent years. Ernisse & McCartney (1995) provide a more detailed summary of these indicators, and Loeblich et al. (1968) provide an annotated index of ebridian taxa.

Morphology, taxonomy and preservation in the sediments

Ebridians do not appear in the fossil record before the Cenozoic, but their fossils have been found in particularly large numbers in Cenozoic sedimentary rocks that are rich in various other siliceous remains (Tappan, 1980). After a phase of rapid diversification in the Paleocene, during which at least eight ebridian genera evolved, they suffered massive extinctions by the close of the Miocene. Although they have a relatively diverse fossil history, ebridians are relatively rare in the sediments of modern seas. Only three (possibly four) species, assigned to two genera (*Ebria* and *Hermesinum*), are known to exist today (Taylor, 1990). The ecological characteristics of the few living ebridians are poorly known, and so they have played a relatively small role in modern research.

Ebridians commonly range from 10 to 60 μm in length and from 20 to 30 μm in width (Lipps, 1979). Their cells are more or less spherical, colourless, and contain a single nucleus and two flagella of different lengths (Lipps, 1979; Tikkanen, 1986). Some ebridian taxa produce cyst-like features at one end of the skeleton that may have served as a protective enclosure during periods of environmental stress (Ernisse & McCartney, 1995). Ebridians are preserved in bottom deposits because of their internal siliceous skeleton, which consists of a framework of solid rods, 2–4 μm in thickness (Preisig, 1994). In *Ebria tripartita*, the system of rods comprises three stems, which branch in a regular manner from an initial branching point, thus forming a triadial, basket-like symmetry (Figs. 1 and 2). In *Hermesinum* four such branches are found. The rods are usually angular and possess many small spines and nodules, as shown by electron microscopy (Fig. 2B). The average length of the siliceous skeleton of *E. tripartita* is 31 μm, and the average width is 24 μm (Tikkanen, 1986). The siliceous skeleton of ebridians is highly resistant both against mechanical abrasion and chemical dissolution.

The taxonomic status of ebridians is presently rather uncertain (Preisig, 1994). On one hand, they resemble silicoflagellates, but have a skeleton of solid tubular rods rather than hollow ones. On the other hand, the structure of their nuclei as well as their endoskeleton has some common features with some dinoflagellates (Loeblich, 1982; Lipps, 1979; Locker & Martini, 1986). Indeed, the evolutionary relations among ebridians, silicoflagellates, and endoskeletal dinoflagellates are in many respects unclear. Today, the ebridians are, nevertheless, often regarded as a separate taxonomic unit (Ebriophyceae/Ebriida) with uncertain phylogenetic affinities (Taylor, 1990; Preisig, 1994).

In the older literature, there have been erroneous identifications concerning these taxa. Most common mistakes include the assignment of the species with other dinoflagellates, silicoflagellates or even Radiolaria. It is perhaps because of these identification problems that these taxa have rarely been recognized and have only been used in a few palaeolimnological studies.

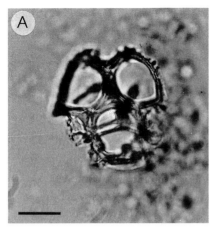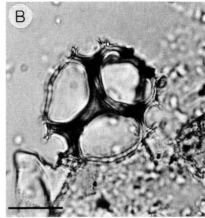

Figure 1. Subfossil *Ebria tripartita* (Schumann) Lemmermann, LM photographs (A-B). Scale bars = 10 μm. Reproduced from Korhola & Grönlund (1999); used with permission.

Methodological aspects

The chemical treatment of subsamples follows the standard methodology developed for diatom sample preparation (Battarbee, 1986 and Battarbee et al., this volume). Naphrax® or Hyrax® or comparable media can be used as mountants. Counting is performed under a light microscope at 1000× magnification. The frequencies of ebridians can be determined relative to diatoms (Korhola & Grönlund, 1999) or the total siliceous microfossil assemblage (Westman, 1999), which can then be based on the sum of total diatoms/siliceous fossils + ebridians. In general, ebridians tend to be rather rare, occurring in low absolute frequencies in an average microfossil preparation (Ernisse & McCartney, 1995). Moreover, their remains are often fragmented (Korhola & Grönlund, 1999). The preservation of the main part of the central skeleton (a complete driadial system of rods) can be used as the criterion for the enumeration of one species.

Brief history of use of ebridians in palaeoecological research

Ebridian research has developed episodically, with an initial phase of interest in the nineteenth century, a quiet interval between about 1910 and 1950, and a second, mostly geologic phase of development since then, in conjunction with renewed interest in subjects such as plate tectonics and palaeoceanography after WWII. Most extensive and reliable data about the biostratigraphic ranges of ancient ebridian species have been generated during the Deep Sea Drilling Project (DSDP, 1968 to 1983) and the Ocean Drilling Project (ODP, 1985 to present). For example, Locker & Martini (1989) established nine ebridian zones that span the interval from the Early Miocene to the early Quaternary, and demonstrated that these organisms might have some value as biostratigraphic markers and environmental indicators. The work dealing with ancient ebridians (i.e., pre-Quaternary biostratigraphy) is referenced and discussed in more detail in Ernisse & McCartney (1995), whereas here

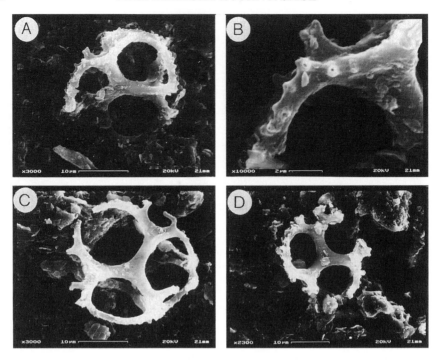

Figure 2. SEM photographs (A-D) of *Ebria tripartita.* B. Detail of photograph A. Reproduced from Korhola & Grönlund (1999); used with permission.

we discuss the few living ebridians and their applicability to more recent environmental and palaeolimnological problems.

Little is known about the present-day ecological preferences of the ebridians, which, of course, hinders palaeoecological interpretations. According to Lipps (1979), they are very abundant in nutrient-rich, cool marine waters at high latitudes. It seems that from the two main living species, *Hermesinun adiraticum* is a stenohaline organism that thrives in warm waters, whereas *E. tripartita* has a wider tolerance to temperature and salinity variations (Rhodes & Gibson, 1981). The latter is the only ebridian species alive in the Baltic Sea today, and intact remains of *E. tripartita* are frequently found in sediments representing the ancient history of the Baltic basin, such as the Eemian Baltic Sea about 120 000 yrs BP (Brander, 1937; Niemelä & Tynni, 1979; Eriksson et al., 1980; Grönlund, 1991).

In the Baltic Sea, increased abundances of this taxon are regularly noted in the uppermost parts of sediment cores and are commonly interpreted as indicating nutrient enrichment and/or upwelling (e.g., Miller & Risberg, 1990; Grönlund, 1991, 1993; Saxon & Miller, 1993; Andrén, 1995). For example, Miller & Risberg (1990) noted increased abundances of *Ebria tripartita* in surface sediments of a core from the northwestern Baltic, and related this change to the acceleration of eutrophication during the last ca. 20 years. Similarly, Korhola & Grönlund (1999) observed in the Gotland basin core a slight increase in the abundance of *E. tripartita* towards the core surface. This increase was also tentatively interpreted as being related to the progressive eutrophication of the basin. Numerous studies (e.g., Rosenberg,

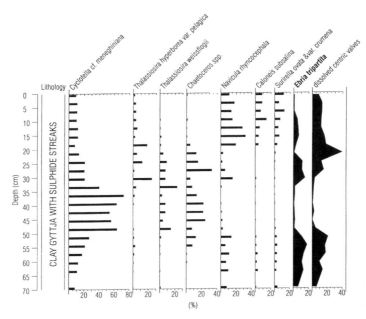

Figure 3. Relative abundance of *Ebria tripartita* in the Töölö Bay core Tol-1, central Helsinki, Finland, plotted together with the frequency distributions of the most common diatoms and partly dissolved valves of centric diatoms found in the same core. The dominance of planktonic diatoms between 50–30 cm (AD 1890-AD 1960) refers to the period when the bay received non-treated or only partially treated sewage effluent from the local sewage works and was therefore heavily polluted (Korhola & Blom, 1996; Håkansson & Korhola, 1998). Reproduced from Korhola & Grönlund (1999); used with permission.

1984; Elmgren, 1989; Baltic Marine Environment Protection Commission, 1990) point to a marked increase in nutrients in the Baltic Sea during the last 200 years.

However, in the sediment core from Töölö Bay (Fig. 3), central Helsinki (Finland), the abundances of *E. tripartita* were particularly low during the most pronounced eutrophication stage, as inferred from diatoms (Korhola & Blom, 1996; Korhola & Grönlund, 1999). One explanation could be that, despite the species' general preference for nutrient-rich waters, it might be less competitive in hypereutrophic waters, especially if the eutrophication is associated with heavy blooms of other planktonic algae (Korhola & Grönlund, 1999). On the other hand, studies of recent blooms in Long Island Sound USA (Ernisse & McCartney, 1993) and the Baltic Sea (Kononen & Niemi, 1984) have observed significant year-to-year fluctuations in the absolute abundance of ebridians, which do not seem to be related to changes in either salinity or trophic status.

In recent years, there has been an increased interest in modern (extant) forms found as fossils in the surface sediment of the shallow seas. Westman (2000) studied four long sediment cores, five short cores, and more than 20 surface-sediment samples representing the years 1993 and 1997 for subfossil ebridians in the Baltic Sea proper. He was unable to distinguish any systematic trend in the abundance of *E. tripartita* during the most recent centuries, although a slight increase in the relative abundance of this taxon was observed in all sub-surface cores. All variation in the ebridian data seemed to be confined to periods

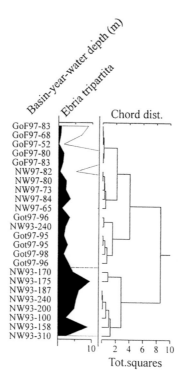

Figure 4. Relative proportion (%) of *Ebria tripartita* in surface-sediment samples from the Baltic Sea collected in 1993 and 1997 plotted against the major diatom algae clusters as distinguished by the CONISS cluster analysis (Grimm, 1987). The column in the left refers to the sampling stations with indications of the sampling year. Reproduced from Westman (2000); used with permission.

when there were also major changes in diatom assemblages—a situation that seems to hold true in other cases as well (e.g., Korhola & Grönlund, 1999). This prompted Westman (1999) to conclude that the changes in the abundance of *E. tripartita* are "most probably attributable to alterations in the species composition of primary production in the Baltic, and hence represent only a secondary effect of environmental change". As *E. tripartita* mainly feeds on diatoms (Ernisse & McCartney, 1993), changes in its relative frequency could simply result from changes in the composition of the diatom flora and/or the contribution of the diatom flora to total primary production (Westman, 1999). This feature is illustrated in Figure 4, where the abundances of *E. tripartita* in the surface-sediments of the Baltic Sea proper sampled in 1993 and 1997 are contrasted against the major differences in diatom assemblages, as revealed by cluster analysis (Westman, 1999). As the abundance of *E. tripartita* in the sediments seems to be reflected in major differencies in the diatom assemblage, it may be assumed that the distribution of the species in the sediments is predominantly controlled by differencies in the composition of the diatom assemblage.

In an ongoing study dealing with the eutrophication history of the shallow Baltic Sea inlets, ebridians have been identified together with diatoms in 45 sampling stations along the coastal area of southern Finland (Weckström & Korhola, in press). Only a few remains

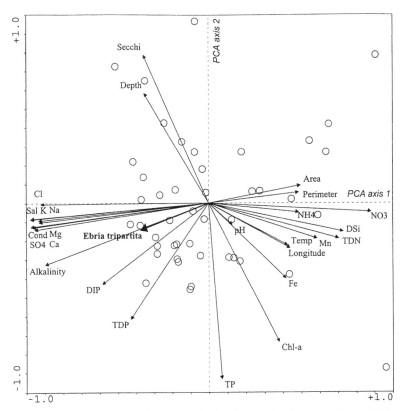

Figure 5. A principal components analysis (PCA) ordination diagram showing the position of *Ebria tripartita* in relation to the measured water quality variables in 45 sampling stations along the southern coast of Finland. *Ebria* was treated as a supplementary variable in the PCA. The PCA plot is from Weckström & Korhola (in press). *Ebria* data are unpublished (Weckström & Korhola). Abbreviations: Sal = salinity, Cond = specific conductance, DIP = dissolved inorganic phosphorus, TDP = total dissolved phosphorus, TP = total phosphorus, TDN = total dissolved nitrogen, DSi = dissolved silica, and temp = water temperature.

of *E. tripartita* were found, with their absolute numbers ranging between 0–12 per 500 diatom frustules in sediment samples. A preliminary principal components analysis (PCA) performed on the measured water chemistry and ebridian data suggests that the ebridian distributions and abundances in this data set are correlated with the PCA axis 1 consisting of salinty and ionic concentration (Fig. 5). This axis explained 46.6% of the variance in the environmental and ebridian data. It thus seems that ebridians in these shallow embayments prefer sites of relatively high salinity (>4 psu), conductivity and, to some extent, PO_4 concentrations. Further studies will focus on the relationships of ebridians in these shallow coastal waters to various diatom species and assemblages.

Indicator value and future research priorities

At present, contradictory data regarding the indicator value of ebridians exist, which are most obviously related to the rudimentary ecological data available for these organisms.

As shown by the examples discussed in this chapter, ebridians clearly show stratigraphic changes in sediment profiles, yet the ecological interpretations of these changes are still speculative. Clearly, more systematic studies on the distribution of these organisms along a range of various ecological gradients are needed in order to achieve a more comprehensive understanding of the palaeolimnological significance of the group.

Future studies should focus on the collection of modern surface-sediment data sets along a wide range of environments and ecological conditions in different marine systems. Some results indicate that the observed patterns in the frequency distributions of ebridians may be partly due to the selective preservation of their siliceous skeletons (Korhola & Grönlund, 1999). Thus, the questions dealing with the preservation and taphonomy of these organisms should be researched more fully. The taxonomy of ebridians is in many respects vague; thus in order for ebridians to become more useful indicators, detailed taxonomic studies using both light and electron microscopy are needed. Moreover, if ebridians are strongly affected by grazing pressure, as suggested by Westman (1999), then the question of which food web changes might affect ebridians would be an interesting research question.

Summary

Ebridians are a group of microscopic, heterotrophic marine plankters. Their siliceous endoskeletons are preserved in sedimentary deposits, and so they can be studied using the same techniques developed for other siliceous indicators, such as diatoms. They are primarily marine, and so are not frequently encountered by palaeolimnologists, but they may be common in brackish waters, such as the Baltic Sea, estuaries, and some lacustrine environments that may have had an influx of marine material. Only a few species are known to exist today. Relatively little is known about the ecological optima and tolerances of taxa, which currently hampers palaeoecological interpretations. However, ongoing research suggests they have some potential in palaeoenvironmental studies.

Acknowledgements

The study was funded by the Academy of Finland Grant 101 7383 to AK. We thank Kaarina Weckström for counting the ebridian remains from the surface-sediments of the Gulf of Finland.

References

Andrén, E., 1995. Recording environmental changes in the Southern Baltic Sea—current results from a diatom study within Project ODER. In Marino, D. & M. Montesor (eds.) Proceedings of the 13th International Diatom Symposium. Maratea, Italy, 1st-7th September 1994: 443–455.

Baltic Marine Environment Protection Commission, Helsinki Commission, 1990. Second Periodic Assessment of the State of the Marine Environment of the Baltic Sea, 1984–1988; Background Document. Baltic Sea Environment Proceedings 35 B, 432 pp.

Brander, G., 1937. Zur Deutung der intramoränen Tonablagerungen an der Mga. unweit von Leningrad. Bull. Comm. Géol. Fin. 119: 93–113.

Edler, L., G. Hällfors & Å. Niemi, 1984. A preliminary check-list of the phytoplankton of the Baltic Sea. Acta Bot. Fenn. 128, 26 pp.

Elmgren, R., 1989. Man's impact on the ecosystem od the Baltic Sea: Energy flows today and at the turn of the century. Ambio 6: 326–332.

Eriksson, B., T. Grönlund & R. Kujansuu, 1980. Interglasiaalikerrostuma Evijärvellä Pohjanmaalla (English summary: An interglacial deposit at Evijärvi in the Pohjanmaa region). Geologi 32: 65–71.

Ernisse, J. & K. McCartney, 1993. Ebridians. In Lipps, J. H. (ed.) Fossil Prokaryotes and Protists. Blackwell Scientific Publication Inc., Boston: 131–140.

Ernisse, J. & McCartney, 1995. Ebridians and endoskeleton dinoflagellates. In Blome, C. D., P. Whalen & K. Reedet (eds.) Siliceous Microfossils. Paleontology Society Short Courses in Paleontology 8: 177–185.

Grimm, E., 1987. CONISS: A Fortran 77 program for stratigraphically constrained cluster analysis by the method of incremental sum of squares. Comp. Geosci. 13: 13–25.

Grönlund, T., 1991. The diatom stratigraphy of the Eemian Baltic Sea on the basis of sediment discoveries in Ostrobothnia. Finland. Geol. Surv. Fin., Report 102, 26 pp.

Grönlund, T., 1993. Diatoms in surface sediments of the Gotland Basin in the Baltic Sea. Hydrobiologia 269/270: 235–242.

Håkansson, H. & A. Korhola, 1998. Phenotypic plasticity in the diatom *Cyclotella meneghiniana* or a new species? Nova Hedwigia 66: 187–196.

Kononen, K. & Å. Niemi, 1984. Long term variation of the phytoplankton composition at the entrance to the Gulf of Finland. Ophelia suppl. 3: 101–110.

Korhola, A. & T. Blom, 1996. Marked early 20th century pollution and the subsequent recovery of Töölö Bay, central Helsinki, as indicated by subfossil diatom assemblage changes. Hydrobiologia 341: 169–179.

Korhola, A. & T. Grönlund, 1999. Observations of *Ebria tripartita* (Schumann) Lemmermann in Baltic sediments. J. Paleolim. 21: 1–8.

Lipps, J., 1979. Ebridians. In Fairbridge, R. W. & D. Jablonski (eds.) Encyclopedia of Earth Sciences. Vol 7. Dowden, Hutchinson & Ross, Stroudsburg, 276 pp.

Locker, S. & E. Martini, 1986. Ebridians and actiniscidians from the southwest Pacific. In Kennett, J. P., et al. (eds.) Initial Reports of the Deep Sea Drilling Project. 90. U.S. Goverment Printing Office, Washington, D.C.: 939–951.

Locker, S. & E. Martini, 1989. Cenozoic silicoflagellates, ebridians, and actiniscidians from the Vøring Plateau (ODP Leg 104). In Eldholm, O., et al. (eds.) Proceedings of the Ocean Drilling Program, Scientific results. 104. Ocean Drilling Program, College Station, Texas: 543–585.

Loeblich, A. R. III, 1982. Dinophyceae. In Parker, S. P. (ed.) Synopsis and Classification of Living Organisms. Vol 1. McGraw-Hill, New York: 101–115.

Loeblich, A. R. III, A. L., Loeblich, H., Tappan & A. R. Loeblich, Jr., 1968. Annotated index of fossil and recent silicoflagellates and ebridians with descriptions and illustrations of validly proposed taxa. Geological Society of America, Memoire 106: 1–319.

Miller, U. & J. Risberg, 1990. Environmental changes, mainly eutrophication, as recorded by fossil siliceous micro-algae in two cores from the uppermost sediments of the north-western Baltic. Nova Hedwigia 100: 237–352.

Niemelä, J. & R. Tynni, 1979. Interglacial and interstadial sediments in the Pohjanmaa region. Finland. Geol. Surv. Fin. Bull. 302, 48 pp.

Preisig, H. R., 1994. Siliceous structures and silicification if flagellated protists. Protoplasma 181: 29–42.

Rhodes, R. G. & V. R., Gibson, 1981. An annual survey of Hermesinum adriaticum and Ebria tripartita, two ebridian algae in the lower Chesepeake Bay. Estuaries 4: 150–152.

Rosenberg, R. (ed.), 1984. Eutrophication in Marine Waters Surrounding Sweden, A review. Statens naturvårdsverk, Solna. PM 1808, 140 pp.

Saxon, M. & U. Miller, 1993. Diatom assemblages in superficial sediments from the Gulf of Riga, eastern Baltic Sea. Hydrobiologia 269/270: 243–249.

Tappan, H., 1980. The Paleobiology of Plant Protists. Freeman, San Francisco, 1028 pp.

Taylor, F. J. R., 1990. Ebridians. In Margulis, L., J. O. Corliss, M. Melkonian & D. J. Chapman (eds.) Handbook of Protoctista. Jones and Bartlett, Boston, 720–721.

Tikkanen, T., 1986. Kasviplanktonopas (Guide to Phytoplankton). Suomen Luonnonsuojelun Tuki Oy, Forssa, 278 pp.

Weckström, K. & A. Korhola. Physical and chemical characteristics of 45 shallow embayments on the southern coast of Finland. Hydrobiologia, in press.

Westman, P., 2000. The Siliceous microalgae *Dictyocha speculum* and *Ebria tripartita* as biomarkers and palaeoecological indicators in Holocene Baltic Sea sediments. GFF 112: 287–292.

11. PHYTOLITHS

DOLORES R. PIPERNO* (Pipernod@stri.org)
Center for Tropical Paleoecology and Archaeology
Smithsonian Tropical Research Institute
Balboa, Republic of Panama
*Mailing address: STRI, Unit 0949, APO AA 34002-0948

Keywords: paleoenvironmental reconstruction, phytoliths, vegetational change, multiproxy analysis

Introduction and history

Phytolith analysis is a relative newcomer to the field of paleolimnology. It has been mainly during the last 10 to 15 years that investigators have developed regional reference collections and keys capable of identifying a wide spectrum of plant taxa, and carried out studies providing basic information on how phytoliths are transported to, and then deposited onto the bottoms of lakes. The limited number of investigations carried out to date suggest that phytolith analysis is an important method of environmental reconstruction in lakes. Herein, I describe what phytoliths are, how they are recovered from lake sediments and identified, and what they potentially contribute to paleolimnological study.

Opaline phytoliths ($SiO_2 \cdot nH_2O$) are created when hydrated silica dissolved in groundwater is absorbed through the roots of a plant and carried throughout its vascular system. This soluble form of silica may then polymerize and be deposited in intercellular spaces, cell walls, and lumina of any plant organ. Aerial structures, including leaves, fruits, seeds, and inflorescence bracts, accumulate this silica more readily than do subterranean organs, although some angiosperms also incorporate solid silica in copious quantities in roots and tubers. After death and decay of the plant, these pieces of silica are ultimately deposited into soils and sediments as discrete particles called phytoliths. Quite literally "plant stones", phytoliths are resistant to oxidation and are well-preserved in many depositional environments for long periods of time. They arguably are the most durable plant fossils known to science.

Phytolith production and taxonomy

Research carried out in many regions of the world during the past two decades has demonstrated that phytoliths are produced in abundant quantities in many different kinds of plants, not just primarily in the Poaceae and Cyperaceae as conventional wisdom predicts (e.g.,

235

J. P. Smol, H. J. B. Birks & W. M. Last (eds.), 2001. *Tracking Environmental Change Using Lake Sediments.*
Volume 3: Terrestrial, Algal, and Siliceous Indicators. Kluwer Academic Publishers, Dordrecht, The Netherlands.

Bozarth, 1992; Piperno, 1988, 1998; Kealhofer & Piperno, 1998; Kondo et al., 1994; Runge, 1995) (Table I). Indeed, tropical evergreen forests probably produce as many, if not more, phytoliths per unit biomass of vegetation than do grassland environments (Piperno, 1988). Environmental factors can cause increases or decreases in the amount of silica absorbed from the soil and subsequently deposited into plants, and may, therefore, result in the formation of different amounts of phytoliths within the same species sampled from different regions. Nevertheless, the basic patterns of phytolith production summarized in Table I, in which families, tribes, and genera tend either to manufacture or not manufacture appreciable amounts of phytoliths, appear to be largely congruent across the world. These factors suggest that certain loci of silicification in plants are under genetic control (see discussion below). Phytolith analysts can, with increasing confidence, predict in which families and genera phytoliths are most likely to be found and in which they are least likely to be found, a factor which contributes to more efficient constructions of modern reference collections. This knowledge is also essential for interpreting paleoecological assemblages because researchers can take into account which taxa are/are not likely to be represented due to variability in phytolith production.

It was once believed that all deposits of opaline silica represented waste products or useless residues produced by plants and, therefore, phytoliths were not of functional utility in many species. This belief was based on the often-repeated observation that silica uptake was a passive process, and that its subsequent deposition in aerial structures was largely controlled by water loss through transpiration. However, Piperno (1988) and Sangster & Hodson (1997) reviewed evidence pointing out that: 1) there were many loci of phytolith deposition that could not be explained by water loss through transpiration; 2) active transport of soluble silica (e.g., expenditure of energy by the plant during silica uptake) had been demonstrated in some species; and 3) incorporation of large amounts of phytoliths clearly benefitted some plants by conferring protection from insects and pathogens. Furthermore, investigators have long noted close links between silicification and lignification in plants, and there is no doubt about the latter substance's important role in plant defence (e.g., Parry & Smithson, 1966; Coley & Barone, 1996).

Dorweiler & Doebley (1997) substantially clarified this picture by showing that phytolith production in teosinte (wild maize) and maize glumes was genetically controlled, and that the gene that controlled silicification was the same one that lignified the glumes. More studies such as these are likely to reveal genetic control over silicification in certain organs and tissues of other species. In view of the facts that toughness is the most effective defence that can be marshaled by plants, chemical compounds are "expensive" for plants to produce (Coley & Barone, 1996), and that it may be "cheaper" for plants to incorporate silica rather than lignin (Piperno, 1988, p. 48), it is likely that a primary function of phytoliths is to deter herbivores and pathogens. Hodson & Evans (1995) and Hodson et al. (1997) provide convincing evidence that silicon dioxide also ameliorates the toxic effects of aluminum in plants, and such "detoxification" is likely to be another important reason why many plants make phytoliths.

Phytolith researchers are now agreed that phytoliths identifiable at precise taxonomic levels (family, subfamily, genus) are produced in a considerable number of angiosperms, not solely in Poaceae and Cyperaceae (Table I), overturning the older notion still prevalent in some of the literature that diagnostic shapes are limited. Gymnosperms and pteridophytes also possess some diagnostic forms. Phytolith shapes are congruent in families, genera, and

Table I. Patterns of phytolith production and taxonomic significance in plants.

Families where production is high, phytoliths specific to family are common, and sub-family and genus-specific forms occur, sometimes widely in the family Monocotyledons: Arecaceae*, Cyperaceae* (*Cyperus/Kyllinga**, *Carex**), Heliconiaceae*● (*Heliconia***), Marantaceae*● (*Maranta**, *Calathea**), Musaceae* (*Musa***), Orchidaceae, Poaceae* (*Chusquea***, *Streptochaeta***, *Tripsacum**, *Zea**), Zingiberaceae*

Dicotyledons: Acanthaceae (*Mendoncia*), Annonaceae, Burseraceae* (*Bursera**, *Canarium**, *Protium**), Chrysobalanaceae*, Asteraceae, Cucurbitaceae* (*Cucurbita**, *Lagenaria**), Dilleniaceae, Fagaceae, Magnoliaceae (*Magnolia***, *Talauma***), Moraceae* (*Morus**), Podostemaceae, Ulmaceae* (*Celtis**), Urticaceae (*Boehmeria***, *Pilea***).

Pteridophytes: Equisetaceae (*Equisetum***), Hymenophyllaceae (*Trichomanes***), Selaginellaceae

Families where production is not high in many species, but where family and genus-specific forms occur
Pinaceae (*Pinus***, *Pseudotsuga***), Chloranthaceae (*Hedyosmum**), Dipterocarpaceae Euphorbiaceae* (*Sapium**), Flacourtiaceae

Examples of families where phytoliths have not been observed, or where production is generally uncommon and of lesser taxonomic utility
Amaranthaceae, Araceae, Araucariaceae, Bignoneaceae, Cactaceae, Chenopodiaceae, Convolvulaceae, Dioscoreaceae, Lecythidaceae, Liliaceae, Melastomataceae, Myrtaceae, Podocarpaceae, Rubiaceae, Sapindaceae, Solanaceae, Taxodiaceae, Tiliaceae

---

*Reproductive structures (fruits and seeds) also produce high amounts of diagnostic phytoliths.
●Underground organs (roots, tubers, and corms) contribute high amounts of diagnostic phytoliths.

Notes: Taxa in parentheses are examples of genus-level discrimination, and in many cases the diagnostic phytoliths were described in plants sampled from at least two different regions of the world; *next to the taxon indicates genus-specific shapes occur in fruits and/or seeds, **indicates shapes occur in leaves. All information is based on phytolith keys developed for the following parts of the world: American tropics (Piperno & Pearsall, 1998; Piperno, 1998; Pearsall, www.missouri.edu/~phyto); mainland southeast Asia (Kealhofer & Piperno, 1998), southern China (Zhao & Piperno, 1999), New Zealand (Kondo et al., 1994), North America (Bozarth, 1992, 1993; Brown, 1984; Hodson et al., 1997; Mulholland, 1989), the Near East (Miller-Rosen, 1993; Albert et al., 1999), equatorial Africa (Runge, 1995, 1998), New Guinea (Boyd et al., 1998), and Australia (Wallis, n.d.).

species across the world (compare descriptions of phytoliths in Piperno, 1988; Pearsall, www.missouri.edu/~phyto; Bozarth, 1992, 1993; Hodson et al., 1997; Sangster et al., 1997; Kealhofer & Piperno, 1998; and Runge, 1995, 1998). In other words, like pollen grains, phytoliths from the Cyperaceae, Asteraceae, Poaceae, Pinaceae, Magnolicaceae, and Ulmaceae, and other cosmopolitan plants have the same basic form regardless of whether they are from an arctic tundra or a tropical rain forest (Figs. 1–4). Phytolith morphology often exhibits a strong correspondence to a plant's taxonomic affiliation. Some examples follow. Spherical "scalloped" phytoliths have been observed in the fruit rinds of a single tribe, Cucurbiteae, of one family, the Cucurbitaceae, where they are differentiable to the level of genus (Bozarth, 1987; Piperno, 1998; Pearsall, www.missouri.edu/~phyto) (Figs. 5–6). Long phytoliths with smooth, conical ends and decorated shafts of various types

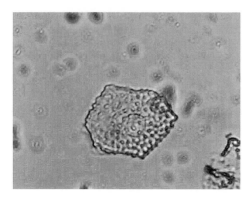

Figure 1. A phytolith from an achene of *Cyperus cayennesis* (Cyperaceae). It is specific to the closely related genera *Cyperus* and *Kyllinga*. Cyperaceae achene phytoliths are often diagnostic to individual genera. The phytolith is 38 μ long.

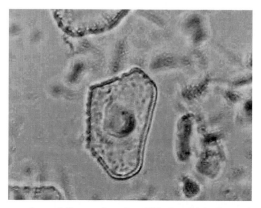

Figure 2. A phytolith from an achene of *Carex polystachya*. This phytolith is diagnostic of the genus *Carex*. Note the differences in the shape, surface sculpture, and protuberance that emanates from the center of the phytolith, compared with the phytolith in Figure 1. It is 38 μ long.

have been observed only in the seeds of the Marantaceae, an important family of tropical forest herbs, and tribal and generic differentiation is possible here too (Piperno, 1998; Pearsall, www.missouri.edu/~phyto; Runge, 1998). Finally, the Bambusoideae produce leaf phytoliths with tribal and genus-level affinities, making it possible to distinguish conclusively bamboos from other grasses in the fossil record (Piperno & Pearsall, 1998; Kealhofer & Penny, 1998).

As mentioned above, phytoliths formed in the epidermal tissue of fruits and seeds of angiosperms seem to be particularly useful. They typically are of a single morphological type and possess distinctive shapes and surface decorations that reveal their origin in reproductive structures. Genus-level identifications are possible with reproductive structure phytoliths from a substantial number of families (Table I) (Fig. 7) (e.g., Piperno, 1989; Tubb et al., 1993; Kealhofer & Piperno, 1998). Species-specific identifications appear to be possible in some domesticated crop species such as maize (*Zea mays* L.) and rice (*Oryza sativa* L.) using phytoliths formed in glumes and seeds, the structures that were manipulated

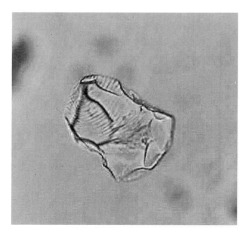

Figure 3. A phytolith from the leaves of *Magnolia* sp. In the Neotropics, this type appears to be diagnostic of the genus *Magnolia*. It is 65 μ long.

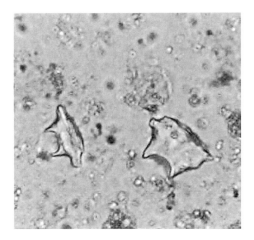

Figure 4. Phytoliths with troughs from the leaves of *Musa* sp. (banana). These are diagnostic of the genus. The largest is 18 μ long.

and genetically altered by prehistoric farmers (Pearsall et al., 1995; Piperno, 1998; Zhao et al., 1998). Phytolith taxonomy is still a young science but large reference collections and keys are available or under construction for many regions of the world (see Table I).

Laboratory methods

Phytoliths are usually isolated from lake and other sediments with heavy-liquid flotation. A number of steps are carried out before phytolith separation is attempted to ensure that phytoliths have been released from the sediment matrix, and that clays, carbonates, and organic materials, all of which may bind to phytoliths and impede their flotation, have been

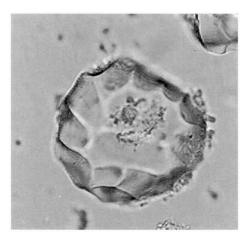

Figure 5. A phytolith from the fruit rind of *Lagenaria siceraria* (bottle gourd). It has elongated surface decorations. Bottle gourd phytoliths appear to be species-specific. The phytolith is hemispherical and is 95 μ long.

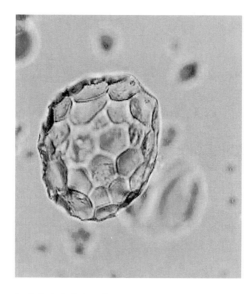

Figure 6. A phytolith from the fruit rind of *Cucurbita moschata*, a domesticated species of squash from Ecuador. It differs from bottle gourd in having regular, rounded surface decorations, and in having a spherical shape. It is 95 μ long. Domesticated squashes, which have much larger fruits that wild squashes, also have significantly larger phytoliths.

removed. Not surprisingly, different techniques for phytolith extraction can produce varying results and, thus, extraction procedures should be modified according to the characteristics of the sediments for which they are used. A flow chart showing the steps, dispersing agents, and chemicals used by various investigators is provided in Figure 8. Comparisons of different methods applied to different sediment types can be found in Zhao & Pearsall (1999) and Lentfer & Boyd (1998). Some points about Figure 1 follow.

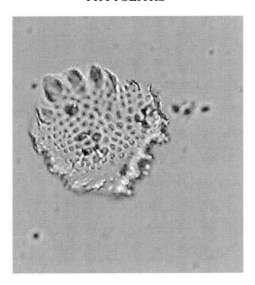

Figure 7. A phytolith from the fruit epidermis of *Protium panamense*, an important species of tree in the lowland Neotropical forest. These phytoliths are specific to at least the level of genus. The phytolith is 75 μ long.

Phytolith analysis of lakes requires more sediment than used in pollen analysis (see Bennet & Willis, this volume), where the standard unit of study is 0.5–1 cm^3. Judging from the (albeit small) number of lakes studied thus far, 5–10 cm^3 of sediment is usually required for non-grass taxa to be represented in quantifiable amounts. It can be imagined that, in order to save time and resources, a technique that separates phytoliths and pollen at the same time is desirable. Figure 1 contains a flow chart for a simultaneous extraction of pollen and phytoliths (Lentfer & Boyd, 1998). A caveat is that this procedure worked well when sediments were highly oxidized and, hence, low in organic matter, and not very well when sediments contained appreciable amounts of organic matter, as lake deposits typically do. The author believes that the optimal phytolith and pollen studies of lacustrine settings will utilize separate extractions for each type of fossil. Because they have nearly identical compositions, phytolith and diatom preparations (see Battarbee et al., this volume) can be made at the same time without problem, although the use of 10 cm^3 of wet sediment will result in the separation of many more diatoms than needed. Microscopic charcoal particles (see Whitlock & Larsen, this volume) are also isolated along with phytoliths using the technique outlined in Figure 8.

Phytoliths are most easily viewed with the microscope when they are mounted on slides in agents with refractive indices of 1.51 to 1.54. These include Permount, Canada Balsam, and Histoclad. Silicone oils, with refractive indices of about 1.4, are generally to be avoided as phytolith mounting media, unless phase contrast is used. Phytoliths not mounted on slides can be dried in acetone or absolute ethyl alcohol and transferred to storage vials. The identification of some plants requires phytolith rotation and the examination of three-dimensional structure. Using the mounting agents listed above, this is not a problem if slides are studied immediately. However, if a repeat examination is desired a few weeks later, phytoliths will have to be mounted again because the medium has likely hardened. This can be avoided by using mounting agents that remain liquid, such as benzyl benzoate or glycerin.

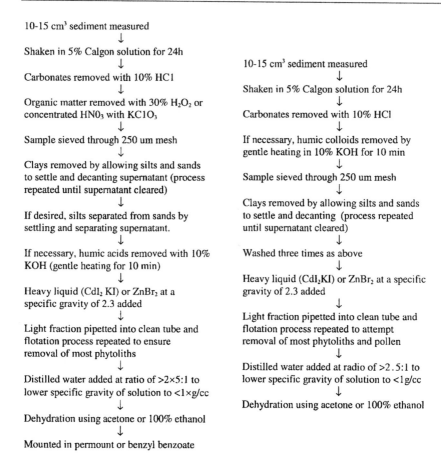

Method for phytolith
extraction

Method for simultaneous
extraction of phytoliths and pollen
(After Lentfer & Boyd, 1998)

10-15 cm³ sediment measured
↓
Shaken in 5% Calgon solution for 24h
↓   10-15 cm³ sediment measured
Carbonates removed with 10% HCl   ↓
↓   Shaken in 5% Calgon solution for 24h
Organic matter removed with 30% H₂O₂ or   ↓
concentrated HNO₃ with KClO₃   Carbonates removed with 10% HCl
↓   ↓
Sample sieved through 250 um mesh   If necessary, humic colloids removed by
↓   gentle heating in 10% KOH for 10 min
Clays removed by allowing silts and sands   ↓
to settle and decanting supernatant (process   Sample sieved through 250 um mesh
repeated until supernatant cleared)   ↓
↓   Clays removed by allowing silts and sands
If desired, silts separated from sands by   to settle and decanting (process repeated
settling and separating supernatant.   until supernatant cleared)
↓   ↓
If necessary, humic acids removed with 10%   Washed three times as above
KOH (gentle heating for 10 min)   ↓
↓   Heavy liquid (CdI₂KI) or ZnBr₂ at a specific
Heavy liquid (CdI₂ KI) or ZnBr₂ at a   gravity of 2.3 added
specific gravity of 2.3 added   ↓
↓   Light fraction pipetted into clean tube and
Light fraction pipetted into clean tube and   flotation process repeated to attempt
flotation process repeated to ensure   removal of most phytoliths and pollen
removal of most phytoliths   ↓
↓   Distilled water added at radio of >2.5:1 to
Distilled water added at ratio of >2×5:1 to   lower specific gravity of solution to <1g/cc
lower specific gravity of solution to <1×g/cc   ↓
↓   Dehydration using acetone or 100% ethanol
Dehydration using acetone or 100% ethanol
↓
Mounted in permount or benzyl benzoate

Figure 8. Flow charts showing the procedures used to extract phytoliths from sediments. The steps using chemicals should be carried out under a fume hood. Gloves should also be warn when handling chemicals.

Once mounted on slides and counted, phytolith quantities can be expressed through percentage, concentration, or absolute (accumulation rate) values. As with pollen studies, calculations of all three, where possible, provide more informed evaluations of the significance of past vegetational changes. At present, no exotic marker exists for a simple estimation of phytolith concentrations, so researchers have used the time-consuming aliquot method (Piperno, 1993), whereby all phytoliths occurring in a known volume of sediment are counted. This is an area of lake phytolith studies that should soon see improvement. For example, use of an exotic marker such as tablets containing *Musa* phytoliths should be possible in New World research.

Applications of phytolith analysis in lake sediments

At present, examples of phytolith applications in lake sediments come from western and eastern equatorial Africa—lakes Guiers, Sinnda and Simba (Alexandre et al., 1997; Mworia-Maitaima, 1997), tropical southeast Asia—Lake Kumphawapi, Thailand (Kealhofer & Penny, 1998), tropical America-lakes La Yeguada, El Valle, and Wodehouse, Panama (Piperno, 1993, 1994), and Lake Ayauch[i], Ecuador (Bush et al., 1989) and southern, subtropical China—Poyang Lake, middle Yangtze River Valley (Zhao & Piperno, 1999). It is clear that four issues conspicuous in the analysis and interpretation of pollen records (see Bennett & Willis, this volume) have relevance for phytolith analysis. These are: 1) differential production and preservation of phytoliths; 2) the definition of a source area; 3) the mode of transport of phytoliths into lakes; and 4) characteristics of phytolith sedimentation. We know little about some of these factors, and thorough study is needed in order for phytolith applications in lake environments to grow and mature. The available phytolith studies of lakes were often carried out in tandem with detailed pollen applications, and they appear to demonstrate the following with regard to these issues.

Phytolith representation and source area

The issue of phytolith production has been reviewed above, and it is clear, that, as in pollen analysis, many taxa will leave "blind spots" in records because they form no to low numbers of phytoliths. Phytolith representation in lake sediments from the many taxa that do produce substantial amounts of phytoliths is good, and phytoliths are present in sufficient numbers and taxonomic diversity to achieve meaningful quantitative and qualitative assessments of climatic and vegetational change. In one of the Panamanian lakes, La Yeguada, located in a lowland, evergreen tropical forest, a 14,000 year-old sequence was dated with sufficient ^{14}C precision to determine phytolith accumulation rates (PARs) (Piperno, 1993). Non-grass PARs per 20 cm^2 per annum ranged from 300 to 13,200, with values usually falling between 500 and 3500 per 20 cm^2 per annum. These were mainly from the trees, shrubs, and herbs of the tropical forest in the lake's watershed. Accumulation rates of short cell phytoliths (bilobates, saddles, etc.) from the Poaceae in the same lake were usually between 8,000 and 18,000 per 20 cm^2 per annum. If the numerous, less diagnostic forms from dicotyledons and grasses had been counted at Lake La Yeguada, phytolith quantities probably would increase by about ten-fold. By comparison, pollen accumulation rates usually varied between 8,000 and 30,000 per cm^2 per annum (Bush et al., 1992). At Poyang Lake, south China, phytolith concentrations typically ranged between 5,000 and 20,000 per cm^3 of sediment, whereas pollen concentrations were usually between 2,000 and 60,000 per cm^3. Hence, in these two lakes, phytolith quantities compared favorably with those of pollen, and typically the highest/lowest concentrations of pollen and phytoliths co-varied at each site.

Arboreal and other taxa important in the regional forests were well-represented in the lakes from China, Thailand, and Panama (Figs. 9 & 10 a,b). Many phytoliths clearly were derived from taxa that do not represent the lake-edge, but which grew on well-drained soils in the watersheds' forests. In all of the sequences, a significant number of taxa were identified to the family and genus levels, unknown (and highly distinctive) phytoliths that usually could be assigned to a habitat type (e.g., mature tropical forest, secondary-growth forest) routinely occurred, and several general phytolith categories encompassing a number of different taxa that share the same growth habits (e.g., trees, herbs) were recorded.

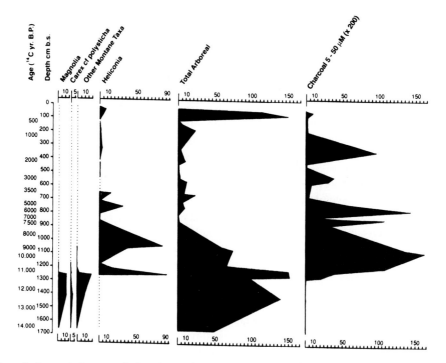

Figure 9. Accumulation rates of selected phytolith types and charcoal into Lake La Yeguada, Panama. The phytolith accumulation rate is expressed as the number per 20 cm^2 sediment yr^{-1}. Depth is expressed in centimeters below surface (cm b.s.). ×200 next to the charcoal category indicates that the actual counts are 200× larger than graphed. Note the presence of *Magnolia*, the sedge *Carex*, and other high elevation forest taxa between ca. 14,000 and 10,500 B.P., indicating much cooler climates at this time, and the decline of arboreal taxa at ca. 7000 B.P. as a result of the onset of slash and burn activity in the watershed (modified from Piperno et al., 1991).

How did these phytoliths get to the lakes? Phytolith transport to a lake reflects a number of different mechanisms, depending on the biome in which the study site is located and past cultural uses of that biome. It appears that if a lake has inflowing rivers and streams, a primary mechanism of phytolith transport will be through river/stream-flow. Lakes La Yeguada, Wodehouse, Poyang, and Kumphawapi are fed by streams and/or rivers. Moreover, in these sequences the presence of phytoliths from the Podostemaceae, a family of plants that grows attached to rocks in swiftly flowing water, seemed to confirm phytolith transport in rivers (Fig. 11). In Africa, phytolith transport was also explained largely through rivers, with perhaps a minor influence from movement in aerosols in this drier environment (Alexandre et al., 1997). In the situations where many phytoliths have arrived through river transport, it should not be concluded that phytolith assemblages are dominated by stream valley vegetation because phytoliths will persist in soils after leaf/seed drop and decay, and should be washed into rivers during those times of the year when sediment erosion is strongest.

On the other hand, in lakes that receive no riverine input, particularly those located in humid and sub-humid regions where the potential vegetation is a dense forest and transport in aerosols is unlikely, the phytolith record probably largely reflects a local source area unless proven otherwise (e.g., if phytolith transport through inwash from surrounding slopes

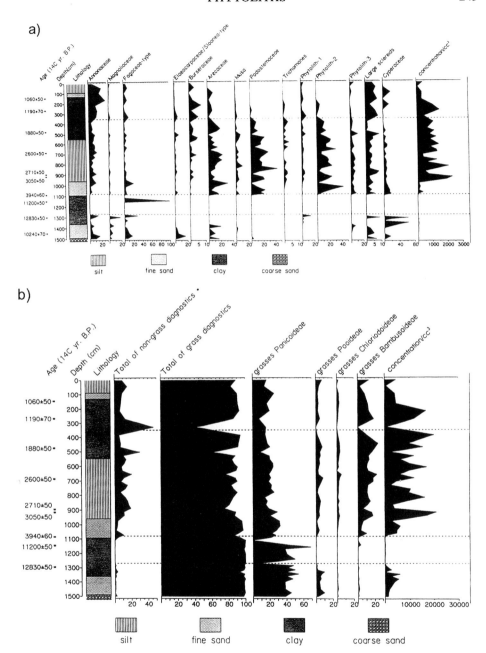

Figure 10. a. A profile of selected non-grass phytolith types from Poyang Lake, south China. Note the good representation of various arboreal taxa from the regional forest Phytoliths 1, 2, and 3 are unknown but highly distinctive phytoliths that should provide more details on vegetational and climatic changes as the modern reference collections are expanded. (Reprinted with permission from Zhao & Piperno, 1999; John Wiley & Sons, Inc.). b. The grass phytolith profile from Poyang Lake, south China. The percentage of non-grass phytoliths is on the left. Note the discrimination of Poaceae subfamilies. High frequencies of Panicoid grasses and low frequencies of bamboos during the Late Pleistocene suggest a cooler and drier climate, and may also reflect limited C3 grass growth as a result of lower atmospheric CO_2 concentrations at that time. (Reprinted with permission from Zhao & Piperno, 1999; John Wiley & Sons, Inc.).

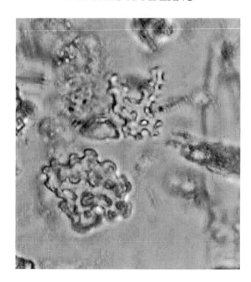

Figure 11. Phytoliths from the leaves of *Tristicha trifaria* (Podostemaceae). They are distinctive at the level of family and can be used to document phytolith arrival to lakes through stream and river input. The largest phytolith is $27\,\mu$ long.

is likely to have been considerable). Researchers should also be aware that large numbers of phytoliths can be injected into the atmosphere from plant tissue during major fires, an effect that is probably greater in open and windy landscapes than in forested ones.

The sedimentation of phytoliths at the bottom of lakes

At present, we know little about the factors that control the sedimentation of phytoliths at the bottoms of lakes. Sedimentation processes are no doubt complex and probably are affected by sediment composition, wave action, and basin structure, among other factors. Another potential source of variability is the relative proportion of phytoliths arriving as disarticulated particles discrete from plant tissue vs. those still embedded in vegetal matter because they may have different settling rates. Since individual phytoliths have a specific gravity very similar to that of pollen grains (1.5 to 2.3 for phytoliths vs. 1.4 to 1.7 for pollen) and they broadly overlap pollen in size (most phytoliths and pollen have diameters of 10–$100\,\mu$m), it is possible, though unproven with experimental study, that settling characteristics of both types of fossils are broadly similar. Some insight into this question comes from existing studies, where trends in concentrations of phytolith and pollen largely co-varied through time, suggesting similar settling rates in the lakes (e.g., Piperno, 1993; Jiang & Piperno, 1998; Zhao & Piperno, 1999).

Summary of the major results

The following aspects of paleovegetational reconstruction using phytolith data are apparent from existing lake records. First, phytolith assemblages from lakes located in the humid tropics and subtropics of the Old and New Worlds discriminate various kinds of forest (e.g., lowland evergreen and semi-evergreen forest; high-elevation montane forest) and grassland formations (those dominated by C4 vs. C3 grasses), and identify forest disturbance resulting from slash and burn agriculture (presence of microscopic fragments of charcoal plus burned weed phytoliths also contributes to such an interpretation) (Figs. 9 & 10 a,b) (Piperno, 1993, 1994; Kealhofer & Penny, 1998; Zhao & Piperno, 1999). Phytolith assemblages from African lakes in regions dominated by tropical savannas discriminate between different grassland types together with the climatic regimes that control their distributions (e.g., cooler, tall grass vs. warmer, short grass savanna), and indicate the relative density of tree and shrub vs. grass cover on the landscape through time (Alexandre et al., 1997).

Phytoliths often document a different spectrum of taxa, and/or contribute to finer discrimination of taxa, and provide more complete and refined reconstructions. For example, in Panama, the presence of phytoliths from *Magnolia*, a plant confined to high elevations in Central America, in late-glacial but not in Holocene deposits gave investigators confidence that high frequencies of *Quercus* pollen recorded between ca. 14,000 B.P. and 10,500 B.P. were actually a result of downward depression of montane forest vegetation caused by climatic cooling, not of an expansion of a lowland, dry-forest oak resulting from the drier late-glacial conditions (Piperno, 1993) (Fig. 9). Similarly, in southern China, greatest frequencies of the Magnoliaceae, rare presence of bamboo phytoliths, and dominance of Panicoid grasses during the late Pleistocene portion of the sequence provided signals that the late-glacial climate there was also cooler in addition to being drier (Zhao & Piperno, 1999) (Figs. 10 a,b). The pollen data registered only drier late-glacial conditions (Jiang & Piperno, 1998). The Poyang Lake phytolith record also demonstrated that *Oryza* (rice) was a native component of the vegetation in the middle Yangtze Valley before its domestication at ca. 10,000 B.P. At Lake Kumphawapi in Thailand, late Pleistocene sediments were oxidized and devoid of pollen (Kealhofer & Penny, 1998). Here, well-preserved phytolith assemblages dominated by grasses provided the data indicating that the late-glacial period was drier than the Holocene. As in southern China, most of the late Pleistocene grass phytoliths were from Panicoid taxa, while bamboo phytolith frequencies were greatly reduced compared with during the Holocene, further indicating drier conditions. In Africa, identification of subfamilies of the Poaceae permitted the Holocene history of different grassland communities and their climatic affiliations to be documented for the first time (Alexandre et al., 1997).

In all of the lake sequences, the changes in climate and vegetational cover that were inferred from the phytolith records (e.g, colder and drier to warmer and wetter climates; mature tropical forest to forest subjected to slash and burn agriculture) were largely concordant with those made from the pollen records. Major changes in phytolith spectra often occurred in synchrony with changes in pollen spectra. Also, when pollen spectra suggested rapid climatic perturbations, as inferred from major oscillations of taxa over short periods of time, phytolith spectra acted in tandem. It would be difficult to explain such findings in the absence of stratigraphically and taxonomically valid phytolith assemblages.

Other potential applications of phytoliths in lake sediments

It has been shown that the residual organic carbon of plant cells that was trapped inside of phytoliths during their formation can be reliably dated by AMS (Wilding, 1967; Wilding et al., 1967; Kelly et al., 1991; Mulholland & Prior, 1993). A phytolith assemblage from about 10 cm³ of phytolith-rich sediment is usually required for a direct ^{14}C determination. Also, the δ^{13}C of the same carbon can be utilized to distinguish vegetation dominated by C3 and C4 plants (Kelly et al., 1991; Fredlund, 1993). These approaches may find substantial applications in lake sediments, particularly where the "hard water error" makes it difficult to obtain reliable dates on the sediment matrix.

Summary

Phytolith analysis of lake records is an important and currently underutilized tool of paleoenvironmental reconstruction. If investigators choose to undertake the time-consuming task of constructing modern phytolith reference collections for their study regions, their efforts are likely to be rewarded because phytoliths can be useful in the identification of specific taxa. Identifications of some key indicator families and genera should be possible utilizing existing published keys and photographs.

The potential advantages of carrying out phytolith and pollen studies of a single lake record as part of a multi-proxy analysis seem to be considerable. For example, where pollen analysis is weak, as in the recognition of herbaceous and arboreal taxa of mature tropical rain forest, phytolith analysis provides significant information (Piperno, 1993). Conversely, in the recognition of woody, secondary tropical forest growth where phytolith analysis may be "silent", pollen data come to the rescue. Phytolith studies have also significantly increased the number of taxa represented in lake profiles from the American tropical forest (e.g., Piperno, 1993), an important improvement in our attempts to decipher the history of species-rich tropical formations. Phytoliths and pollen grains are complementary avenues of paleoenvironmental reconstruction and they should be studied in tandem whenever possible.

Acknowledgments

The author's research was supported by the Smithsonian Tropical Research Institute (STRI) and a grant to the STRI by the Andrew W. Mellon Foundation.

References

Alexandre, A., J.-D. Meunier, A.-M. Lezine, A. Vincens & D. Schwartz, 1997. Phytoliths: indicators of grassland dynamics during the late Holocene in intertropical Africa. Palaeogeo., Palaeoclim., Palaeoecol. 136: 213–229.

Albert, R. M., O. Lavi, L. Estroff & S. Weiner, 1999. Mode of occupation of Tabun Cave, Mt Carmel, Israel during the Mousterian period: A study of the sediments and phytoliths. J. Arch. Sci. 26: 1249–1260.

Boyd, W. E., C. J. Lentfer & R. Torrence, 1998. Phytolith analysis for a wet tropics environment: Methodological issues and implications for the archaeology of Garua Island, West New Britain, Papua New Guinea. Palynology 22: 213–228.

Bozarth, S. R., 1987. Diagnostic opal phytoliths from rinds of selected Cucurbita species. Amer. Anti. 52: 607–615.

Bozarth, S. R., 1992. Classification of opal phytoliths formed in selected dicotyledons native to the great plains. In Rapp G., Jr. & S. C. Mulholland (eds.) Phytolith Systematics: Emerging Issues, Plenum Press, New York, pp. 193–214.

Bozarth, S. R., 1993. Biosilicate assemblages of boreal forests and Aspen parklands. In Pearsall, D. M. & D. R. Piperno (eds.) Current Research in Phytolith Analysis: Applications in Archaeology and Paleoecology. The University Museum of Archaeology and Anthropology, Philadelphia, pp. 95–108.

Brown, D. A., 1984. Prospects and limits of a phytolith key for grasses in the central United States. J. Arch. Science 11: 345–368.

Bush, M. B., D. R. Piperno & P. A. Colinvaux, 1989. A 6,000 year history of Amazonian maize cultivation. Nature 340: 303–305.

Bush, M. B., D. R. Piperno, P. A. Colinvaux, P. E. DeOliveira, L. A. Krissek, M. C. Miller & W. E. Rowe, 1992. A 14,300 Yr paleoecological profile of a lowland tropical lake in Panama. Ecol. Mono. 62: 251–275.

Coley, P. D. & J. Barone, 1996. Herbivory and plant defenses in tropical forests. Ann. Rev. Ecol. Syst. 27: 305–335.

Dorweiler, J. & J. Doebley, 1997. Developmental analysis of *Teosinte Glume Architecture* 1: A key locus in the evolution of maize (Poaceae). Amer. J. Bot. 84: 1313–1322.

Fredlund, G. G., 1993. Paleoenvironmental intepretations of stable carbon, hydrogen, and oxygen isotopes from opal phytoliths, Eustis Ash Pit, Nebraska. In Pearsall, D. M. & D. R. Piperno (eds.) Current Research in Phytolith Analysis: Applications in Archaeology and Paleoecolog. The University Museum of Archaeology and Anthropology, Philadelphia, pp. 37–46.

Hodson, M. J. & D. E. Evans, 1995. Aluminum/silicon interactions in higher plants. J. Exp. Bot. 46: 161–171.

Hodson, M. J., S. E. Williams & A. G. Sangster, 1997. Silica deposition in the needles of the gymnosperms. I. Chemical analysis and light microscopy. In Pinilla, A., J. Juan-Tresserras & M. J. Machado (eds.) The State of the Art of Phytoliths in Soils and Plants, Mongrafias del Centro de Ciencias Medioambientales, Madrid, pp. 123–133.

Jiang, Q. & D. R. Piperno, 1998. Late Quaternary pollen sequence from Poyang Lake. southern China, and its environmental and archaeological implications. Quat. Res. 52: 250–258.

Kealhofer, L. & D. Penny, 1998. A combined phytolith and pollen record for 14,000 years of vegetation change in northeast Thailand. Rev. Palaeobot. Paly. 103: 83–93.

Kealhofer, L. & D. R. Piperno, 1998. Phytoliths in the modern southeast Asia and Thai flora. Smith. Contri. Bot. No. 88.

Kelly, E. F., R. G. Amundson, B. D. Marino & M. J. Deniro, 1991. Stable isotope ratios of carbon in phytoliths as a quantitative method of monitoring vegetation and climate change. Quat. Res. 35: 222–233.

Kondo, R., C. W. Childs & I. A. E. Atkinson, 1994. Opal Phytoliths of New Zealand. Manaaki Whenua Press, Lincoln, 344 pp.

Lentfer, C. J. & W. E. Boyd, 1998. A comparison of three methods for the extraction of phytoliths from sediments. J. Arch. Sci. 25: 1159–1183.

Miller-Rosen, A., 1993. Phytolith evidence for early cereal cultivation in the Levant. In Pearsall, D. M. & D. R. Piperno (eds.) Current Research in Phytolith Analysis: Applications in Archaeology and Paleoecology. The University Museum of Archaeology and Anthropology, Philadelphia, pp. 160–171.

Mulholland, S., 1989. Phytolith shape frequencies in North Dakota grasses: A comparison to general patterns. J. Arch. Sci. 16: 489–511.

Mulholland, S. C. & C. Prior, 1993. AMS radiocarbon dating of phytoliths. In Pearsall, D. M. & D. R. Piperno (eds.) Current Research in Phytolith Analysis: Applications in Archaeology and Paleoecology. The University Museum of Archaeology and Anthropology, Philadelphia, pp. 21–23.

Mworia-Maitima, J., 1997. Prehistoric fires and land-cover change in western Kenya: evidences from pollen, charcoal, grass cuticles, and grass phytoliths. The Holocene 7: 409–417.

Parry, D. W. & F. Smithson, 1966. Opaline silica in the inflorescences of some British grasses and cereals. Ann. Bot. 30: 525–539.

Pearsall, D. M., D. R. Piperno, E. H. Dinan, M. Umlauf, Z. Zhao & R. A. Benfer, Jr., 1995. Distinguishing rice (Oryza sativa Poaceae) from wild Oryza species through phytolith analysis: Results of Preliminary Research. Econ. Bot. 49: 183–196.

Piperno, D. R., 1988. Phytolith Analysis: An Archaeological and Geological Perspective. Academic Press, San Diego, 280 pp.

Piperno, D. R., 1989. The occurrence of phytoliths in the reproductive structures of selected tropical angiosperms and their significance in tropical paleoecology, paleoethnobotany, and systematics. Rev. Palaeobot. Paly. 61: 147–173.

Piperno, D. R., 1991. The status of phytolith analysis in the American tropics. J. World Pre. 5: 155–191.

Piperno, D. R., 1993. Phytolith and charcoal records from deep lake cores in the American tropics. In Pearsall, D. M. & D. R. Piperno (eds.) Current Research in Phytolith Analysis: Applications in Archaeology and Paleoecology. The University Museum of Archaeology and Anthropology, Philadephia, pp. 58–71.

Piperno, D. R., 1994. Phytolith and charcoal evidence for prehistoric slash and burn agriculture in the Darien rain forest of Panama. The Holocene 4: 321–325.

Piperno, D. R., 1998. Paleoethnobotany in the Neotropics from microfossils: New insights into ancient plant use and agricultural origins in the tropical forest. J. world Pre. 12: 393–449.

Piperno, D. R. & D. M. Pearsall, 1998. The silica bodies of tropical American grasses: Morphology, Taxonomy, and Implications for Grass Systematics and Fossil Phytolith Identification. Smith. Cont. Bot. No. 85.

Piperno, D. R., M. B. Bush & P. A. Colinvuax, 1991. Paleoecological perspectives on human adaptation in Central Panama. II. The Holocene. Geoarch. 6: 227–250.

Runge, F., 1995. Potential of opal phytoliths for use in paleoecological reconstruction in the humid tropics of Africa. Zeitschrift fhr Geomorphologie, N.F., Supplement 99: 53–64.

Runge, F., 1998. The opal phytolith inventory of soils in central Africa-quantities, shapes, classification, and spectra. Rev. Palaeobot. Palyn. 107: 23–53.

Sangster, A. G. & M. J. Hodson, 1997. Botanical studies of silicon localization in cereal roots and shoots, including cryotechniques: a survey of work up to 1990. In Pinilla, A. F., J. Juan-Tresserras & M. J. Machado (eds.) The State of the Art of Phytoliths in Soils and Plants, Monografias del Centro de Ciencias Medioambientales, Madrid, pp. 113–121.

Sangster, A. G., S. E. Williams & M. J. Hodson, 1997. Silica deposition in the needles of the gymnosperms, II. Scanning electron microscopy and x-ray analysis. In Pinilla, A., J. Juan-Tresseras & M. J. Machado (eds.) The State of the Art of Phytoliths in Soils and Plants. Monografias del Centro de Ciencias Medioambientales, Madrid, pp. 123–133.

Tubb, H. J., M. J. Hodson & G. C. Hodson, 1993. The inflorescence papillae of the Triticeae: A new tool for taxonomic and archaeological research. Ann. Bot. 72: 537–545.

Wallis, L. (n.d.). Phytolith Analysis in the Kimberley Region, Australia. Manuscript in possession of the author.

Wilding, L. P., 1967. Radiocarbon dating of biogenetic opal. Science 156: 66–67.

Wilding, L. P., R. E. Brown & N. Holowaychuk, 1967. Accessibililty and properties of occluded carbon in biogenic opal. Soil Sci. 103: 56–61.

Zhao, Z. & D. M. Pearsall, 1998. Experiments for improving phytolith extraction from soils. J. Arch. Sci. 25: 587–598.

Zhao, Z., D. M. Pearsall, R. A. Benfer, Jr. & D. R. Piperno, 1998. Distinguishing Rice (Oryza sativa Poaceae) from wild Oryza species through phytolith analysis, II: Finalized Method. Econ. Bot. 52: 134–135.

Zhao, Z. & D. R. Piperno, 1999. Late Pleistocene/Holocene environments in the middle Yangtze River Valley, China and rice (*Oryza sativa* L.) domestication: The phytolith evidence. Geoarchaeology 15: 203–222.

12. FRESHWATER SPONGES

THOMAS M. FROST*
Trout Lake Station
Center for Limnology
University of Wisconsin-Madison
10810 County N
Boulder Junction, WI 54512, USA

*deceased

Keywords: freshwater sponges, gemmoscleres, gemmules, megascleres, microscleres, paleolimnology, Porifera, spicules, silica, *Spongilla lacustris*

Introduction

The skeletons of freshwater sponges (Porifera: Demospongiae) are comprised of siliceous structures termed spicules. These spicules are well preserved in most sedimentary environments. They are quite similar chemically to diatom frustules. The taxonomy of freshwater sponges is based upon the structure of some kinds of spicules, which are the fundamental, identifying features of species (Penney & Racek, 1968). As such, the occurrence of sponge species is recorded in the sediments of many freshwater habitats (Harrison, 1988). Sponges generally require clean-water habitats and some species may be restricted to a limited range of chemical conditions (Frost et al., 2001). The presence of any sponge spicules, or evidence of the occurrence of particular sponge species, may serve as indicators of past environmental conditions (Harrison, 1974). In addition, some characteristics of sponge spicules can respond quantitatively to habitat silica concentrations and provide a paleolimnological measure of past water-chemistry conditions (Kratz et al., 1991). Despite some work with sponges, however, the full potential of sponge spicules in paleolimnological research has not been explored effectively. In this chapter, I describe the elements of sponge biology that are relevant to paleolimnological research, review a range of previous papers that have used sponges in paleolimnological work, outline the techniques necessary for processing sponge spicules in sediments, and discuss the potential for further work with sponges in paleolimnology.

It is important to note that this chapter is focused on the use of spicules of freshwater sponges in paleolimnological investigations. Many marine sponges also produce spicules and these microfossils can serve in paleontological research (Rigby, 1995). Marine fossils

J. P. Smol, H. J. B. Birks & W. M. Last (eds.), 2001. *Tracking Environmental Change Using Lake Sediments.*
Volume 3: Terrestrial, Algal, and Siliceous Indicators. Kluwer Academic Publishers, Dordrecht, The Netherlands.

are not considered here because such materials are associated with lake sediments only under rare circumstances and are unlikely to be indicators of past limnological conditions.

Sponge species and their distribution

Although the entire phylum Porifera consists of more than 5000 species (Berquist, 1978), there are probably fewer than 300 species that occur in freshwater habitats. In the most comprehensive source available, Penney & Racek (1968) list 95 freshwater species world-wide. In America north of Mexico and the Caribbean, there are 27 freshwater species in 12 genera in two families (Frost et al., 2001).

The presence of sponges across freshwater habitats is influenced on the broadest scale by biogeographic patterns. Some species in the largest freshwater sponges family, the Spongillidae, are cosmopolitan and truly worldwide in their distribution (Penney & Racek, 1968). On the opposite extreme, some species have been reported from a single lake (e.g., Poirrier, 1978). The distributions of most species lie between these two extremes with many species, for example, restricted to limited portions of North America (Frost et al., 2001).

Within biogeographic constraints, the presence of a species within a particular habitat is controlled by physical, chemical, and, possibly, biological factors. For physical conditions, some species, such as *Ephydatia fluviatilis* Linnaeus, appear to occur primarily in lotic habitats but the majority favor, or at least tolerate, still-water conditions (Frost et al., 2001). Chemically, sponges generally require relatively clean-water conditions in order to maintain the filtering of their water processing system and/or the functioning of their algal symbionts, which require access to some light (Frost et al., 2001). Some sponges appear to be restricted to a limited range of chemical conditions. These restricted ranges are summarized for North American species in Harrison (1988). In the most extensive surveys of sponge species distributions among lakes that has been published, Jewell (1935, 1939) reported that several species appeared to be restricted to a subset of the chemical conditions that she found in her sampled lakes. Other sponges appear to be tolerant of a wide range of chemical conditions. In general, I would recommend that any paleolimnological surveys should conduct a careful calibration of present-day sponge distribution versus chemical conditions within a region before drawing any inferences about past chemical conditions. With this caveat, however, sponge fossils have the potential to provide some useful insights into past limnological conditions particularly when considered in conjunction with other paleolimnological indicators.

Sponge life history

Two features of freshwater sponge life history influence the forms and relative abundances of the spicules that sponges produce: the formation of gemmules and the extent of active growth. Sponges usually grow actively during only some seasons, although some species maintain an active growth form throughout the year (Frost et al., 2001). Most sponges undergo a change of body form when active growth is limited by adverse environmental conditions. During stressful periods, sponges produce resting stages termed gemmules (Fig. 1), which are resistant forms that are produced asexually and capable of generating active sponge tissue after adverse environmental conditions have diminished. Gemmules

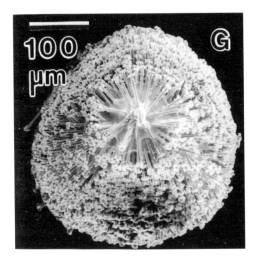

Figure 1. A nearly intact, single gemmule (G) of *Radiospongilla crateriformis* (Potts) showing the typical arrangement of gemmoscleres in the outer, protective layer of the gemmule. The photograph is used with permission from Ricciardi & Reiswig (1993).

consist of undifferentiated sponge cells that have low rates of metabolism and are encased in a resistant membrane. This membrane is, in most cases, embedded with specialized spicules, which provide some of the gemmule's protective structure. In colder regions, gemmules are produced during winter periods when most active sponge tissue is absent. Gemmulation can also be used to avoid situations that are too hot for sponges to survive or when a habitat undergoes drying. The extent of active growth during periods of suitable environmental conditions relative to the occurrence of gemmules controls the proportional growth of the different forms of sponge spicules and their occurrence in lakes sediments. The spicules of gemmules are critical for the identification of sponge species. For some species, the structure of the intact gemmules is also used in taxonomic identifications.

Sponge spicules

The past record of sponge occurrence in a habitat is provided primarily by siliceous spicules that are readily preserved in aquatic sediments. Freshwater sponges produce spicules of two or three basic forms depending upon the species: megascleres, gemmoscleres, and in some cases microscleres.

Megascleres

Megascleres are produced by all actively growing sponges. They are usually the largest skeletal components of sponges and provide their primary structure. Megascleres are needle-like in form and are either smooth or spined (Figs. 2 and 3). They vary in length depending upon the species. In North America, lengths can range from 100 to 450 μm (Frost

Figure 2. A scanning electron micrograph of megascleres from *Spongilla lacustris* (Linnaeus). One spined microsclere is also shown near the center of the illustration. The scale bar is $10\,\mu$m in length. The photograph is used with permission from Frost et al. (2001).

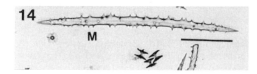

Figure 3. A megasclere (M) of *Trochospongilla horrida* Weltner showing the occurrence of spines that are typical for this species. The scale bar is $50\,\mu$m in length. The photograph is used with permission from Ricciardi & Reiswig (1993).

et al., 2001). Length is fairly constant for a species but width can vary systematically with habitat chemical conditions (Kratz et al., 1991). Megasclere form does not vary substantially from species to species and, by itself, is of little use in identifying species.

Gemmoscleres

Gemmoscleres are skeletal elements associated with sponge gemmules (Frost et al., 2001). They are usually embedded in the protective membrane of the gemmules. Gemmoscleres vary substantially in form among sponge species. Some are needle-like (Fig. 4) while others are dumbbell shaped with considerable variation in the formation of their ends, which are termed rotules when gemmoscleres are rounded (Figs. 5 and 6). Some gemmoscleres are spined while others are smooth. Differences in form are generally much more substantial among sponge species than within a species. Gemmoscleres range considerably in size, from as little as 8 to more than $300\,\mu$m in length. Overall, gemmosclere differences are essential in taxonomic identifications of freshwater sponges (Penney & Racek, 1968).

Microscleres

Microscleres are skeletal elements that are widely distributed throughout the body of some sponge species. They are not responsible for overall structure, which is provided by megascleres, but seem to be responsible for some stiffening of the sponge body, particularly for

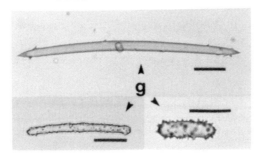

Figure 4. Needle-shaped gemmoscleres (g) of *Eunapius fragilis* (Leidy) showing smooth and spined forms in the same species. The scale bars are 25 μm in length. The photograph is used with permission from Ricciardi & Reiswig (1993).

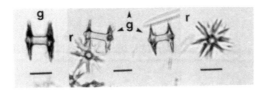

Figure 5. Short, dumbbell-shaped gemmoscleres (g) of *Ephydatia muelleri* (Lieberkühn) with some views of the rotules (r) of these spicules. The scale bars are 10 μm in length. The photograph is used with permission from Ricciardi & Reiswig (1993).

the dermis (Koehl, 1982). They occur in only some species. For example, they are absent in half of the freshwater sponge genera that occur in North America (Frost et al, 2001). Their presence or absence and their form can be useful in taxonomic work. Some are needle-like (Fig. 7) while others are dumbbell shaped (Fig. 8) with variation also in the presence and forms of spines. They range in length from 8 to 130 μm.

Paleolimnological studies using freshwater sponges

Spicules that are distributed in lake sediments or soils provide the fossil record for freshwater sponges in most situations. In some cases, in addition, intact gemmules can be preserved in aquatic sediments and can be used in identifying the species that have occurred during the history of a habitat (Turner, 1985). Deposited sponge spicules can be used in several different ways to make inferences about past habitat conditions.

The abundance of some sponge spicules, irrespective of their species, has been used as a general indication of suitable habitat conditions. An overall decline in sponge spicules has been interpreted, in conjunction with diatom-species changes, to represent a deterioration in water quality conditions (Yang et al., 1993). Information on declines in sponge spicules have been proposed in combination with information on diatom fossils along with chrysophyte cysts and scales to make inferences about past salinity conditions in western Canadian lakes (Cumming et al., 1993). In an analysis of soils, Schwandes & Collins (1994) suggested that the presence of sponge spicules can be used to identify past aquatic habitats or terrestrial

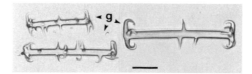

Figure 6. Elongate, dumbbell-shaped, spined gemmoscleres (g) of *Anheteromeyenia argyrosperma* (Potts). The scale bar is 25 μm in length. The photograph is used with permission from Ricciardi & Reiswig (1993).

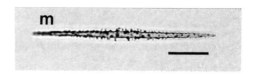

Figure 7. Needle-shaped, spined microsclere (m) of *Heteromeyenia tubisperma* (Potts). The scale bar is 25 μm in length. The photograph is used with permission from Ricciardi & Reiswig (1993).

areas that had been wet, as well as to serve as an indicators of a soil's genesis. Conley & Schelske (1993) suggested that silica in sponge spicules may represent an important component of the total silica in some lake sediments. Because sponge spicules are on average larger than diatom frustules and are slower to dissolve, they may constitute an important sink for silica in some habitats.

Other analyses have been based on species-related inferences drawn from sponge spicules. Information on the distribution of sponge species across regions has been based on evaluations of sediments collected in surveys of lakes. A total of 13 sponge species were recorded in Connecticut throughout the state's five distinct geologic regions (Paduano & Fell, 1997). *Spongilla lacustris* Linnaeus occurred in all of the 28 lakes surveyed while spicules of 3 species occurred in only one lake. Four other sponge species occurred in many of the surveyed lakes and spicules were found regularly in sediment strata representing time periods from approximately 1875 to 1990. Other studies have considered the histories of particular habitats. A sediment core from a montane Colorado lake suggested that *Eunapius fragilis* Leidy and *Ephydatia muelleri* Lieberkühn occurred nearly constantly during that last 7000 years but that *S. lacustris* was present only intermittently during this period and appears to have been absent for the last thousand years (Hall & Herrmann, 1980). Hall & Herrmann (1980) suggest that high levels of total dissolved solids, as indicated by concentrations of calcium and magnesium, and a high acid neutralizing capacity, may have had adverse effects on the occurrence of *S. lacustris* in the lake. Spicular remains in the sediments of Lake Okeechobee, Florida, USA indicate that the lake has been moderately eutrophic and probably turbid for the past 3000 to 4000 years (Harrison et al., 1979). Wilkins et al. (1991) combined information on spicules from three sponge species in a central Kentucky pond with information on plant microfossils to make inferences about past regional conditions and on the development of the pond itself. In general, techniques are readily available to quantify the distribution of sponge spicules with depth in lake sediments (e.g., Harrison & Warner, 1986) and studies have evaluated past sponge species

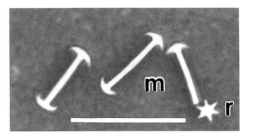

Figure 8. Dumbbell-shaped microscleres (m) of *Corvomeyenia everetti* (Mills) with one view of a rotule (r). The scale bar is 25 μm in length. The photograph is used with permission from Ricciardi & Reiswig (1993).

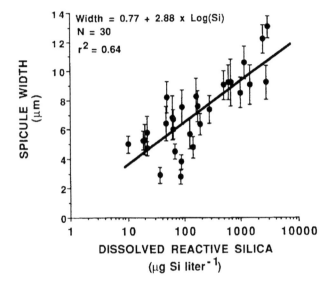

Figure 9. Relationship between megasclere widths in surface sediments and dissolved reactive silica in lake water of 30 lakes in the Northern Highland Lake District in Wisconsin, USA. Error bars indicate 95% confidence intervals for spicule width. Used with permission from Kratz et al. (1991).

abundances in Lago di Monterosi in Italy (Racek, 1970), Lake Baikal in Siberia (Stoermer et al., 1995), and other lakes throughout the world (Harrison, 1988).

Quantitative assessments of past lake chemical conditions have also been made from measurements of the dimensions of sponge spicules preserved in lake sediments (Kratz et al., 1991). Across a series of lakes in the Northern Highland Lake District, Wisconsin, USA, comparisons of spicules in surface sediments revealed substantial differences in the width of megascleres that showed a strong correlation with habitat silica concentrations (Fig. 9). Sediment cores taken in 8 lakes in the Northern Highland each showed systematic and substantial declines in the concentrations of dissolved reactive silica during the histories of the lakes (from 4.1 to 270 fold, Fig. 10). Although these cores were not dated, their deepest

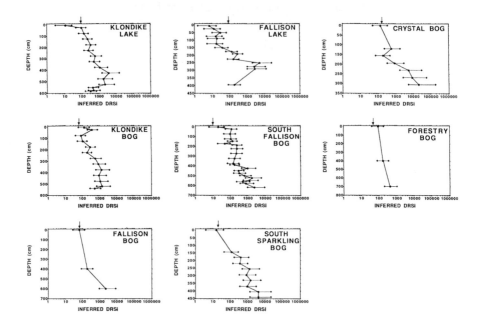

Figure 10. Relationship between reconstructed concentration of dissolved reactive silica (DRSi) and depth in cores from two clearwater lakes and six bog lakes. Present-day DRSi concentrations are indicated by arrows. Inferred dissolved reactive silica is in units of μg/L. Error bars indicate 95% confidence intervals, which were computed based on regression calculations with the inverse regression procedure described by Draper & Smith (1981). Used with permission from Kratz et al. (1991).

portions included materials that were deposited shortly after deglaciation in the region, about 12,000 B.P. (Pielou, 1991). The observed decline most likely occurred due to changes in the rates of silica release from minerals in the region over the 12,000 year period recorded in the lake sediments. With appropriate calibration, quantitative paleo-assessements of spicules may be useful for determining long-term silica dynamics in other regions.

Techniques for assessing sponge spicules in sediments

Methods for processing samples for sponge spicules are very similar to those for diatom frustules (see Battarbee et al., this volume). Once sediment samples or cores have been collected, material can be digested with concentrated nitric acid heated in a boiling water bath for approximately one hour (Frost et al., 2001) or ashed at 550 °C to remove organics (Kratz et al., 1991), carefully washed, mounted permanently on glass microscope slides using a medium such as Permount®, and examined microscopically. (Nitric acid is extremely dangerous and should be handled very carefully. Protective eyeware is absolutely required. Protective gloves and clothing are also essential.) These digestion processes will not eliminate resistant minerals and a process to concentrate the spicules will be necessary if such minerals are abundant. Harrison (1988) provides a detailed report of a technique for quantifying spicule deposition rates. Sources of information for identifying sponge species from spicules include Ricciardi & Reiswig (1993) for eastern Canada, Frost et al. (2001) for habitats in America north of Mexico and the Caribbean or Penney & Racek (1968) for habitats worldwide.

Future applications of sponges in paleolimnology

The potential uses of fossils from freshwater sponges have not been explored fully in paleolimnological research. Given the detailed record that sponge spicules provide in some lake sediments, their limited use may be related to the lack of information available on the distributional ecology of sponges. The detailed assessments of chemical factors influencing the occurrence and distribution of freshwater sponges that are necessary to calibrate paleolimnological research are fairly limited (Harrison, 1988; Frost et al., 2001). Careful evaluation of the factors controlling the distribution of freshwater sponges would certainly expand their potential use in paleolimnological investigations.

Expanded use of sponge spicules could provide valuable insights into past lake conditions. General declines of sponges in a lake as evidenced by decreased spicules could be an indication of declining, water-quality conditions (Harrison, 1974) and could be applied to a broad range of situations with proper calibration or in conjunction with separate measures. Some sponge species appear to be restricted to a limited range of environmental conditions (Harrison, 1988). More detailed evaluations of the distributions of a wider diversity of sponge species, particularly those with limited distributions, are likely to reveal a broad range of paleo-indicators. Quantitative assessment of spicule morphology has the potential of broad application as an indicator of past, habitat-silica conditions (Kratz et al., 1991). It may also be possible to evaluate additional past habitat conditions using other minerals or the proportions of stable isotopes fossilized in sponge spicules. Sponge spicules have the potential to serve as another useful tool on the workbench of paleolimnologists.

Summary

Freshwater sponges have skeletal elements, spicules, that consist of biogenic silica and are well preserved in most aquatic sediments. The taxonomy of freshwater sponges is based primarily on the structures of some spicules, particularly the gemmoscleres but also characteristics of megascleres and microscleres, which taken together are generally quite specific to species. There are as many as 300 species of freshwater sponges world-wide and the past presence of sponges in general and of particular species in many cases can be inferred from examinations of sediments. The collection and identification of sponge spicules in aquatic sediments is relatively straightforward.

Sponge spicules can serve several roles in paleoenvironmental investigations. The general presence or absence of sponges, or their overall abundance, can reflect the presence of previously wet conditions in some locations and the salinity or overall water quality of past conditions in aquatic habitats. The distribution of individual sponge species can be controlled by their chemical, physical, and possibly biological environment such that the presence of a particular taxon in sediments can be serve as an indicator of past habitat conditions. The width of megascleres has been shown to be related to habitat concentrations of dissolved reactive silica in a systematically quantitative fashion. The width of spicules in sediments can serve as a reliable indicator of past, habitat-silica availability. The reliable use of sponges as paleo-indicators typically requires a calibration of the occurrence of sponges and present-day habitat conditions.

Sponge spicules have not routinely played a role in paleoenvironmental investigations. Their preservation in sediments and their specificity to environmental conditions, however, suggest that they may be quite useful in future paleo-research. They may be particularly helpful when combined with other paleo-indicators of past habitat conditions.

Acknowledgements

The preparation of this chapter was supported by several grants from the National Science Foundation. I particularly appreciate the constructive suggestions of the editors and a reviewer that allowed improvements on an earlier draft of this chapter.

References

Berquist, P. R., 1978. Sponges. University of California Press, Berkeley, California, U.S.A., 268 pp.

Conley, D. J. & C. L. Schelske, 1993. Potential role of sponge spicules in influencing the silicon biogeochemistry of Florida lakes. Can. J. Fish. Aquat. Sci. 50: 296–302.

Cumming, B. F., S. E. Wilson & J. P. Smol, 1993. Paleolimnological potential of chrysophyte cysts and scales and of sponge spicules as indicators of lake salinity. Int. J. Salt Lake Res. 2: 87–92.

Draper, N. R. & H. Smith, 1981. Applied Regression Analysis, 2nd edition. Wiley, 709 pp.

Frost, T. M., H. M. Reiswig & A. Ricciardi, 2001. Porifera. Pages 97–133 in Thorp, J. H. & A. P. Covich (eds.) Ecology and Classification of North American Freshwater Invertebrates (Second Edition). Academic Press, New York, New York, USA.

Hall, B. V. & S. J. Herrmann, 1980. Paleolimnology of three species of fresh-water sponges (Porifera: Spongillidae) from a sediment core of a Colorado semidrainage mountain lake. Trans. am. Microscop. Soc. 99: 93–100.

Harrison, F. W., 1974. Sponges (Porifera: Spongillidae). Pages 29–66 in Hart, C. W. Jr. & S. L. H. Fuller (eds.) Pollution Ecology Freshwater Invertebrates. Academic Press, New York, Y.Y., 389 pp.

Harrison, F. W., 1988. Utilization of freshwater sponges in paleolimnological studies. Palaeogeogr. Palaeoclim. Palaeoecol. 62: 387–397.

Harrison, F. W., P. J. Gleason & P. A. Stone, 1979. Paleolimnology of Lake Okeechobee, Florida: An analysis utilizing spicular components of freshwater sponges (Porifera: Spongillidae). Not. Nat. Acad. Nat. Sci., Philadelphia 454: 1–6.

Harrison, F. W. & B. G. Warner, 1986. Fossil fresh-water sponges (Porifera: Spongillidae) from Western Canada: An overlooked group of Quaternary paleoecological indicators. Trans. am. Microscop. Soc. 15: 110–120.

Jewell, M. E., 1935. An ecological study of the fresh-water sponges of northern Wisconsin. Ecological Monographs 5: 461–504.

Jewell, M. E., 1939. An ecological study of the fresh-water sponges of northern Wisconsin, II. The influence of calcium. Ecology 20: 11–28.

Koehl, M. A. R., 1982. Mechanical design of spicule-reinforced connective tissues: Stiffness. J. exp. Biol. 98: 239–267.

Kratz, T. K., T. M. Frost, J. E. Elias & R. B. Cook, 1991. Reconstruction of a regional, 12,000-year silica decline in lakes using fossil sponge spicules. Limnol. Oceanog. 36: 1244–1249.

Paduano, G. M. & P. E. Fell, 1997. Spatial and temporal distribution of freshwater sponges in Connecticut lakes based upon analysis of siliceous spicules in dated sediment cores. Hydrobiologia 350: 105–121.

Penney, J. T. & A. A. Racek, 1968. Comprehensive revision of a worldwide collection of freshwater sponges (Porifera: Spongillidae). United States National Museum Bulletin 272, 184 pp.

Pielou, E. C., 1991. After the Ice Age: The Return of Life to Glaciated North America. The University of Chicago Press, Chicago, IL, USA, 366 pp.

Poirrier, M. A., 1978. *Corvospongilla becki* n. sp., a fresh-water sponge from Louisiana. Trans. am. Microscop. Soc. 97: 240–243.

Racek, A. A., 1970. The Porifera. In Hutchinson, G. E. (ed.) Ianula: An Account of the History and Development of the Lago di Monterosi, Latium, Italy. Trans. am. Phil. Soc. 60: 143–149.

Ricciardi, A. & H. M. Reiswig, 1993. Freshwater sponges (Porifera, Spongillidae) of eastern Canada: taxonomy, distribution, and ecology. Can. J. Zool. 71: 665–682.

Rigby, J. K., 1995. Sponges as microfossils. In Blome, C. D. et al. (convenors). Siliceous Microfossils. Paleontological Society Short Course in Paleontology, 8, 185 pp.

Schwandes, L. P. & M. E. Collins, 1994. Distribution and significance of freshwater sponge spicules in selected Florida soils. Trans. am. Microscop. Soc. 113: 242–257.

Stoermer, E. F., M. B. Edlund, C. H. Pilskaln & C. L. Schelske, 1995. Siliceous microfossil distribution in the surficial sediments of Lake Baikal. J. Paleolim. 14: 69–82.

Turner, J., 1985. Sponge gemmules from lake sediments in the Puget Lowland, Washington. Quat. Res. 24: 240–243.

Wilkins, G. R., P. A. Delcourt, H. R. Delcourt, F. W. Harrison & M. R. Turner, 1991. Paleoecology of central Kentucky since the last glacial maximum. Quat. Res. 36: 224–239.

Yang, J. -R., H. C. Duthie & L. D. Delorme, 1993. Reconstruction of the recent environmental history of Hamilton Harbour (Lake Ontario, Canada) from analysis of siliceous microfossils. J. Great Lakes Res. 19: 55–71.

13. SILICEOUS PROTOZOAN PLATES AND SCALES

MARIANNE S. V. DOUGLAS
(msvd@geology.utoronto.ca)
Paleoecological Assessment Laboratory (PAL)
Department Geology
University of Toronto
22 Russell St., Toronto, Ontario
M5S 3B1, Canada

JOHN P. SMOL (smolj@biology.queensu.ca)
Paleoecological Environmental Assessment
and Research Lab (PEARL)
Department Biology
Queen's University
Kingston, Ontario
K7L 3N6, Canada

Keywords: protozoan plates, Rhizopoda, siliceous, testate amoeba, thecamoebians, heliozoans, peat, *Sphagnum*

Introduction

The testate amoebae (subphylum: Sarcodina; superclass: Rhizopoda) are a group of pre-dominantly freshwater protozoans protected by an external test or shell composed of proteinaceous, calcareous, agglutinated, or siliceous material (Beyens & Meisterfeld, this volume). Some genera's tests consist of small, internally metabolized siliceous plates; these genera and their plates are the focus of this paper. Testate amoebae are often the most common protists living in bog vegetation, with concentrations as high as 16×10^6 individuals per m^2, or a biomass of 1 g/m^2 (Heal, 1962). However, they are also found in a wide variety of other habitats ranging from desert soils to streams and lakes (Tolonen, 1985). Their hard, outer tests ensure that they are often well preserved in sediments where they can be isolated and identified, and then used in paleoenvironmental assessments. Several reviews summarize the many applications of testate amoebae for paleoecological studies (e.g., Frey, 1964; Tolonen, 1985; Warner, 1988; Beyens & Meisterfeld, this volume).

Although considerable paleoenvironmental work has been completed on testate amoe-bae (also called testaceans), several genera of siliceous testate amoebae can be underrep-resented in paleolimnological studies, as taphonomic processes and laboratory procedures often separate their tests into the individual plates making up their test (Tolonen, 1985),

J. P. Smol, H. J. B. Birks & W. M. Last (eds.), 2001. *Tracking Environmental Change Using Lake Sediments.*
Volume 3: Terrestrial, Algal, and Siliceous Indicators. Kluwer Academic Publishers, Dordrecht, The Netherlands.

and they are therefore no longer easily identified using the typical microscopic techniques used by paleo-protozoologists. Experimental data collected by Lousier & Parkinson (1981) have shown that tests of siliceous plate-bearing testaceans disassociate at an exponential rate versus those of agglutinated tests, which decompose at a linear rate. However, because these plates are siliceous and of similar size to diatoms and other microfossils, disassociated plates are easily studied using standard diatom (see Battarbee et al., this volume) or chrysophyte (see Zeeb & Smol, this volume) microscope-slide preparation techniques. Although often ignored, it is our view that protozoan plates and scales from related microorganisms (e.g., heliozoans) can provide useful paleoenvironmental information, and because they can be studied using standard diatom/chrysophyte preparations, they can easily be included in ongoing paleolimnological studies. We limit our discussion to disassociated plates and scales. A thorough review of intact protozoan tests and their use in paleolimnological studies, including discussion of intact siliceous protozoan tests, is presented in Beyens & Meisterfeld (this volume). As well, a comprehensive thecoamoebae bibliography has been compiled by Mediloli et al. (1999) and can be accessed on the Internet.

History and taxonomy

The siliceous nature of certain protozoan plates was first noted by Dujardin in 1841 (Bovée, 1981), but it was not until the early 1900's that the systematics of these rhizopods was studied in detail (Cash & Hopkins, 1905; Cash et al., 1915, 1919). Siliceous tests are formed in one of two ways: they are composed either of ingested and subsequently agglutinated siliceous material such as sand grains and diatom frustules, or from ingested dissolved silica that is internally metabolized into uniformly shaped plates and held together by organic cement. Taxonomic identification of testaceans is based on apertural details and the shape and other morphometric characteristics of the intact test. The identification of individual siliceous plates, however, can only be based on the shape and size of each disassociated plate.

Two basic plate types are produced: thecal (body) plates and apertural (mouth) plates (Figs. 1 and 2). Apertural plates generally make up less than 5% of the plate complement and may be different within a single organism (Figs. 1 and 2). The morphologies of these plates are often generic and sometimes species-specific (Dogiel, 1965; Bovée, 1981). The fine structure and a description of how the plates are held together are described by Hedley & Ogden (1974) who describe the reproductive process involving binary fission.

Siliceous plate-bearing genera include *Assulina, Campascus, Corythion, Cyphoderia, Euglypha, Heleopera, Lesquereusia, Nebela, Paulinella, Placocista, Quadrullela, Sphenoderia, Tracheleuglypha* and *Trinema. Paraeuglypha* and *Paraquadrula* are also plate-bearing genera although many studies may include them as *Euglypha* and *Quadrullela* genera, as they are rarely reported on.

Some of the most comprehensive early taxonomic classification studies were conducted by Cash et al. (1909, 1915, 1919), who included descriptions of thecal and apertural plates. More recent works by Ogden & Hedley (1980), Ellison & Ogden (1987), and Charman et al. (2000) have continued this trend, although these taxonomic treatments are based on intact tests.

The general morphology of individual plates is described in Tables I and II. Included in these tables are some key taxonomic and morphological references, as well as some

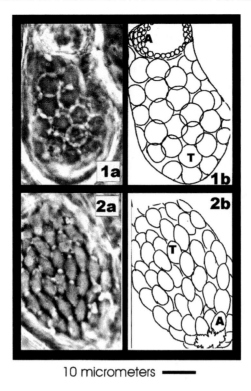

10 micrometers ━━━

Figures 1 and 2. 1a, photograph of intact *Trinema* test; 1b, line drawing of intact *Trinema* test. Note the circular plates. 2a, light micrograph of collapsed *Euglypha* test; 2b, line drawing of *Euglypha* test. Note oval thecal plates and dentured apertural plates. T = thecal plate, A = apertural plate.

ecological and geographical details. The reader is also referred to the major synthesis by Ogden & Hedley (1980). Light micrographs of some of these plates are shown in Figures 3–38. Given the rather plain ornamentation of protozoan plates, scanning electron micrographs of the plates are not included here, but some are presented in Douglas & Smol (1987).

Protozoan plates of each genus have broadly characteristic shapes and sizes; however, in certain taxa, there is some overlap in morphological characteristics and taxonomic identifications may sometimes have to be grouped. For example, it might be impossible to assign a generic label to an individual plate as it could belong to either the genus *Assulina* or *Euglypha*. (e.g., Tables I, II).

The general morphological characteristics of plates from most genera are summarized in Tables I and II. A few additional comments, however, are warranted for some taxa. The test of the genus *Euglypha* is made up of two basic kinds of plates: apertural and thecal (Figs. 2a and 2b). A third kind of plate, a plate-bearing spine (Fig. 29), is present in only a few species. Thecal plates are generally oval of varying dimensions (e.g., Figs. 7, 8, 13–18), making it difficult to match positively these plates to a particular species. As mentioned above, overlaps in plate morphology between two or more genera might exist. However, the presence of distinctive toothed apertural plates (Figs. 31–38) confirms the presence of

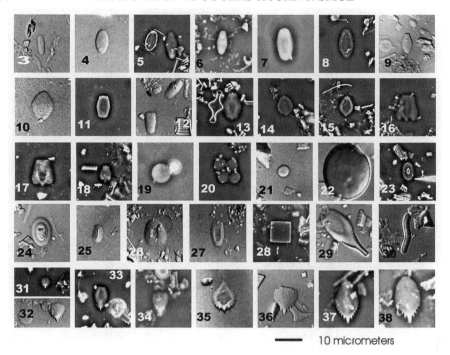

—— 10 micrometers

Figures 3–38. Thecal plates: Figures 3–30: 3, *Euglypha/Placocista*; 4, *Euglypha*; 5, *Assulina* sp.; 6, *Assulina* sp.; 7, *Euglypha* sp.; 8, *Euglypha* sp.; 9, *Assulina* sp.; 10, *Tracheleuglypha/Sphenodeia*; 11, Type "B"; 12, *Assulina* sp.; 13–18, *Euglypha acanthophora*; 19, *Trinema* sp.; 20, *Trinema* sp.; 21, *Trinema* sp.; 22, *Tracheleuglypha/Sphenodeia*; 23–27, unknowns, possibly spine-bearing thecal plates belonging to *Euglypha*; 28, *Quadrulella* sp.; 29, *Euglypha* sp. elongated plate; 30, *Lesquereusia* sp.; Apertural plates. 31–38, *Euglypha* spp.

this genus in a sample. In some *Euglypha* species, it is possible to identify the thecal plates to the species level, e.g., *E. acanthophora* (e.g., Figs. 13–18) and the potential exists, with further study, to identify other plates to the species level.

Thecal plates belonging to the genus *Nebela* are polygonal of various sizes (Table I). However, this cannibalistic genus also ingests and uses the siliceous plates of other genera, making it nearly impossible to track its presence based upon analyses of this eclectic collection of plates.

Ecology

The main purpose of this chapter is to provide a means of recognizing and identifying individual protozoan plates in standard diatom microscope slide preparations (Battarbee et al., this volume). A comprehensive overview of their ecology would require an additional chapter and many of these details are included in Beyens and Meisterfeld (this volume). Instead, we simply provide reference to some key papers that include information on ecological and geographical distributions (Tables I, II).

Table I. Morphological plate characteristics and selected publications pertaining to each genus. Morphological characteristics of protozoan plates (i.e., plate shape and dimensions) with cross-references to light micrographs included in this chapter are also provided. Literature references are sorted according to taxonomy, ecological affiliation, and geographical distribution. Numbered references are listed in the References in **bold**.

Genus or Generic Grouping	Plate Shape	Plate Dimensions (microns)	Figure example	Taxonomic References	Ecological Affiliations	Geographic Distributions
Assulina		4–10	5, 6, 9, 12	12, 23, 31	3, 4, 7, 8, 10, 12, 15, 17, 18, 19, 23, 30, 32, 34, 35, 50	3, 4, 7, 8, 10, 12, 19, 23, 35, 40, 50
Campascus	unknown			46	18, 46	46
Corythion	See Table II			12, 21, 31	3, 4, 6, 7, 9, 10, 12, 15, 17, 19, 21, 30, 32, 34, 35, 48, 50,	3, 4, 6, 7, 9, 10, 19, 21, 35, 40, 47, 48, 50
Cyphoderia	See Table II			12, 27, 46	1, 12, 18, 34, 46	46, 12
Euglypha	Apertural		31–38	12, 15, 20, 22, 23, 31, 33, 36, 38, 41, 42, 43, 44, 46, 47	1, 3, 4, 5, 6, 7, 8, 9, 10, 12, 15, 17, 18, 19, 22, 23, 28, 30, 32, 34, 35, 36, 38, 46, 47, 48, 50	3, 4, 5, 6, 7, 8, 9, 10, 12, 16, 19, 22, 23, 28, 35, 36, 40, 46, 48, 50
	Thecal	5–10	4, 7, 8, 13–18, 23–27, 29			
Helopera	See Table II			11, 13, 31, 41, 45, 46, 47	1, 3, 4, 6, 7, 8, 11, 13, 15, 18, 19, 30, 32, 34, 46, 47, 50	3, 4, 6, 7, 8, 11, 12, 13, 19, 46, 47, 50,
Lesquereusia		Varying sizes	30	11, 31, 35, 45	1, 3, 6, 11, 15, 18, 25, 26, 28, 32	3, 6, 11, 25, 26, 28, 49
Nebela		Varying sizes	28	11, 13, 15, 23, 31, 33, 41, 45, 46, 47, 52	1, 2, 3, 4, 5, 6, 7, 8, 9, 11, 13, 15, 16, 17, 18, 19, 23, 26, 28, 30, 32, 33, 34, 35, 46, 47, 48, 50, 52	2, 3, 4, 5, 6, 7, 8, 9, 11, 13, 16, 23, 28, 35, 40, 46, 47, 48, 50, 51, 52
Paraeuglypha				31	28	28

Table I. (continued) Morphological plate characteristics and selected publications pertaining to each genus. Morphological characteristics of protozoan plates (i.e., plate shape and dimensions) with cross-references to light micrographs included in this chapter are also provided. Literature references are sorted according to taxonomy, ecological affiliation, and geographical distribution. Numbered references are listed in the References in bold.

Genus or Generic Grouping	Plate Shape	Plate Dimensions (microns)	Figure example	Taxonomic References	Ecological Affiliations	Geographic Distributions
Paraquadrula					2, 4, 5, 8, 10, 16, 19	2, 4, 5, 8, 10, 16, 19
Paulinella		5–9		12, 29, 31, 39	1, 12, 18, 29	12
Placocista	See Table II	5–10		12	12	12, 29
Quadrulella	▢	2–15	28	13, 23, 24, 45, 47 Quadrula: 1, 31, 34, 46	3, 4, 11, 13, 18, 23, 28, 46, 47, 50	3, 4, 11, 13, 23, 25, 28, 46, 50
Sphenoderia	See Table II			12, 23, 31	3, 6, 12, 15, 18, 19, 23, 30, 34, 46	3, 6, 12, 19, 23, 46,
Traecheleuglypha	See Table II			31	6, 10, 15, 18, 19, 34	6, 10, 19,
Trinema	◯	3–7.5	19, 20, 21	12, 14, 23, 31, 37, 41	1, 2, 3, 4, 5, 6, 7, 8, 9, 10, 12, 15, 16, 17, 18, 19, 23, 28, 30, 34, 35, 37, 46, 47, 48	2, 3, 4, 5, 6, 7, 8, 9, 10, 12, 16, 19, 23, 28, 35, 37, 46, 47, 48

Siliceous testaceans have a world-wide distribution in freshwater habitats (Bovée, 1981), but they are most abundant in swamps and shallow bodies of waters (Grell, 1953), with especially high concentrations recorded in wet bog vegetation (e.g., Heal, 1962; Tolonen, 1985). A collation of the available habitat preferences shows that *Sphagnum* moss is often the preferred substrate for many taxa (e.g., Cash et al., 1915, 1919; Couteaux, 1969; Decloitre, 1962; Heal, 1964). In a study of 163 lakes in Finland, Douglas (unpublished data) showed that the ratio of siliceous protozoan plates to diatom frustules was correlated to the percent of peatland within a lake's catchment. Nonetheless, testaceans have been documented in many other microhabitats, such as soils (Bonnet, 1964; Smith & Headland, 1983), deep lakes (Penard, 1899), and on mosses and lichens (Couteaux, 1969). For example, Beyens et al. (1986a) noted four distinct assemblages growing on mosses that were influenced by moisture content, pH, and increasing continentality with latitudes. In particular, *Assulina* showed an increase in abundance on lichens as sampling proceeded northwards, but a reverse trend was noted for protozoa living on mosses. *Trinema lineare* is characteristic of oligotrophic, alkaline conditions (Beyens et al., 1991). Beyens (1985)

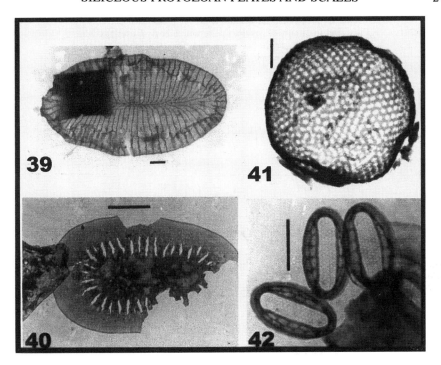

Figures 39–42. Transmission electron micrographs of heliozoan scales. 39, *Raphidiophyrys elegans,* scale bar = 0.5 μm; 40, *Acanthocystis pectinata,* scale bar = 0.5 μm; 41, *Pompholyxophrys* cf. *exigua,* scale bar = 0.25 μm; 42, *Gyromitus disomatus,* scale bar = 0.25 μm. From Douglas & Smol (1987); used with permission.

used the presence of *Assulina* species to confirm wet paleohydrological conditions during the Sub Boreal in Belgium.

Although moisture gradients have been shown to be especially important in influencing assemblages (Warner, 1987; Charman & Warner, 1992), acidity levels are also important (Charman & Warner, 1992; Costan & Planas, 1986), as are food sources (Heal, 1964; Yeates & Foissner, 1995). Substrate preferences have been observed by Bérzinš & Stensdotter (1990). Moreover, siliceous plate-bearing protozoans have been recorded in estuarine-marine waters. For example, Golemansky & Ogden (1980) observed *Cyphoderia* taxa in the littoral of the Black Sea, and Hannah et al. (1996) described a new species of *Paulinella* from subtidal coastal sediments.

Paleoecological potential

We first explored the potential paleolimnological usefulness of siliceous protozoan plates in a surface-sediment study of 38 lakes from the Adirondack Mountains in New York State (Douglas & Smol, 1987). Protozoan plates fulfilled the five criteria outlined by Bradbury (1975) which are required for effective indicators: (1) Identification: the thick, unornamented siliceous plates characteristic of testaceans are easily recognized in sedimentary

Table II. Morphological characteristics of plates that can only be assigned to groups of genera, due to morphological overlaps. Morphological characteristics of protozoan plates (i.e., plate shape and dimensions) are provided, with cross-reference to light micrographs included in this chapter.

Genus or Generic Grouping	Plate Shape	Plate Dimensions (microns)	Figure
Traecheleuglypha/ Sphenoderia		6–15	10, 22
Sphenoderia/ Trinema/ Euglypha		8–13	
Euglypha/ Helopera/ Paulinella		3–11	
Corythion/Trinema		3–5	
Euglypha/Placocista		5–10	3
Assulina/Euglypha		3–10	
Cyphoderia (Trinema may be included)		1–2.5	
Type A		3–8	
Type B		4–13	11

deposits and can be readily identified. (2) Preservation: plates are relatively heavily silicified and are typically well preserved. Even in cores where most other siliceous microfossils were relatively poorly preserved (i.e., diatom valves and chrysophyte scales and cysts showed evidence of dissolution), the more heavily silicified protozoan plates remained intact. (3) Abundance: the abundance of plates ranges from total absence to trace levels, to representing the dominant microfossil group in sedimentary deposits. (4) Quantification: plate morphotypes are easily enumerated and quantified using standard counting procedures. (5) Ecological diversity: siliceous testaceans have a world-wide distribution in freshwater habitats (Bovée, 1981), but they appear to be most abundant in swamps and shallow bodies of waters (Grell, 1953), with especially high concentrations recorded in wet bog vegetation (e.g., Heal, 1962; Tolonen, 1985). It was concluded that protozoan plates are useful indicators, and warrant inclusion in paleolimnological investigations.

Laboratory methods

Protozoan plates can be extracted from a variety of deposits (e.g., peats, lake sediments, soils). Due to their siliceous composition, preparation techniques are identical to those used for other siliceous microfossils, such as diatoms or chrysophycean scales and cysts (see Battarbee et al., this volume; Zeeb & Smol, this volume). Briefly, organic matter is removed from a sample by using a strong oxidizer such as sulphuric and/or nitric acids, or hydrogen peroxide. Digested slurries can then be plated onto glass cover slips and mounted in a permanent mounting medium with a high refractive index, e.g., Naphrax®. The microfossils can then be observed and enumerated under high magnification (i.e., 1000X) using a light microscope. Phase contrast and differential interference contrast (DIC) optics yield the best resolution of plates with texture and thickness being most evident using such optics.

Data presentation

Plate morphotypes are easily identified and quantified using standard diatom and chrysophyte counting procedures, as they are often enumerated alongside these indicators. Typical forms of data presentation include the relative percentage of each plate morphotype expressed as the total plate count, or less frequently as a concentration or accumulation rate of plates (e.g., Douglas & Smol, 1988).

Another common form of data presentation is to express the ratio of protozoan plates to the other siliceous microfossils being enumerated. For example, the ratio of protozoan plates to diatom frustules has been used as an easily calculated estimate of the relative importance of these two indicator groups (Douglas & Smol, 1987, 1988; Brown et al., 1994; Rühland et al., 2000). This ratio is often calculated as the number of protozoan plates/(number of plates + frustules) and expressed as a percentage.

Similar to scaled chrysophytes (Siver, 1991), the relationship between the number of plates to the organisms that produce them is not a linear function, as the average number of plates comprising an amoeba's test varies between species and, to a lesser extent, within species. Consequently, taxa with more plates/test will be overestimated in percentage diagrams. A similar problem is faced by palynologists (see Bennett & Willis, this volume), as different plant species produce variable amounts of pollen grains, or with scaled chrysophytes (see Zeeb & Smol, this volume), as different taxa have different average numbers of scales per cell (Siver, 1991). Further studies on living testaceans can be used to estimate the average number of plates/organism, and so that plate counts can be standardized in order to estimate protozoan numbers more accurately. However, as with the other indicators mentioned above, overall trends are still evident by counting individual plates.

Paleolimnological applications

Only a few paleolimnological studies have thus far been published on protozoan plates. However, this is not due to the paucity of plates, as they are often present, at least in small numbers, on most diatom and chrysophyte slides, and at times are the dominant siliceous microfossils. Douglas & Smol (1987) identified and enumerated protozoan plates in the surface sediments of 38 lakes from Adirondack Park (New York, U.S.A) and found that plates far outnumbered diatom frustules and chrysophyte scales and cysts in some lakes. The distribution of protozoan taxa seemed to track more closely catchment features (e.g., presence of semi-aquatic vegetation and bog development) than lakewater chemical characteristics. Douglas & Smol (1987) also calculated the protozoan plate : diatom frustule ratio for a profile from Leech Fen (Labrador, Canada), and showed that a dramatic increase in the ratio of plates occurred at the transition from a limnic environment to a fen with the development of *Sphagnum* communities. A subsequent paleolimnological study of Lake Colden, in the Adirondack Mountains, NY, used protozoan plates to infer and confirm the long-term presence of a *Sphagnum* mat within the lake catchment, which was a natural source of acidity (Douglas & Smol, 1988). These data helped to explain the historic fishless state of the lake.

Protozoan plates are especially common in peat cores and other cores associated with bogs. Brown et al. (1994) analysed the siliceous microfossils from a high arctic peat deposit on a small island off the northwest coast of Greenland. Although diversity was low (only four plate morphotypes were identified), protozoan plates were often the dominant siliceous microfossils. The relative numbers of plates to diatom frustules and to chrysophyte cysts also fluctuated considerably. Brown et al. (1994) hypothesized that these changes were related to past moisture differences. In a similar study of a peat/lake core from north-central Siberia, Rühland et al. (2000) included protozoan plates in their interpretation of past hydrology and moisture, as well as fire episodes, in this lake/bog sequence. They found that the protozoan plates provided an independent line of evidence from the diatoms in inferring wet and dry periods. The appearance of protozoan plates in the sediments of Taylor Lake, Nova Scotia, marked the development of bog vegetation at that site (Spooner et al., unpublished). Meanwhile, Reavie et al. (2001) used protozoan plates to help assess the paleolimnological history of a groundwater spring in southern Ontario. The testate amoeba community shifted from a *Lesquereusia* to a *Trinema* dominated assemblage over the course of ~150 years (AD 1840–1993). This shift reflected a change from deeper waters to a macrophyte-dense spring. Although protozoan plates have been recorded across all latitudes, Pienitz et al. (1995) did not track any noticeable trends in assemblage composition in relation to changing latitude in Finland and northern Norway.

In some instances, heavily silicified microfossils, such as protozoan plates and sponge spicules, are better preserved than diatoms. This is especially true of ombrotrophic bogs, which receive little ground-water inflow. Douglas & Harington (unpublished) found this to be the case in a Pliocene-age, high arctic beaver dam deposit in which diatoms were scarce. The presence of protozoan plates, mostly dominated by *Trinema*, was used to infer shifts in environmental conditions. The site shifted from an alkaline site to a more acidic site containing mosses.

Other related siliceous indicators

We have restricted our discussion thus far to the siliceous plates of testate amoebae. However, other protozoa can be tracked in sediments using other siliceous microfossils. For example, several heliozoan taxa are covered by siliceous scales that are species-specific.

Nicholls (1983a), Dürrschmidt (1985) and Vigna (1988) described 13 species, 12 species, and three species of the centrohelid genus *Acanthocystis* from Ontario lakes, Chilean and Argentinian sites, respectively. Siemensma & Roijackers (1988a) proposed further division of this genus into three generic groups. The structure and taxonomy of scales belonging to the genera *Raphidocystis, Raphidoiophyrys*, and *Pompholyxophyrys* were described by Nicholls & Dürrschmidt (1985). *Raphidiophyrys* taxa have been described by Siemensma & Roijackers (1988b) and Vigna (1988). Belcher & Swale (1978) reported on *Pinaciophora fluviatilis* and its unusual scales. Nicholls (1983b) described two *Pinaciophora* species new to North America. This genus is among several taxa (i.e., *Rabdiophrys* and *Pompholyxophrys*) that Roijackers & Siemensma (1988) argue should belong to the Crisidiscoidid amoeba, distinct from the Heliozoea. These taxa occur at all latitudes. For example, Hawthorn & Ellis-Evans (1984) described the benthic protozoa from maritime Antarctica, including thecamoeba and heliozoans, and Green (1994) described the planktonic temperate-tropical gradient of protozoa, including heliozoans. As all of these scales typically preserve in lake sediments (Figs. 39–42), they warrant further research.

Summary

Siliceous protozoan plates and scales are commonly observed on diatom and chrysophyte microscopic slide preparations, however they are rarely included in paleolimnological interpretations. A major deterrent to their full exploitation is that plates can rarely be identified to the species level, and at times even generic-level identifications are tentative. This relatively coarse taxonomic resolution discourages detailed paleoeoenvironmental interpretations. Hopefully, further research on disassociated plates, as well as continued ecological studies, may fine-tune these interpretations. However, even at these relative coarse levels of data presentation and analysis, valuable paleolimnological information can still be inferred (e.g., the succession and development of bog formation in and around the lake). In samples where preservation of siliceous microfossils is poor, the use of protozoan plates becomes more important as they are often more resistant to dissolution than other siliceous proxies. A major advantage of using siliceous protozoan plates and scales is that no additional preparation procedures are needed if, for example, diatoms are already being considered in the study.

Acknowledgements

Paleolimnological work in our labs is primarily funded by grants from the Natural Sciences and Engineering Research Council of Canada. We thank K. Rühland, I. Gregory-Eaves, A. Betts-Piper, T. Karst, W. Last, and H. J. B. Birks for comments on an earlier draft.

References

N.B. The **bold** numbers in parentheses refer to the reference numbers used in Table I.

Belcher, J. H. & E. M. F. Swale, 1978. Records from England of the heliozoan-like organism *Pinaciophora fluviatilis* Greefe and of its scales, "*Potamodiscus kalbei*" Gerloff. Arch. Protistenk. 120: 367–370.

Bérzinš, B. & U. Stensdotter, 1990. Ecological studies of freshwater rhizopods. Hydrobiologia 202: 1–11. **(1)**

Beyens, L., 1985. On the subboreal climate of the Belgian Campine as deduced from diatom and testate amoebae analyses. Rev. Palaeobot. Palyn. 46: 9–31.

Beyens, L. & D. Chardez, 1986. Some new and rare testate amoebae from the Arctic. Acta Protozoologica 25: 81–91. **(2)**

Beyens, L., D. Chardez & D. De Baere, 1988. Some data on the testate amoebae from the Shetland Islands and the Faeröer. Arch. Protistenkd. 136: 79–96. **(3)**

Beyens, L., D. Chardez & D. De Baere, 1990. Ecology of terrestrial testate amoebae assemblages from coastal lowlands on Devon Island (NWT, Canadian Arctic). Polar Biol. 10: 431– 440. **(4)**

Beyens, L., D. Chardez & D. De Baere, 1991. Ecology of aquatic testate amoebae in coastal lowlands of Devon Island (Canadian High Arctic). Arch. Protistenkd. 140: 23–33. **(5)**

Beyens, L., D. Chardez & D. De Baere, 1992. The testate amoebae from the Søndre Strømfjord region (West-Greenland): their biogeographic implications. Arch. Protistenkd. 142: 5–13. **(6)**

Beyens, L., D. Chardez & R. De Landtsheer, 1986a. Testate amoebae populations from moss and lichen habitats in the Arctic. Polar Biol. 5: 165–173. **(7)**

Beyens, L., D. Chardez & R. De Landtsheer, 1986b. Testate amoebae communities from aquatic habitats in the Arctic. Polar Biol. 6: 197–205. **(8)**

Beyens, L., D. Chardez, D. De Baere & C. Verbruggen, 1995. The aquatic testate amoebae fauna of the Strømness Bay area, South Georgia. Antarctic Sci. 7: 3–8. **(9)**

Bonnet, L., 1964. Le peuplement thécamoebien des sols. Rev. Écol. Biol. Sol. I. 2: 123– 408. **(10)**

Bovée, E. C., 1981. Distribution and forms of siliceous structures among Protozoa. In Simpson, T. & B. E. Volcanni (eds.) Silicon and Siliceous Structures in Biological Systems. Springer-Verlag, New York (N.Y.): 233–280.

Bradbury, J. P., 1975. Diatom stratigraphy and human settlement in Minnesota. Geol. Soc. Spec. Pap. 171: 1–74.

Brown, K., M. S. V. Douglas & J. P. Smol, 1994. Siliceous microfossils in a Holocene, high arctic peat deposit (Nordvestö, northwestern Greenland). Can. J. Bot. 72: 208–216.

Cash, J. & J. Hopkins, 1909. The British freshwater Rhizopoda and Heliozoa, 2. The Ray Society, London, 166 pp. **(11)**

Cash, J., G. H. Wailes & J. Hopkinson, 1915. The British freshwater Rhizopoda and Heliozoa, 3. The Ray Society, London, 156 pp. **(12)**

Cash, J., G. H. Wailes & J. Hopkinson, 1919. The British freshwater Rhizopoda and Heliozoa, 4. The Ray Society, London, 130 pp. **(13)**

Chardez, D., 1956. Variations morphologiques et teratologiques chez quelques rhizopodes testaces. Biologisch Jaarboek, pp. 265–276. Dr. W. Junk. Den Haag. **(14)**

Chardez, D., 1958. Etude sur les thécamoebiens d'une petite pièce d'eau. Hydrobiologia 10: 293– 304. **(15)**

Chardez, D. & L. Beyens, 1988. *Centropyxis gasparella* sp. nov. and *Parmulina louisi* sp. nov., new testate amoebae from the Canadian High Arctic (Devon Island, NWT). Arch. Protistenkd. 136: 337–344. **(16)**

Charman, D. J., D. Hecton & W. A. Woodland, 2000. The identification of testate amoebae (Protozoa: Rhizopoda) in peats. QRA Technical Guide No. 9, Quaternary Research Association, London, 147 pp.

Charman, D. J. & B. G. Warner, 1992. Relationship between testate amoebae (Protozoa: Rhizopoda) and microenvironmental parameters on a forested peatland in northeastern Ontario. Can. J. Zool. 70: 2474–2482. **(17)**

Costan, G. & D. Planas, 1986. Effects of a short-term experimental acidification on a microinvertebrate community: Rhizopoda, Testacea. Can. J. Zool. 64: 1224–1230. **(18)**

Coûteaux, M.-M., 1969. Thécamoebiens muscicoles de Gaume et de Moyenne-Belgique. Rev. Écol. Biol. Sol 6: 413–428. (**19**)

Coûteaux, M.-M., 1979. L'effet de la déforestation sur le peuplement thécamoebien en Guyane française: étude préliminaire: Rev. Ecol. Biol. Sol. 16: 403–413.

Coûteaux, M.-M., A. Munsch & J. -F. Ponge, 1979. Le genre *Euglypha*: essai de taxinomie numérique. Protistologica 15: 565–579. (**20**)

Cowling, A. J., 1986. Culture methods and observations of *Corythion dubium* and *Euglypha rotunda* (Protozoa: Rhizopoda) isolated from maritime Antarctic moss peats. Protistologica XXII: 181–191. (**21**)

Decloitre, L., 1962. Le genre *Euglypha* Dujardin. Arch. Protistenk. 106: 51–100. (**22**)

Decloitre, L., 1964. Thécamoebiens de la XIIeme expedition Antarctique Francaise en Terre Adelie. No. 259. Expeditions polaires françaises (Missions Paul-Emile Victor) Paris. (**23**)

Deflandre, G. & M. Deflandre-Rigaud, 1959. *Difflugia? marina* Bailey, une espèce oubliée, synonyme de *Quadrulella symmetrica* (Wallich), Rhizopode testacé d'eau douce. Remarques sur la systématique des Nebelidae. Hydrobiologia 12: 299–307. (**24**)

Dektyar, M. N., 1994. Shelled amoebas (Testacea, Rhizopoda) in the Dnieper Reservoirs. Hydrobiological J. 30: 34–41. (**25**)

Dogiel, V. A., 1965. General protozoology. Oxford University Press, Oxford, 747 pp.

Douglas, M. S. V. & J. P. Smol, 1987. Siliceous protozoan plates in lake sediments. Hydrobiologia 154: 13–23.

Douglas, M. S. V. & J. P. Smol, 1988. Siliceous protozoan and chrysophycean microfossils from the recent sediments of *Sphagnum* dominated Lake Colden, N.Y., U.S.A. 1988. Verh. int. Ver. Limnol. 23: 855–859.

Dürrschmidt, M., 1985. Electron microscopic observations on scales of species of the genus *Acanthocystis* (Centrohelidia, Heliozoa) from Chile. I. Arch. Protistenk. 129: 55–87.

Ellison, R. L., 1995. Paleolimnological analysis of Ullswater using testate amoebae. J. Paleolim. 13: 51–63. (**26**)

Ellison, R. L. & C. G. Ogden, 1987. A guide to the study and identification of fossil testate amoebae in Quaternary lake sediments. Int. Revue ges. Hydrobiol. 72: 639–652.

Frey, D. G., 1964. Remains of animals in Quaternary lake and bog sediments and their interpretation. Arch. Hydrobiol. Beitrg. 2: 1–114.

Golemansky, V. & C. G. Ogden, 1980. Shell structure of three littoral species of testate amoebae from the Black Sea (Rhizopodea: Protozoa). Bull. Br. Mus. Nat. Hist. (Zool.) 38:1–6. (**27**)

Green, J., 1994. The temperate-tropical gradient of planktonic protozoa and rotifera. Hydrobiologia 72: 13–26.

Green, J., 1996. Associations of testate rhizopods (Protozoa) in the plankton of a Malyasian estuary and two nearby ponds. Zool. Lond. 239: 485–506. (**28**)

Grell, K. G., 1953. Protozoology. Springer-Verlag, Berlin, 554 pp.

Hannah, F., A. Rogerson & O. R. Anderson, 1996. A description of *Paulinella indentata* N. Sp. (Filosea: Euglyphina) from subtidal coastal benthic sediments. J. Euk. Microbiol. 43: 1–4. (**29**)

Harnisch, O., 1937. Neue Daten zur testaceen Rhizopdenfauna nicht moorbildender Sphagnete. Zool. Anz. Bd. 120: 129–137. (**30**)

Harnisch, O., 1958. II. Klasse: Wurzelfüssler, Rhizopóda: In Brohmer, P., P. Ehrmann & G. Ulmer (eds.) Die Tierwelt Mitteleuropas, Band 1: Urtiere-Hohltiere-Würmer, Lieferung 1b, p. 1–75, pls. 1–26; Quelle and Meyer, Leipzig. (**31**)

Hawthorn, G. R. & J. C. Ellis-Evans, 1984. Benthic protozoa from maritime Antarctic freshwater lakes and pools. Br. Antarct. Surv. Bull. 62: 67–81.

Heal, O. W., 1962. The abundance and micro-distribution of testate amoebae (Rhizopoda: testacea) in *Sphagnum*. Oikos 13: 35–47. (**32**)

Heal, O. W., 1963. Morphological variation in certain Testacea (Protozoa: Rhizopoda). Arch. Protistenk. 106: 351–368. (**33**)

Heal, O. W., 1964. Observations on the seasonal and spatial distribution of testacea (Protozoa: Rhizopoda) in *Sphagnum*. J. Anim. Ecol. 33: 395–412. (**34**)

Heal, O. W., 1965. Observations on testate amoebae (Protozoa: Rhizopoda) from Signy Island, South Orkney Islands. Br. Antar. Surv. Bull. 6: 43–47. (**35**)

Hedley, R. H. & C. G. Ogden, 1973. Biology and fine structure of *Euglypha rotunda* (Testacea: Protozoa). Bull. Br. Mus. Nat. Hist. (Zool.) 25: 121–137. (**36**)

Hedley, R. H. & C. G. Ogden, 1974a. Observations on *Trinema lineare* Penard (Testacea: Protozoa). Bull. Br. Mus. Nat. Hist. (Zool.) 26: 187–199. (**37**)

Hedley, R. H. & C. G. Ogden, 1974b. Adhesion plaques associated with the production of a daughter cell in Euglypha (Testacea; Protozoa). Cell. Tiss. Res. 153: 261–268.

Hedley, R. H., C. G. Ogden & J. I. Krafft, 1974. Observations on clonal cultures of *Euglypha acanthophora* and *Euglypha strigosa* (Testacea: Protozoa). Bull. Br. Mus. Nat. Hist. (Zool.) 27: 103–111. (**38**)

Kies, L., 1974. Elektronenmikroskopische Untersuchungen an *Paulinella chromatophora* Lauterborn, einer Thekamöbe mit blau-grünen Endosymbionten (Cyanellen). Protoplasma 80: 69–89. (**39**)

Laminger, H., 1972. Notes on some terrestrial testacea (Protozoa, Rhizopoda) from Nepal, Himalaya (Lhotse Shar). Arch. Protistenk. Bd. 114: 486–488. (**40**)

Lousier, J. D. & D. Parkinson, 1981. The disappearance of the empty tests of litter- and soil-testate amoebae (Testacea, Rhizopoda, Protozoa). Arch. Protistenk. 124: 312–326.

Medioli, F. S., D. B. Scott, E. Collins, S. Asioli & E. G. Reinhard, 1999. Paleontologia Electronica 2(1) http://www-odp.tamu.edu/paleo/1999_1/biblio/issue1_99.htm

Meisterfeld, R., 1979. Contribution to the systematic of testacea (Rhizopoda,Testacea) in *Sphagnum*. A SEM-investigation. Arch. Protistenk. 121: 246–269. (**41**)

Netzel, H., 1972. Morphogenese des Gehäuses von *Euglypha rotunda* (Rhizopoda, Testacea). Z. Zeliforsch. 135: 63–69. (**42**)

Netzel, H., 1977. The structure of the theca in *Euglypha rotunda* (Rhizopoda, Testacea). Arch. Protistenk. 119: 191–216. (**43**)

Nicholls, K. H., 1983a. Little-known and new heliozoans: the centrohelid genus *Acanthocystis*, including descriptions of nine new species. Can. J. Zool. 61: 1369–1386.

Nicholls, K. H., 1983b. Little-known and new heliozoans: *Pinaciophora triangula* Thomsen new to North America and a description of *Pinaciophora pinea* sp. nov. Can. J. Zool. 61: 1387–1390.

Nicholls, K. H. & M. Dürrschmidt, 1985. Scale structure and taxonomy of some species of *Raphidocystis, Raphidiophrys*, and *Pompholyxophrys* (Heliozoea) including descriptions of six new taxa. Can. J. Zool. 63: 1944–1961.

Ogden, C. G., 1981. Observations of clonal cultures of Euglyphidae (Rhizopoda, Protozoa). Bull. Br. Mus. Nat. Hist. (Zool.) 41: 137–151. (**44**)

Ogden, C. G., 1979. Siliceous structures secreted by members of the subclass Lobosia (Rhizopodea: Protozoa). Bull. Br. Mus. Nat. Hist. (Zool.) 36: 203–207. (**45**)

Ogden, C. G. & R. H. Hedley, 1980. An atlas of freshwater testate amoebae. British Museum (Natural History), Oxford University Press, Oxford, 222 pp.

Penard, E., 1899. Les Rhizopodes de faune profonde dans le lac Léman. Rev. Suisse de Zool. T. 7., 142 pp + 9 plates. (**46**)

Penard, E., 1903. Notice sur les Rhizopodes du Sptizberg. Arch. Protistenkd 2: 239–282. (**47**)

Pienitz, R., M. S. V. Douglas, J. P. Smol, P. Huttunen & J. Meriläinen, 1995. Diatom, chrysophyte and protozoan distributions along a latitudinal transect in Fennoscandia. Ecography 18: 429–439.

Reavie, E. D., M. S. V. Douglas & N. E. Williams, 2001. Paleoecology of a groundwater outflow using siliceous microfossils. Écoscience 8: 239–246.

Roijackers, R. M. M. & F. J. Siemensma, 1988. A study of Cristidiscoidid amoebae (Rhizopoda, Filosea), with descriptions of new species and keys to genera and species. Arch. Protistenkd. 135: 237–253.

Rühland, K., J. P. Smol, J. P. P. Jasinski & B. G. Warner, 2000. Response of diatoms and other siliceous indicators to the developmental history of a peatland in the Tikski Forest, Siberia. Arctic, Antarctic, Alpine Res. 32: 167–178.

Siemensma, F. J. & R. M. M. Roijackers, 1988a. A study of new and little-known Acanthocystid heliozoans, and a proposed division of the genus Acanthoscystis (Actinopoda, Heliozoea). Arch. Protistenkd. 135: 197–212.

Siemensma, F. J. & R. M. M. Roijackers, 1988b. The genus *Raphidiophrys* (Actinopoda, Heliozoea): scale morphology and species distinctions. Arch. Protistenkd. 136: 237–248.

Siver, P., 1991. Implications for improving paleolimnological inference models utilizing scale-bearing siliceous algae: transforming scale counts to cell counts. J. Paleolim. 5: 219–225.

Smith, H. G. & R. K. Headland, 1983. The population ecology of soil testate rhizopods on the sub-Antarctic island of South Georgia. Rev. Ecol. Biol. Sol. 20: 269–286. **(48)**

Stager, J. C., 1988. Environmental changes at Lake Cheshi, Zambia since 40,000 years BP. Quat. Res. 29: 54–65. **(49)**

Tolonen, K., 1985. Rhizopod analysis. In Berglund, B. E. (ed.) Handbook of Holocene Palaeoecology and Palaeohydrology. John Wiley & Sons, New York (N.Y.): 645–666.

Vigna, M. S., 1988. Ultraestructura y taxonomia de algunos heliozoa dulceacuicolas interesantes o poco conocidos para Argentina. Physis B, 46: 11–16.

Warner, B. G., 1987. Abundance and diversity of testate amoebae (Rhizopoda, Testacea) in *Sphagnum* peatlands in Southwestern Ontario, Canada. Arch. Protistenkd. 133: 173–189. **(50)**

Warner, B. G., 1988. Methods in Quaternary Ecology #5. Testate Amoebae (Protozoa). Geoscience Canada 15: 251–260.

Wilkinson, D. M., 1990. Multivariate analysis of the biogeography of the protozoan genus *Nebela* in southern temperate and Antarctic zones. Europ. J. Protistol. 26: 117–121. **(51)**

Yeates, G. W. & W. Foissner, 1995. Testate amoebae as predators of nematodes. Biol. Fertil. Soils 20: 1–7. **(52)**

14. BIOGENIC SILICA

DANIEL J. CONLEY (dco@dmu.dk)
Department of Marine Ecology
National Environmental Research Institute
P.O. Box 358 DK-4000 Roskilde, Denmark

CLAIRE L. SCHELSKE
Department of Geological Science
Land Use and Environmental Change Institute
University of Florida
Gainesville, FL 32611
USA

Keywords: biogenic silica, amorphous silica, sediments, diatoms, diatom abundance, dissolved silicate, eutrophication, climate

Introduction and history

Siliceous microfossils, particularly diatoms, are commonly used as proxies in paleolimnology for studies related to eutrophication and acidification of surface waters, and hydrologic and climate change on a regional scale (see Stoermer & Smol, 1999 for further references). Biogenic silica (BSi), a chemically determined proxy, measures the amorphous Si content of sediments. It has been shown to be a good proxy for diatom abundance (Conley, 1988) and for other siliceous microfossils including sponges (Conley & Schelske, 1993) and phytoliths (Schwandes, 1998). Therefore, as a single proxy, BSi provides an index of both diatom abundance and, in many systems, of diatom productivity (Ragueneau et al., 1996).

Techniques to estimate the BSi content of sediments can be classified into 5 broad categories: 1) X-ray diffraction (Goldberg, 1958), 2) point counting of diatoms (Pudsey, 1992), 3) infrared analysis (Fröhlich, 1989), 4) a normative calculation technique whereby BSi is estimated by difference from mineral silicates (Leinen, 1977), and 5) the wet-alkaline digestion techniques (Hurd, 1972; DeMaster, 1979, 1981, 1991; Eggiman et al., 1980; Mortlock & Froelich, 1989; Müller & Schneider, 1993). Wet-alkaline techniques are most often used because of their ease of use and reliability.

Although early measurements of alkali soluble Si had been made in paleolimnological studies (Hansen, 1956; Diggerfeldt, 1972; Renberg, 1976), reliable wet chemical methods were not available to measure BSi until the early 1980s (DeMaster, 1979, 1981; Eggiman et al., 1980). While other methods had been used in the literature, none provided the ease of use or the theoretical underpinning that DeMaster (1979) provided. Consequently,

J. P. Smol, H. J. B. Birks & W. M. Last (eds.), 2001. *Tracking Environmental Change Using Lake Sediments.*
Volume 3: Terrestrial, Algal, and Siliceous Indicators. Kluwer Academic Publishers, Dordrecht, The Netherlands.

widespread paleolimnological applications began in the 1980s (e.g., Schelske et al., 1983) and determination of BSi in sediments is currently one of the routine measurements made in paleolimnological studies.

Methods

We recommend that investigators employ one of the wet chemical digestion techniques for the measurement of biogenic silica (BSi) in sediments. Three basic modifications of this technique are commonly used: 1) A moderate strength base is used ($2\,M\ Na_2CO_3$) and the sample is digested for 5 hours after which the amount of Si extracted is measured (Mortlock & Froelich, 1989); 2) A weak base is used for the extraction solution ($1\%\ Na_2CO_3$) and the aliquots are withdrawn at selected intervals (3, 4, and 5 h) and the amount of Si extracted is measured on these sub-samples. A least-squares linear regression is made to correct for mineral dissolution during the digestion and the amount of BSi in the sample is determined by extrapolation to the intercept (DeMaster, 1979, 1981); 3) a strong base is used ($0.5–1\,M$ NaOH) with the amount of Si extracted continuously measured and a mineral correction made from the slope of the increase in the amount of Si extracted with time (Müller & Schneider, 1993). Comparison of chemical estimates of BSi with diatom microfossil point counts demonstrate (Conley, 1988) that these wet chemical extraction techniques for the measurement of BSi in sediments provides a valid proxy for the abundance of diatom microfossils in sediments.

We recommend that the specific technique or techniques selected by an investigator should be tested to confirm its reliability for a specific application. There are a variety of amorphous siliceous components in sediments, including chrysophycean cysts (Newberry & Schelske, 1986), diatoms (Conley, 1988), phytoliths (Schwandes, 1998), radiolarians (Eggiman et al., 1981), silicoflagellates, and sponge spicules (Conley & Schelske, 1993). In addition, there is a wide range of BSi concentrations observed in nature, from high BSi concentrations found in diatomites (Mortlock & Froelich, 1989), to low BSi concentrations found in nearshore sediments (Gehlen & van Raaphorst, 1993). Therefore, one must optimize the selected technique for the type of sample analyzed.

In a recent comparison of laboratories that included a broad representation from the aquatic science community, e.g. paleolimnologists, paleoceanographers, limnologists, estuarine scientists and oceanographers, 23 of 30 participating laboratories analyzed all samples with acceptable precision in an interlaboratory comparison of variability in measurement of BSi in sediments (Conley, 1998). This comparison demonstrated that the X-ray diffraction technique overestimated BSi relative to methods based on wet chemical digestion. In addition, neither digestion solution (Na_2CO_3 or NaOH) nor methodology (DeMaster, 1981 versus Mortlock & Froelich, 1989) was a significant factor in the measurement of BSi in these samples from contemporary environments that did not contain sponge spicules nor radiolarians.

Theoretical considerations

The wet chemical digestion techniques rely on the ability of a weak base solution to quantitatively dissolve all amorphous Si components of the sediments, while dissolving only a small fraction of the mineral silicates. Either a "mineral correction" is made for

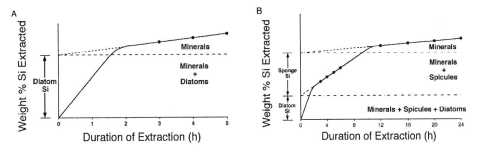

Figure 1. Hypothetical diagram describing the extraction of Si from sediments during digestion in 1% Na$_2$CO$_3$ at 85 °C as a function of time. The solid line depicts the hypothetical increase in the weight % of Si extracted with time in sediments with different amorphous Si components, the points represent the weight % Si extracted at different sampling times during the digestion process, and the dotted lines represent extrapolation to the intercept to determine the amorphous Si fraction. A) Sediments containing only diatoms and silicate minerals (modified from DeMaster, 1979) and B) sediments containing diatoms, sponge spicules and silicate minerals (modified from Conley & Schelske, 1993).

separation between the two components (e.g., DeMaster, 1981) or in sediment samples with high BSi concentrations the relative difference between Si extracted from amorphous compounds and from mineral silicates is large so that the "mineral correction" is ignored (Mortlock & Froelich, 1989; Conley, 1998). During a typical extraction (1% Na$_2$CO$_3$, 85 °C) in sediments where only diatoms and mineral silicates are present, the diatoms are rapidly dissolved (<2 h) and any increases after that time period are due strictly to digestion of mineral silicates (Fig. 1A). In this case, a least-squares linear regression is made and the extrapolation to the intercept gives the BSi concentration (DeMaster, 1979). Continuous measurement of the amount of Si extracted (Müller & Schneider, 1993) provides the best estimate of the contribution of mineral silicates (Conley, 1998), although this technique requires a level of sophistication beyond the capability of most paleolimnological laboratories. Mineral corrections can also be made by simultaneous measurement of Al concentrations during the digestion (Kamatani & Oku, 2000).

During a typical extraction (1% Na$_2$CO$_3$, 85 °C) in sediments where sponge spicules as well as diatoms and mineral silicates are present, the diatoms are rapidly dissolved (<2 h), sponge spicules are dissolved during the first 8–12 h of the digestion, and increases in Si extracted after that time period are due to digestion of mineral silicates (Fig. 1B). In this case, two least-squares linear regressions are made to the data giving the total amorphous Si extracted, the diatom Si extracted and the sponge Si by difference (Conley & Schelske, 1993). The procedure for separating diatom BSi from sponge BSi probably underestimates sponge BSi because smaller and/or lightly silicified sponges can be completely dissolved early during the digestion process (Conley & Schelske, 1993). Sponge BSi can comprise a significant portion of the total amorphous Si extracted from sediments and may act as a significant biogeochemical sink of amorphous Si in sediments (Conley & Schelske, 1993; Bavestrello et al., 1996).

Other amorphous Si compounds in sediments

While some success has been achieved separating diatom BSi from sponge BSi, it is difficult to separate chemically the various amorphous Si components of sediments. For example, the Si contained in radiolarians is typically dissolved by using a strong base solution (Mortlock & Froelich, 1989) and is measured as part of total BSi. Marsh sediments can have a significant contribution of phytoliths to the total BSi content of sediments (Norris & Hackney, 1999), e.g., the amorphous Si structures found in many plants (Wilding et al., 1977). Schwandes (1998) has shown that phytoliths (Piperno, this volume) from terrestrial plants extract more rapidly into a base than diatom BSi, suggesting that some component of the BSi measured in aquatic sediments can potentially be comprised of phytoliths. In addition, there are other forms of amorphous Si that are present in the environment, e.g., duripans (Wilding et al., 1977), precipitation from hydrothermal spring emissions (Juniper et al., 1995), and volcanic glasses (Chadwick et al., 1989), but they, in general, do not comprise a significant portion of BSi measured in most sediment types. Inorganically precipitated Si deposits have been reported in sediments from an African soda lake (Verschuren, 1999).

Sample preparation

Sediment samples should preferably be freeze-dried and not oven dried because better sample integrity is obtained with freeze-drying. In addition, one does not have to worry about loss of volatile organic compounds or other components with oven drying in case sediment samples are used for other purposes. Samples also should be homogenized with a mortar and pestle. Samples should not be mill ground because higher BSi concentrations will be obtained from the exposure of new surfaces available for dissolution (Krausse et al., 1980) and samples should not be combusted before analysis (Conley, unpub. data).

Sediment weight

A 60% reduction in BSi extracted was reported by Flower (1993) with increasing sample weight, e.g. from 2 to 100 mg. Conley (1998) argued that the results can largely be explained by a methodological flaw, with the likely explanation being sorption losses back onto sediments (Mortlock & Froelich, 1989). No significant differences in BSi extracted has been found for sample weights between 20 and 40 mg of dry sediments (Conley, unpub. data). However, because of the potential for weight effects, it is recommended that a consistent weight be used for all samples, e.g. ca. 30 mg.

Methodology

We use a modification of the DeMaster (1979, 1981) technique in our laboratories. We believe this is the simplest and most robust method to use for sediments with a large range in BSi concentrations. Weighed (30 mg), freeze-dried sediments are placed into 125 ml polypropylene round bottles. Immediately prior to digestion 40 mL of 1% Na_2CO_3 solution is added to the bottles. The bottles are placed in a covered shaking bath at 85 °C and 100 rpm with caps slightly loosened to vent gases. After 3 h the bottles are removed, placed into a room temperature water bath and cooled for 5 minutes to retard further dissolution reactions by bringing temperatures down to near room temperature. The bottles are sub-sampled (1 mL) and the sub-sample pipeted into plastic bottles containing 9 mL of 0.021 N HCl to neutralize the Na_2CO_3 digestion solution. After sub-sampling the bottles are placed back into the heating-shaking bath and the sub-sampling procedure is repeated at 4 and

5 h of digestion time. The original technique as published by DeMaster (1979) called for sub-sampling at 2, 3, 4 and 5 h. We have statistically analyzed the sub-sampling strategy and have found out that only three samples are necessary and that sampling at 3, 4 and 5 h provides the best replicability (Newberry, Schelske and Conley, unpub. data).

The sub-samples are analyzed for dissolved silicate by any standard accepted analytical method for the measurement of dissolved silicate. In our laboratories we use variations of the molybdate blue methodology (Technicon Industrial Method 186-72W-Modified) with ascorbic acid as the reductant on a Technican AutoAnalyzer II system. A least-squares regression analysis is made on the increase in Si extracted versus time (taking into account the various dilution steps and the amount of sediment used) and extrapolation to the intercept is made to estimate BSi concentration. If there is no consistent increasing trend in Si extracted with time, then a mean concentration is taken of the 3 sub-samples (Conley, 1998).

Strength of digestion solution

We strongly recommend that the remaining sediment in the digestion bottle of selected samples be examined under the microscope to ensure complete dissolution of all amorphous silica components in the sediments. Numerous studies have clearly shown that when BSi concentrations are high (for example, Mortlock & Froelich, 1989) or if sponge spicules (Conley & Schelske, 1993) or radiolarians (Mortlock & Froelich, 1989; Müller & Schneider, 1993) are present in the sample, a stronger base may be required for complete dissolution. On the other hand, using a stronger base than required to dissolve all the amorphous Si may overestimate BSi concentrations in sediments with low BSi concentrations (Gehlen & van Raaphorst, 1993).

Examination of the slope of increases of Si extracted during the digestion

One also can tell if dissolution is complete from examination of the slope of the increase in the amount of Si extracted with time. If the concentrations of the Si extracted are rapidly increasing it is likely that the amorphous Si in sediments is not all dissolved because increases in dissolved silicate concentration from the dissolution of minerals should be relatively small. The presence of other forms of amorphous Si in a sample such as sponge spicules, and spicules being significantly larger than diatoms takes longer to dissolve (Conley & Schelske, 1993), can contribute to increases in the amount of Si extracted with time.

At low BSi concentrations, slope corrections are absolutely necessary so that BSi concentrations are not overestimated (Conley, 1998). At high BSi concentrations, slope corrections are often not necessary as the error associated during the extraction of BSi can be larger than increases in dissolved silicate concentration from mineral dissolution (Conley, 1998). In these cases, the mean concentration from the three sub-samplings is used instead of the intercept to estimate the BSi concentration.

Additional considerations

It is well known that the use of plastic ware is essential when working with Si mea-surements (Tarapchak et al., 1983). In addition, since dissolved silicate is only a weakly charged species, contaminated water may be found in laboratories without a good water purification system. The widespread use of digestion of the sample in conical centrifuge tubes is strongly discouraged. Conical tubes focus sediments into the container bottom removing the solids from contact with the digestion solution. Conical centrifuge tubes also must be intermittently shaken, which subsequently can influence the bath temperature, and differential expansion of the cap and bottle may cause digestion solution to be lost upon shaking. It is recommended, therefore, that a flat-bottom digestion vessel be used while gently shaken in a heating-shaking bath to ensure continuous contact of sediments with digestion solution throughout the incubation.

Applications

Silica depletion and productivity

Measurements of sedimentary BSi in the Laurentian Great Lakes provide a signal of anthropogenic nutrient enrichment resulting from historic, nutrient-driven changes in BSi production by siliceous phytoplankton (Schelske, 1999), with changes in BSi accumulation used to infer increased diatom production and sedimentation resulting from nutrient enrichment (Schelske et al., 1983). Increased nutrient loading initially results in increased diatom production and accelerated sedimentation of BSi (Schelske & Stoermer, 1971), but ultimately results in decreased diatom abundance after the reservoir of dissolved silicate in the water mass is depleted and diatom production is limited by dissolved silicate. A hypothetical model proposed by Schelske et al. (1983) is supported by BSi profiles in sediment cores from these lakes (see Fig. 4.2 in Schelske, 1999). A BSi peak that is found in the late 1800s in the sediment record of Lake Ontario is inferred to represent increased phosphorus loading resulting from European settlement and forest clearance in the drainage basin (Schelske et al., 1983). This model predicts that a peak in BSi accumulation in the sediment record is a proxy for dissolved silicate depletion in the water column. Two peaks, one resulting from epilimnetic dissolved silicate depletion (Schelske et al., 1983) and a second resulting from water-column dissolved silicate depletion (Schelske et al., 1986; Stoermer et al., 1985b) are found in the sediment record (Schelske, 1999).

The sedimentary BSi record is a sensitive indicator of anthropogenic nutrient enrichment because it integrates seasonal uptake of dissolved silicate utilized in diatom production (Schelske, 1999). Paleolimnological results show that BSi production increased historically in Lake Superior and Lake Huron, lakes with mean TP concentrations ranging from 4–5 μg P/L. Historic BSi production in Lake Michigan with a slightly higher mean TP concentration (8 μg/L) also increased, but decreased as predicted by the hypothetical model to reflect the onset of summer epilimnetic dissolved silicate depletion in the 1970s. Due to the integrative effect, sedimentary BSi provides a record of nutrient enrichment even though a distinguishable change in the TP record, a proxy for the forcing variable, is not apparent (Taylor, 1999). That BSi provides a sensitive measure of nutrient enrichment in these lakes is supported by parallel studies based on diatom microfossils (see Stoermer et al., 1993). Nutrient enrichment effects were found in Lake Superior, using diatom microfossils (Stoermer et al., 1985a) and BSi as proxies of anthropogenic nutrient enrichment (Schelske, 1999). In addition, inferences from both diatom microfossils and BSi show nutrient enrichment effects in another large lake, Great Slave Lake in the Northwest Territories, Canada

(Stoermer et al., 1990). These results are significant because Lake Superior and Great Slave Lake were considered to be pristine lakes with little or no known impact from anthropogenic nutrient loading with the postulated source of nutrient enrichment in Great Slave Lake being atmospheric deposition (Stoermer et al., 1990).

Paleolimnological records from the North American Great Lakes (Schelske et al., 1983) provided the first evidence that small increases in nutrient loading in these P-limited lakes could increase diatom production and result in epilimnetic depletion of dissolved silicate on a time scale of one or two decades (see Schelske, 1999). Varved sediments in Lake Zurich, Switzerland provide evidence of an even more rapid biogeochemical response. A known increase in production of *Tabellaria fenestrata* resulting from increased loading of domestic sewage only lasted three years (from 1896–1898), but this increased production was recorded in the varved sediments both as an increased concentration and sedimentation of BSi (Schelske et al., 1987). Nuisance blooms of *Oscillatoria rubescens* began after the bloom of *T. fenestrata*, a response that would be predicted if diatom production was limited by dissolved silicate (Schelske & Stoermer, 1971). Lake-wide depletion of dissolved silicate resulting from nutrient enrichment has been reported in recent studies from other lakes (Verschuren et al., 1998; Wessels et al., 1999).

The use of BSi as index of paleoproductivity is probably restricted to oligotrophic systems and to systems with long biological residence times. As a paleoproductivity proxy, the use of BSi in most lakes may be confounded by biogeochemical depletion of dissolved silicate or by other system characteristics that control this integrative measure of system responses to forcing functions. In some systems, large inputs of dissolved silicate are utilized to sustain high diatom production over decades. The discharge from such systems may be largely depleted of dissolved silicate (Conley et al., 1993). For example, the large outflow of relatively dissolved silicate rich water from Lake Huron provides a dissolved silicate source for high BSi production in the phosphorus-enriched waters of the central basin of Lake Erie (Schelske, 1999). As a consequence the outflow through the Niagara River to Lake Ontario is low in dissolved silicate. We, therefore, would advise that researchers consider implied assumptions in using BSi as an indicator of paleoproductivity, the primary assumption being that historic supplies of dissolved silicate were replete and not limiting for diatom production.

The depletion of dissolved silicate occurs over the course of river systems as has been shown for the Mississippi River (Turner & Rabalais, 1991). These decreases in dissolved silicate from the Mississippi River basin were hypothesized to occur by nutrient-driven increases in diatom production and sedimentation in the watershed. Recent results suggest that simply the building of dams (Conley et al., 1993) and providing a sedimentation area where none previously existed can also account for decreases in dissolved silicate from the Danube River (Humborg et al., 1997) and from Swedish rivers (Humborg et al., 2000). The interacting roles of dam building and eutrophication are certainly both responsible for the reported declines in dissolved silicate and further research is needed to disentangle the mechanisms behind watershed changes in dissolved silicate concentrations.

Eutrophication is undoubtedly one of the causes of the depletion of dissolved silicate in coastal marine systems (Conley et al., 1983). Increases in the sedimentary signal in BSi concentrations have also been reported in a variety of coastal areas (Turner & Rabalais, 1994; Cornwell et al., 1996). Ecological consequences of reductions in dissolved silicate concentrations are severe with large-scale changes in food webs as has been shown in the

Gulf of Mexico (Turner et al., 1998), in the Baltic Sea (Rahm et al., 1996; Humborg et al., 2000) and the Black Sea (Humborg et al., 1997).

Biogeochemistry/sedimentation/preservation

Whether or to what degree diatom microfossils or BSi are preserved in the sediment record is often questioned in paleoecological studies. Undoubtedly heavily silicified frustules are preserved better than lightly silicified frustules and constitute the largest fraction of recon-structed BSi in sediments (Conley, 1988). Therefore, certain lightly silicified species may be poorly represented in the microfossil assemblage. In addition, the degree of silicification may change in response to variable environments; for example, frustules of *Aulacoseira italica* became lightly silicified in Lake Ontario as historical dissolved silicate supplies in the water column decreased (Stoermer et al., 1985b). Theoretically, preservation of such lightly silicified forms is poorer than for more heavily silicified forms. The implicit assumption in paleoenvironmental studies is that the sediment record is not confounded by time-varying changes in preservation of specific microfossil populations. The sediment record of BSi provides one means of testing whether sedimentary amorphous silica is preserved and whether the quantity preserved varies over time.

By contrast, conclusions from ecological data on diatom productivity stated that the depletion of dissolved silicate would not occur because 80–100% of BSi production by diatoms is recycled (see Schelske, 1988). Depletion of dissolved silicate on a short time scale, however, is possible as shown in a Si mass-balance for Lake Michigan even when as much as 95% of the dissolved silicate is recycled (Schelske, 1985). High rates of sedimen-tation and recycling of Si were reported for three reservoirs in Seine River, France (Garnier et al., 1999). In Lake Vesijarvi, Finland the spring diatom bloom depleted dissolved silicate to 0.10 mg/L, sedimented diatom cells then disappeared from the sediment surface at a rate of 5% d^{-1}, or on a scale of a few weeks (Tallberg, 1999). Without such high rates of Si recycling, dissolved silicate supplies in overlying waters would be severely depleted in the absence of large inputs from other sources. Therefore, high rates of dissolution are to be expected; but even with such high rates, the BSi sedimentary record and diatom microfossils can be preserved.

For geochemical purposes, the quantity of sedimented amorphous Si can be estimated directly from measurements of BSi in the sediment record. Such measurements provide a quantitative estimate of the sedimentary BSi sink when combined with mass sedimenta-tion rate determined from dated cores. It is also possible to infer relative changes in BSi sedimentation from relative changes in BSi concentration in sediments. Such inferences are made with the implicit assumption that relative differences in BSi concentration are not affected significantly by changes in mass sedimentation rate. Using the BSi record in the sediments of the lower Great Lakes to demonstrate biogeochemical silica depletion as discussed above is one example. Quantification of the BSi sedimentary sink is based on changes in the water-column dissolved silicate reservoir that can be described using a mass balance approach (Schelske, 1999).

Climate change

The BSi sedimentary record has been used to study climate change. In Lake Baikal, diatom productivity inferred from BSi varied over the past 5 million years in response to orbital insolation (Colman et al., 1995; Williams et al., 1997). This signal is very strong with BSi varying from <5–50% sediment dry weight. BSi accumulation and paleoproductivity during the Holocene were also studied in Lake Baikal by Qiu et al. (1993). BSi has been shown to provide a record of changes in paleoproductivity with changes in paleoclimatic conditions in Lake Biwa over the past 145,000 years (Xiao et al., 1999).

Oxygen isotopes in BSi have recently been used as hydrologic proxies in studies of climate change (Proft, 1994), and as climatic proxies to study the Allerød-Younger Dryas temperature shift (Shemesh & Peteet, 1998), and the climatic record in lacustrine sediments in other systems (Rietti-Shati et al., 1998; Rosqvist et al., 1999). The temperature effect on the oxygen isotope signal in diatoms was studied experimentally (Brandiriss et al., 1998). The oxygen isotopic composition of opaline phytoliths also records variation in environmental oxygen isotopic composition, providing a potential tool for climatic reconstruction using terrestrial plants (ShahackGross et al., 1996). Using oxygen isotopes extracted from BSi to study climate change (Shemesh et al., 1992) is a very promising technique and undoubtedly will be applied in many future studies.

Other BSi techniques used in marine research may have potential applications in freshwater paleoenvironmental research. Measuring the silicon isotopic composition of diatoms can provide a record of environmental change (De La Rocha et al., 1998). The $^{30}Si : ^{28}Si$ ratio provides a proxy for diatom productivity in that fractionation of the heavier isotope increases with increasing productivity. Because the silicon isotopic ratio in BSi is not affected by dissolution, it may provide a better proxy for some purposes than direct measurements of BSi. The ratio, however, is dependent on the fraction of BSi preserved in the sediment record and, thus, may be affected by differential dissolution among species. In addition, cosmic-ray produced ^{32}Si can be measured with accelerator mass spectrometry and the 178 y half-life of this radionuclide provides the potential for extending radiochronologies to roughly 1000 years (Nijampurkar et al., 1998). This is important because chronologies obtained with ^{210}Pb are realistically no greater than 150 years.

Future directions

Reliable, easy-to-use wet-alkaline techniques for the measurement of BSi in sediments have been developed over the last 20 years and have been shown to have great utility in paleolimnological studies. BSi is an important, valid, and useful proxy in paleoenvironmental studies. As a single proxy, it provides an index of diatom abundance and can provide information on productivity and climate change. Combined with diatom microfossil abundance, inferences about system-wide changes, especially those resulting from eutrophication, are possible. New studies combining measurements of BSi with analysis of stable isotopes of oxygen and radiometric isotopes of Si in BSi show great promise for paleoenvironmental research.

Summary

The measurement of the biogenic silica (BSi) content of sediments is a chemical estimate of the siliceous microfossil abundance. Briefly, sediments are leached with a weak base, usually Na_2CO_3, for a period of time (2–5 hours), and aliquots withdrawn over time. The aliquots are then measured for the amount of Si extracted and a least-squares regression is made on the increase in concentration with time to separate the Si extracted from amorphous Si compounds, e.g. diatoms, sponges, etc., from that of mineral silicates. Comparison of chemical estimates of BSi with diatom microfossil point counts demonstrate that the extraction techniques provide a valid proxy for the abundance of diatom microfossils in sediments. However, the exact choice of methodology will depend upon the type of siliceous components in the sediments and the ability of the digestion solution to dissolve those components. Therefore, both the strength of the digestion solution used and the time over which subsamples are taken should be adjusted for depending upon the type of sediment used. Application of these techniques as a proxy for siliceous microfossil abundance have been instrumental in unraveling the response of aquatic systems to nutrient enrichment and has provided important information on paleoproductivity in particular in studies of paleoclimate.

Acknowledgements

Much of the research regarding biogenic silica in our laboratories has been supported by the U.S. National Science Foundation, the U.S. Environmental Protection Agency and the Carl S. Swisher Endowment, University of Florida Foundation. Selected sediment samples used in the interlaboratory comparison for the measurement of biogenic silica in sediments (Conley, 1998) are available upon request from the first author.

References

Bavestrello, G., R. Cattaneo-Vietti, C. Cerrano, S. Cerutti & M. Sara, 1996. Contribution of sponge spicules to the composition of biogenic silica in the Ligurian Sea. P.S.Z.N. I. Mar. Ecol. 17: 41–50.

Brandriss, M. E., J. R. O'Neil, M. B. Edlund & E. F. Stoermer, 1998. Oxygen isotope fractionation between diatomaceous silica and water. Geochim. Cosmochim. Acta 62: 1119–1125.

Chadwick, O. A., D. M. Henricks & W. D. Nettleton, 1989. Silicification of Holocene soils in Northern Monitor Valley. Nevada. Soil Soc. Amer. J. 53: 158–164.

Colman, S. M., J. A. Peck, E. B. Karabanov, S. J. Carter, J. P. Bradbury, J. W. King & D. F. Williams, 1995. Continental climate response to orbital forcing from biogenic silica records in Lake Baikal. Nature 378: 769–771.

Conley, D. J., 1988. Biogenic silica as an estimate of siliceous microfossil abundance in Great Lakes sediments. Biogeochemistry 6: 161–179.

Conley, D. J., 1998. An interlaboratory comparison for the measurement of biogenic silica in sediments. Mar. Chem. 63: 39–48.

Conley, D. J. & C. L. Schelske, 1993. Potential role of sponge spicules in influencing the silicon biogeochemistry of Florida lakes. Can. J. Fish. Aquat. Sci. 50: 296–302.

Conley, D. J., C. L. Schelske & E. F. Stoermer, 1993. Modification of the biogeochemical cycle of silica with eutrophication. Mar. Ecol. Prog. Ser. 101: 179–192.

Cornwell, J. C., J. C. Stevenson, D. J. Conley & M. Owens, 1996. A sediment chronology of Chesapeake Bay eutrophication. Estuaries 19: 488–499.

De La Rocha, C. L., M. A. Brzezinski, M. J. DeNiro & A. Shemesh, 1998. Silicon isotope composition of diatoms as an indicator of past oceanic change. Nature 395: 680–683.

DeMaster, D. J., 1979. The marine budgets of silica and ^{32}Si. Ph.D. Dissertation, Yale University, 308 pp.

DeMaster, D. J., 1981. The supply and accumulation of silica in the marine environment. Geochim. Cosmochim. Acta 45: 1715–1732.

DeMaster, D. J., 1991. Measuring biogenic silica in marine sediments and suspended matter. In Marine Particles: Analysis and Characterization, Hurd, D. C. & D.W. Spenser (eds.) Geophysical Monograph 63. American Geophysical Union, Washington, D.C., pp. 363–367.

Diggerfeldt, G., 1972. The post–glacial development of Lake Trummen. Folia Limnol. Scand. 16: 1–104.

Eggiman, D. W., F. T. Manhiem & P. R. Betzer, 1980. Dissolution and analysis of amorphous silica in marine sediments. J. Sed. Petrol. 50: 215–225.

Flower, R. J., 1993. Diatom preservation-experiments and observations on dissolution and breakage in modern and fossil material. Hydrobiologia 269: 473–484.

Fröhlich, F., 1989. Deep-sea biogenic silica: new structural and analytical data from infrared analysis—geological implications. Terra Res. 1: 267–273.

Garnier, J., B. Leporcq, N. Sanchez & X. Philippon, 1999. Biogeochemical mass balances (C, N, P, Si) in three large reservoirs of the Seine Basin (France). Biogeochemistry 47: 119–146.

Gehlen, M. & W. van Raaphorst, 1993. Early diagenesis of silica in sandy North Sea sediments: quantification of the solid phase. Mar. Chem. 42: 71–83.

Goldberg, E. D., 1958. Determination of opal in marine sediments. J. Mar. Res. 17: 71–83.

Hansen, K., 1956. The profundal bottom deposits of Gribsø. In Berg, K. & I. C. Petersen (eds.) Studies on Humic, Acid Lake Gribsø. Folia Limnol. Scand. 8, Copenhagen.

Humborg, C., V. Ittekkot, A. Cociasu & B. V. Bodungen, 1997. Effect of Danube River dam on Black Sea biogeochemistry and ecosystem structure. Nature 386: 385–388.

Humborg, C., D. J. Conley, L. Rahm, F. Wulff, A. Cociasu & V. Ittekkot, 2000. Silica retention in river basins: far-reaching effects on biogeochemistry and aquatic food webs in coastal marine environments. Ambio in press.

Hurd, D. C., 1972. Factors affecting solution rate of biogenic opal in seawater. Earth Planet. Sci. Lett. 15: 411–417.

Juniper, S. K., P. Martineu, J. Sarrazin & Y. Gélinas, 1995. Microbial-mineral floc associated with nascent hydrothermal activity on CoAxial Segment, Juan de Fuca Ridge. Geophys. Res. Lett. 22: 179–182.

Kamatani, A. & O. Oku, 2000. Measuring biogenic silica in marine sediments. Mar. Chem. 68: 219–229.

Krausse, G. L., C. L. Schelske & C. O. Davis, 1983. Comparison of three wet-alkaline methods of digestion of biogenic silica in water. Freshwater Biol. 13: 1–9.

Leinen, M., 1977. A normative calculation technique for determination of biogenic opal in deep sea sediments. Geochim. Cosmochim. Acta 40: 671–676.

Mortlock, R. A. & P. N. Froelich, 1989. A simple and reliable method for the rapid determination of biogenic opal in pelagic sediments. Deep-Sea Res. 36: 1415–1426.

Müller, P. J. & R. Schneider, 1993. An automated leaching method for the determination of opal in sediments and particulate matter. Deep-Sea Res. 40: 425–444.

Newberry, T. & C. L. Schelske, 1986. Biogenic silica record in the sediments of Little Round Lake, Ontario. Hydrobiologia 143: 293–300.

Nijampurkar, V. N., D. K. Rao, F. Oldfied & I. Renberg, 1998. The half-life of Si-32: A new estimate based on varved lake sediments. Earth Planet. Sci. Lett. 163: 191–196.

Norris, A. R. & C. T. Hackney. 1999. Silica content of a mesohaline tidal marsh in North Carolina. Estuar. Coastal Shelf Sci. 49: 597–605.

Proft, G., 1994. Biogenic silica (BSi) in sediments of the Mecklenburgian lake district (Germany) and the calcite-silica relation as indicator for trophy and water level. Acta Hydroch. Hydrob 22: 177–184.

Pudsey, C. J., 1992. Calibration of a point-counting technique for estimation of biogenic silica in marine sediments. J. Sediment Petrol. 63: 760–762.

Qiu, L., D. F. Williams, A. Gvorzdkov, E. Karabanov & M. Shimaraeva, 1993. Biogenic silica accumulation and paleoproductivity in the northern basin of Lake Baikal during the Holocene. Geology 21: 25–28.

Ragueneau, O., A. Leynaert, P. Tréguer, D. J. DeMaster & R. F. Anderson, 1996. Opal studied as a marker of paleoproductivity. EOS 77: 491, 493.

Rahm, L., D. J. Conley, P. Sandén, F. Wulff & P. Stålnacke, 1996. Time series analysis of nutrient inputs to the Baltic Sea and changing DSi:DIN ratios. Mar. Ecol. Prog. Ser. 130: 221–228.

Renberg, I. 1976, Palaeolimnological investigations in Lake Prästsjön. Early Norrland 9: 113–119.

Rietti-Shati, M., A. Shemesh & W. Karlen, 1998. A 3000-year climatic record from biogenic silica oxygen isotopes in an equatorial high-altitude lake. Science 281: 980–982.

Rosqvist, G. C., M. Rietti-Shati & A. Shemesh, 1999. Late glacial to middle Holocene climatic record of lacustrine biogenic silica oxygen isotopes from a Southern Ocean island. Geology 27: 967–970.

Schelske, C. L., 1985. Biogeochemical silica mass balances in Lake Michigan and Lake Superior. Biogeochemistry 1: 197–218.

Schelske, C. L., 1988. Historic trends in Lake Michigan silica concentrations. Int. Revue ges. Hydrobiol. 73: 559–591.

Schelske, C. L., 1999. Diatoms as mediators of biogeochemical silica depletion in the Laurentian Great Lakes. In Stoermer, E. F. & J. P. Smol (eds.) The Diatoms: Applications for the Environmental and Earth Sciences. Cambridge University Press, pp. 73–84.

Schelske, C. L. & E. F. Stoermer, 1971. Eutrophication, silica depletion, and predicted changes in algal quality in Lake Michigan. Science 173: 423–424.

Schelske, C. L., E. F. Stoermer, D. J. Conley, J. A. Robbins & R. M. Glover, 1983. Early eutrophication in the lower Great Lakes: New evidence from biogenic silica in sediments. Science 222: 320–322.

Schelske, C. L., E. F. Stoermer, G. L. Fahnenstiel & M. Haibach, 1986. Phosphorus enrichment, silica utilization, and biogeochemical silica depletion in the Great Lakes. Can. J. Fish. Aquat. Sci. 43: 407–415.

Schelske, C. L., H. Zullig & M. Boucherle, 1987. Limnological investigation of biogenic silica sedimentation and silica biogeochemistry in Lake St. Moritz and Lake Zurich. Schweiz Z. Hydrol 49: 42–50.

Schwandes, L. P., 1998. Environmental durability of biogenic opal. Soil Crop Sci. Soc. Florida Proc. 57: 36–39

ShahackGross, R., A. Shemesh, D. Yakir & S. Weiner, 1996. Oxygen isotopic composition of opaline phytoliths: Potential for terrestrial climatic reconstruction. Geochim. Cosmochim. Acta 60: 3949–3953.

Shemesh, A., L. H. Burckle & J. D. Hays, 1995. Late Pleistocene oxygen isotope records of biogenic silica from the Atlantic sector of the Southern Ocean. Paleoceanography 10: 179–196.

Shemesh, A., C. D. Charles & R. G. Fairbanks, 1992. Oxygen isotopes in biogenic silica: Global changes in ocean temperature and isotopic composition. Science 256: 1434–1436.

Shemesh, A. & D. Peteet, 1998. Oxygen isotopes in fresh water biogenic opal: Northeastern US Allerød-Younger Dryas temperature shift. Geophys. Res. Lett. 25: 1935–1938.

Stoermer, E. F., J. A. Wolin & C. L. Schelske, 1993. Paleolimnological comparison of the Laurentian Great Lakes based on diatoms. Limnol. Oceanogr. 38: 1311–1316.

Stoermer, E. F., C. L. Schelske & J. A. Wolin, 1990. Siliceous microfossil succession in the sediments of McLeod Bay, Great Slave Lake, Northwest Territories. Can. J. Fish. Aquat. Sci. 47: 1865–1874.

Stoermer, E. F., J. P. Kociolek, C. L. Schelske & D. J. Conley, 1985a. Siliceous microfossil succession in the recent history of Lake Superior. Proc. Acad. Nat. Sci., Philadelphia 137: 106–118.

Stoermer, E. F. & J. P. Smol (eds.), 1999. The Diatoms: Applications for the Environmental and Earth Sciences. Cambridge University Press, 469 pp.

Stoermer, E. F., J. A. Wolin, C. L. Schelske & D. J. Conley, 1985b. Variations in Melosira islandica valve morphology in Lake Ontario sediments related to eutrophication and silica depletion. Limnol. Oceanogr. 30: 414–418.

Tallberg, P., 1999. The magnitude of Si dissolution from diatoms at the sediment surface and its potential impact on P mobilization. Archiv. Hydrobiol. 144: 429–438.

Tarapchak, S. J., D. R. Slavens, M. A. Quigley & J. S. Tarapchak, 1984. Silicon contamination in diatom nutrient enrichment experiments. Can. J. Fish. Aquat. Sci. 40: 657–664.

Taylor, C. M., 1999. Recent changes in silica availability after implementation of phosphorus abatement in Lake Ontario. M. S. Thesis, University of Florida, Gainesville, FL, 60 pp.

Turner, R. E. & N. N. Rabalais, 1991. Changes in Mississippi River water quality this century. Implications for coastal food webs. BioScience 41: 140–147.

Turner, R. E. & N. N. Rabalais, 1994. Coastal eutrophication near the Mississippi River delta. Nature 368: 619–621.

Turner, R. E., N. Qureshi, N. N. Rabalais, Q. Dortch, D. Justic, R. F. Shaw & J. Cope, 1998. Fluctuating silicate:nitrate ratios and coastal plankton food webs. Proc. Natl. Acad. Sci. USA 95: 13048–13051.

Verschuren , D. 1999. Influence of depth, mixing regime on sedimentation in a small, fluctuating tropical soda lake. Limnol. Oceanogr. 44: 1103–1113.

Verschuren, D., D. N. Edgington, H. J. Kling & T. C. Johnson, 1998. Silica depletion in Lake Victoria: Sedimentary signals at offshore stations. J. Great Lakes Res. 24: 118–130.

Wessels, M., K. Mohaupt, R. Kummerlin & A. Lenhard, 1999. Reconstructing past eutrophication trends from diatoms and biogenic silica in the sediment and the pelagic zone of Lake Constance. Germany. J. Paleolim. 21: 171–192.

Wilding, L. P., N. E. Smeck & L. R. Drees. 1977. Silica in soils: quartz, cristobalite, tridymite, and opal. In Minerals in Soil Environments. Soil Sci. Soc. Amer., Madison, pp. 471–552.

Williams, D. F, J. Peck, E. B. Karabanov, A. A. Prodopenko, V. Kravchinsky, J. King & M. I. Kuzmin, 1997. Lake Baikal record of continental climate response to orbital insolation during the past 5 million years. Science 278: 1114–1117.

Xiao, J., Y. Inouchi, H. Kumai, S. Yoshikawa, Y. Kondo, T. Liu & Z. An, 1997. Biogenic silica record in Lake Biwa of central Japan over the past 145,000 years. Quat. Res. 47: 277–283.

15. SEDIMENTARY PIGMENTS

PETER R. LEAVITT (Peter.Leavitt@uregina.ca)
Limnology Laboratory
Department of Biology
University of Regina
Regina, SK
S4S 0A2, Canada

DOMINIC A. HODGSON (daho@pcmail.nerc-bas.ac.uk)
British Antarctic Survey
Cambridge, CB3 0ET
United Kingdom

Keywords: pigment, carotenoid, chlorophyll, HPLC, chromatography, mass spectrometry, sediment, fossil, methods

Introduction

Chlorophylls (Chl), carotenoids and their derivatives have been isolated and identified from aquatic sediments for over 50 years (Fox, 1944; Fox et al., 1944; Vallentyne, 1954, 1956). As Brown (1969) pointed out, such pigments from algae, phototrophic bacteria and higher plants often preserve long after all morphological structures have disappeared. Unmodified algal carotenoids have been recovered from 56,000 year-old marine sediments (Watts & Maxwell, 1977), while perhydrocarotene, a fully saturated carotenoid derivative, has been isolated from Green River shale deposited in shallow lakes over 50 million years ago (Murphy et al., 1967). Similarly, preservation of pigments throughout the Holocene history has been recorded for many northern hemisphere lakes (Sanger, 1988) and reflects both the water-insoluble nature of these lipophilic molecules and the widespread occurrence of suitable sedimentary environments (organic, anoxic, aphotic).

The primary role of pigments in early paleolimnological studies was as a biochemical marker for the presence of former populations of phototrophic prokaryotes (Brown & Colman, 1963; Brown, 1968) or for estimates of historic changes in lake production (e.g., Vallentyne, 1957; Fogg & Belcher, 1961; Belcher & Fogg, 1964). However, these innovative early studies were constrained by relatively coarse analytical techniques and understanding of pigment biogeochemistry. Fortunately, with the development of more sophisticated methods of chemical identification (e.g., Liaaen-Jensen, 1971; Davies, 1976; Britton et al., 1995a, 1995b), better determinations of compound distribution and taxonomic specificity (Goodwin, 1980a, 1980b) and detailed studies of pigment taphonomy (reviewed

J. P. Smol, H. J. B. Birks & W. M. Last (eds.), 2001. *Tracking Environmental Change Using Lake Sediments.*
Volume 3: Terrestrial, Algal, and Siliceous Indicators. Kluwer Academic Publishers, Dordrecht, The Netherlands.

in Leavitt, 1993; Cuddington & Leavitt, 1999), the range of ecological (reviewed in Millie et al., 1993; Jeffrey et al., 1997a) and paleolimnological applications (Brown, 1969; Swain, 1985; Sanger, 1988; Cohen, 2002) has expanded greatly. Fossil pigments are now used as indicators of algal and bacterial community composition (Züllig, 1981; Yacobi et al., 1990), food-web interactions (Leavitt et al., 1989, 1994a, 1994b), lake acidification (Guilizzoni et al., 1992), changes in the physical structure of lakes (Hurley & Watras, 1991; Hodgson et al., 1998), mass flux within lakes (Carpenter et al., 1988; Ostrovsky & Yacobi, 1999), past UV radiation environments (Leavitt et al., 1997, 1999) and scales of lake variation (Carpenter & Leavitt, 1991). As well, pigments have been used as indicators of a wide variety of anthropogenic impacts on aquatic ecosystems, including eutrophication, acidification, fisheries management, land-use practices and climate change (Leavitt et al., 1994c; Hall et al., 1999). In effect, analyses of fossil pigments are being applied to any investigation in which the abundance, production and gross composition of phototrophic communities in lakes are important response variables.

Accurate reconstruction of aquatic communities and processes from lake sediments requires reliable methodology for the identification, isolation and quantification of fossil pigments. Fortunately, there are several detailed monographs which provide a thorough consideration of the wide variety of modern analytical techniques available for analyzing pigment abundance and composition (Liaaen-Jensen, 1971; Davies, 1976; Goodwin, 1980a, 1980b; Scheer, 1991; Britton et al., 1995a, 1995b; Jeffrey et al., 1997a). Rather than attempt to summarize these authoritative sources, we present a model analytical system suitable for a typical palaeoecological inquiry, and consider the basic pitfalls and practical considerations needed to acquire reliable, interpretable fossil time series. By necessity, our treatment will be synthetic rather than detailed, and we will attempt to direct potential investigators to specialist references. Finally, we provide a brief review of developing protocols in mass spectrometric analysis of pigments, which, in our view, represents the next significant analytical advance in biogeochemical paleoecology.

This chapter contains several sections. First we summarize the suite of fossil pigments commonly encountered in lake sediments and outline the main controls of pigment biogeochemistry and taphonomy. Next we present a basic protocol for isolating and quantifying fossil pigment abundance and composition using high performance liquid chromatography (HPLC). Simple techniques to tentatively identify these compounds are presented in the third section. Finally, we conclude with a review of advanced techniques in mass spectroscopy. While these latter protocols are often beyond the reach of many investigators, they are increasingly essential for rigorous compound identification and have been applied recently in innovative studies of historical changes in the physical environment of lakes (mixing regime, oxygen content, source of sedimentary material; e.g., Villaneuva et al., 1993; Koopmans et al., 1996; Hodgson et al., 1997, 1998).

Pigments in lake sediments

Lake sediments often preserve a wide range of carotenoids, chlorophylls, photoprotectant compounds and other lipid-soluble pigments produced by phototrophic organisms in both the lake and its surrounding catchment (Table I). Sediments also contain derivatives of most common pigments, including isomers, allomers, epimers, pyrolysis products, and various pigment-specific structural modifications of associated oxygen, methyl, glycosidic

Table I. Pigments commonly recovered from lake sediments and their taxonomic affinities. Quantitatively important sources indicated in *italics*. Distributions modified for lake ecosystems from Goodwin (1980a), Jeffrey et al. (1997a). Predominant source identified as planktonic (P), littoral (L), terrestrial (T) or sedimentary (S, post-depositional derivatives); upper case letters indicate quantitatively more important sources. Relative degree of chemical stability and preservation is ranked from most (1) to least (4) stable based on mesocosm experiments and mass-balance studies (see Leavitt, 1993). - = uncertain stability.

Pigment	Source[1]	Stability	Affinity
β, β-carotene	P,L,t	1	*Plantae, Algae*, some phototrophic bacteria
β, α-carotene	P,l	2	*Cryptophyta, Chrysophyta, Dinophyta*, some *Chlorophyta*
β-isorenieratene[2]	P	1	*Chlorobiaceae* (green sulphur bacteria)
isorenieratene[2]	P	1	*Chlorobiaceae* (brown varieties)
alloxanthin	P	1	*Cryptophyta*
fucoxanthin	P,L	2	*Dinophyta[3], Bacillariophyta, Chrysophyta*
diatoxanthin	P,L,s[4]	2	*Bacillariophyta, Dinophyta, Chrysophyta*
diadinoxanthin	P,L,s[4]	3	*Dinophyta, Bacillariophyta, Chrysophyta, Cryptophyta*
dinoxanthin	P	-	*Dinophyta*
peridinin	P	4	*Dinophyta*
echinenone	P,l	1	*Cyanobacteria*
zeaxanthin	P,l	1	*Cyanobacteria*
canthaxanthin	P,l	1	colonial *Cyanobacteria*, herbivore tissues
myxoxanthophyll	P,l	2	colonial *Cyanobacteria*
scytonemin[5]	p,L	-	colonial *Cyanobacteria*
oscillaxanthin	P,l	2	*Cyanobacteria (Oscillatoriaceae)*
aphanizophyll[6]	P,l	2	N_2-fixing *Cyanobacteria* (Nostocales)
lutein	P,L,t	1	*Chlorophyta, Euglenophyta, Plantae*
neoxanthin	l	4	*Chlorophyta, Euglenophyta, Plantae*
violaxanthin	l	4	*Chlorophyta, Euglenophyta, Plantae*
okenone[2]	P	1	purple sulphur bacteria
astaxanthin	P,l	4	invertebrates, N-limited *Chlorophyta*
chlorophyll *a*	P,L	3	*Plantae, Algae*
chlorophyll *b*	P,L	2	*Plantae, Chlorophyta, Euglenophyta*
pheophytin *a*	P,L,t,s	1	Chl *a* derivative (general)
pheophytin *b*	P,L,t,s	2	Chl *b* derivative (general)
pheophorbide *a*	P,l,s	3	Chl *a* derivative (grazing, senescent diatoms)
pyro-pheo(pigments)	L, S**	2	derivatives of *a* and *b*-phorbins
Chl *c*	P,l	4	*Dinophyta, Bacillariophyta, Chrysophyta*

[1] importance of given source also depends on lake morphometry and physical properties (e.g., clarity).

[2] mainly found in sediments of meromictic lakes, along with bacteriochlorophylls and derivatives (e.g., Hodgson et al., 1998).

[3] some dinoflagellates contain peridinin as their main carotenoid, but this very labile compound is rarely recovered from lake sediments.

[4] Diatoxanthin may interconvert with diadinoxanthin in lake sediments. Both are elements of the xanthophyll cycle of chromophyte algae.

[5] Cyanobacteria, particularly benthic colonial forms, may produce and deposit scytonemin and it's derivatives in some lakes which are transparent to UV radiation (Leavitt et al., 1999).

[6] *Nostocales* and allied cyanobacteria have a highly diverse complement of similar, glycosidic carotenoids. Many of these pigments are not easily separated by HPLC (cf. Leavitt & Findlay, 1994).

and complex hydrocarbon side chains and functional groups. Although we do not discuss them specifically, some groups of water-soluble pigments might also preserve in native or degraded forms within the bulk material or interstitial waters of lake sediments (e.g., phycobiliproteins, mycosporine-like amino acids, flavenoids).

Sedimentary pigments differ widely in their taxonomic specificity (Table I and references therein). In general, pigments have five properties which make them useful chemosystematic compounds and therefore of interest to paleoecologists: common initial biosynthetic pathways within each class of molecule; diverse chemical structures arising from specific terminal steps during biosynthesis; ubiquitous distribution among phototrophic organisms, with some taxonomic specificity; functions essential to survival, and; relative ease of analysis (Liaaen-Jensen, 1979). Overall, these features are more representative of the carotenoids (>600) than Chls (<20) or other groups. For example, while broadly-distributed pigments (β-carotene, Chl a, pheophytin a) are valuable indicators of total algal abundance, unaltered marker carotenoids (e.g., alloxanthin, lutein, echinenone, fucoxanthin, peridinin) are more useful when investigations focus on historical changes in algal classes (divisions) or functional groups (flagellates, colonial, N_2-fixing, etc.). Similarly, derivatives of both Chls and carotenoids are proving to be valuable indicators of sedimentary and water-column characteristics that regulate pigment transformations (grazing, anoxia, stratification, light, etc.) and, consequently, are key indicators of changes in the biotic and physical environment of lakes (Koopmans et al., 1996; Hodgson et al., 1998).

The main sources of pigments in lakes include planktonic and benthic algal communities (Carpenter et al., 1986; Steinman et al., 1998), phototrophic bacterial populations (Yacobi et al., 1990; Hurley & Garrison, 1993; Overmann et al., 1993) and aquatic higher plants, or macrophytes (Bianchi et al., 1993). In addition, water and sediments contain pigments associated with detrital material which may become temporarily resuspended or transported from the terrestrial environment or bottom deposits (Carpenter et al., 1988; Leavitt & Carpenter, 1989; Winfree et al., 1997). In all cases, pigments degrade both in the water column (Fig. 1; Leavitt, 1993; Steenbergen et al., 1994; Cuddington & Leavitt, 1999), and following deposition in the sediments (Hurley & Armstrong, 1990; Damste & Koopmans, 1997; Hodgson et al., 1998). Pigment degradation in the water column is usually very rapid and extensive (>95% of all compounds; half-life of days) and includes chemically- or microbially-mediated oxidation (Leavitt & Carpenter, 1990a; Louda et al., 1998); grazing by invertebrates (McElroy-Etheridge & McManus, 1999; Descy et al., 1999; Poister et al., 1999); bacterial degradation (Leavitt & Carpenter, 1990b), cell lysis and enzymatic metabolism during senescence (Louda et al., 1998). These processes interact in complex manners to regulate the quantity and composition of pigment assemblages in surficial sediments (Fig. 1; Cuddington & Leavitt, 1999). Pigment degradation in sediments is often less rapid, particularly under anoxic conditions (Tett, 1982; Hurley & Armstrong, 1990; Yacobi et al., 1991), but also depends on the presence of light and burrowing invertebrates (Leavitt & Carpenter, 1989). The molecular characteristics of degradation processes differ between chlorophylls and carotenoids, with carotenoids being converted to cis-carotenoids and then to colourless compounds via cleavage of the conjugated (diene) system so that fewer than seven conjugated double bonds remain (Leavitt, 1993). In contrast, chlorophylls degrade to pheophytins via the loss of Mg^{2+} atoms from the tetrapyrole ring, and pheophorbides, following loss of the phytol chain and various side groups. The chemistry of chlorophyll degradation is discussed in Hendry (1987) and Scheer (1991), whereas that of carotenoids is presented in Britton et al. (1995a) and Damste & Koopmans (1997).

Rates of degradation differ greatly among pigments (Table I; Leavitt, 1993; Steenbergen et al., 1994) and between habitats (Hurley & Armstrong, 1990). While pigments are usually protected from net loss via compensatory biosynthesis in living organisms, detrital pigments are rapidly and differentially degraded in most aquatic habitats. The most rapid losses occur when pigments are exposed simultaneously to high irradiance,

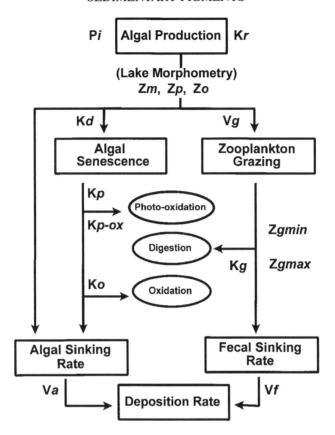

Figure 1. Pathways of pigment production, transformation and degradation within lakes. The main controls of pigment deposition at lake-bottom include: algal cell recruitment rate (K_r), senescence rate (K_d) and population size (P_i); rates of photo-oxidation (K_p), oxidation (K_o), and ingestion (V_g) and digestion (K_d) by herbivores; algal (V_a) and fecal (V_f) sinking rates; the minimum (Z_{gmin}) and maximum (Z_{gmax}) depths of grazers; and, morphometric features of the lake related to the maximum depth (Z_m), depth of light penetration (Z_p) and its extinction coefficient (K_p), depth of oxygen penetration (Z_o). See Cuddington & Leavitt (1999) and text for additional details. Reproduced with permission.

temperature and oxygen—conditions characteristic of the epilimnion of lakes. In addition, recent theoretical and empirical studies suggest that grazing by invertebrates and cell-sinking velocity also constitute important processes regulating deposition rates (Poister et al., 1999; Cuddington & Leavitt, 1999). In general, compounds with complex functional groups and structures degrade more rapidly than less modified compounds (Leavitt & Carpenter, 1990a, 1990b; Hurley & Armstrong, 1990; Steenbergen et al., 1994). However, substantial variability in pigment loss rates also occur because of differences in suscepti-bility of source populations to senescence and decomposition, either because of inherent biochemical characteristics of cells or because of differences in species habitat preferences, selective grazing by invertebrates or other ecological factors (Leavitt & Carpenter, 1989).

Development of a more sophisticated understanding of the patterns of pigment pro-duction, deposition and degradation has significantly improved the reliability of historical reconstructions based on fossil pigment analyses. Rather than a deterrent to paleoecological applications, improved definition of pigment biogeochemistry allows investigators to identify more clearly the scope, limitations and applications of pigment analyses. For example, Cuddington & Leavitt (1999) present a first-generation model which quantifies the relationship between algal production and fossil deposition while accounting for variability in lake morphometry and stratification, algal growth and sinking characteristics, and pigment degradation due to grazing, chemical oxidation and photooxidation. Comparison of model predictions to results from mass balance studies, whole-lake experiments, surveys, empirical models and fossil analyses suggest that algal production and pigment deposition are correlated well for short-term (<millennia) historical reconstructions within single lakes, particularly when lake depth varies little in comparison to the magnitude of fossil signals (e.g., Leavitt & Findlay, 1994; Leavitt et al., 1999). In contrast, the model demonstrates that comparisons of fossil abundance and pigment composition among sites may be complicated by high variance in the physical structure of lakes (transparency, stratification, depth, herbivore communities) overriding the production-deposition relationships. Additionally, while historical trends in abundance of individual taxa and ecological interactions among these groups may be each inferred accurately from stratigraphic variations in pigment concentration or accumulation rates (Leavitt, 1993; Leavitt et al., 1999; Hall et al., 1999), differential preservation of biomarkers prevents quantitative determination of algal community composition based on ratios of indicator pigments, such as is used in oceanographic and limnological studies (Jeffrey et al., 1999).

Sediment collection, transportation and storage

Sediments for pigment analyses can be collected using any standard paleoecological technique including gravity, piston and freeze coring. However, subsequent treatment of the sample must consider the labile nature of such biochemical fossils and take appropriate measures to prevent pigment destruction or transformation. Exposure to direct light, warm temperatures ($>0\,°C$), pH extremes, oxygen, and enzymes should be avoided. Conditions which most closely mimic those of deeply-buried sediments will promote pigment stability and include temperatures $<4\,°C$, absence of light and oxygen, and limited biological activity (grazers, microbes). This is best accomplished by freezing sediment samples in blackened, air-tight containers under an inert atmosphere (argon, N_2, etc.) immediately following sediment collection. Because this is rarely practical under field conditions, we briefly outline some procedures which minimize artefacts.

Collection of sediments by freeze-coring can minimize pigment degradation by freezing samples before their removal from the aquatic environment. Unfortunately, such cores may require considerably more protection from oxidation in subsequent stages of sample isolation. While in the field, frozen cores should be wrapped first with heavy black (>4 mil) black plastic sheeting, then industrial-grade aluminum foil. Both can be strongly secured with cloth-based tape, but not metallic or plastic-based tapes which will not adhere properly when cold. We often tape the entire core in a manner similar to applying a plaster cast in order to strengthen the core for transportation, minimize penetration of light, and because we think it looks cool. Because these precautions will not prevent dehydration of sediments (e.g., freezer-burn) and pigment oxidation to colourless derivatives, we do not recommend storage of frozen cores for long periods of time (>1 month). If storage is unavoidable, efforts should be made to store cores in black, sealed tubes flushed regularly with inert gases at $-20\,°C$ or colder. Similarly, the surface 5 mm of the frozen core should be removed with a wood-plane prior to sectioning for stratigraphic analysis (Leavitt et al., 1989).

Often cores are sectioned in the field and individual samples isolated from bulk sediments. In general, the relative rates of pigment loss from samples will be inversely related to sediment mass isolated and directly proportional to sample:container volume ratios due to enhanced dehydration and oxidation rates of samples with high surface area:volume ratios or in large spaces. Similarly, oxidation rates will be correspondingly great for dewatered (deep) sediments and those with low organic matter content, fine particle size or high redox potential (i.e., low reducing capacity). Oxidative losses can be reduced somewhat by filling containers fully to eliminate headspace for gases. If logistic constraints dictate the use of porous sample bags (whirl-pak, zip-lock, etc.), samples should be transferred as soon as possible to air-tight, darkened containers such as thick-walled vials with gas-tight screwcaps, black plastic 35 mm photographic film canisters, or ampules which can be evaculated and sealed.

Pigment losses can be great during sediment storage prior to isolation and quantification of pigments. Comparisons of the rates of pigment oxidation in algal samples collected on glass-fibre filters demonstrate that substantial destruction can occur in a matter of days unless steps are taken to eliminate exposure to light, oxygen and ambient temperatures (Mantoura et al., 1997a). In particular, freeze-drying of sediments (see below) should be avoided if samples will not be analyzed within a week, as dry samples have a powdery texture which maximizes exposure to oxygen. In most cases, the key to proper storage requires low oxygen tension and biological activity. As in field storage, these requirements can be met by matching volumes of sample and container (of a non-porous, inert medium), sealing samples under an atmosphere of high-purity inert gas, and storing in extreme cold (e.g., $-80\,°C$ ultrafreezer). While storage can also be effective if isolated crude or purified sediment extracts are placed under similar environments, we do not recommend final extraction of sediments until pigments are to be completely isolated and quantified. Even under ideal conditions, it is advised that sediment samples not be stored for more than 6–12 months prior to analysis. In theory, this storage period can be extended through the use of anti-oxidants.

Extraction of sedimentary pigments

The choice of extraction procedure for carotenoids and Chls depends upon several factors including the characteristics of sediments, the ease of solvent extraction, and the properties and relative quantities of the pigments present (Liaaen-Jensen, 1971; Wright et al., 1997; Schiedt & Liaaen-Jensen, 1995). The main problems associated with isolation and extraction relate to preventing cleavage of the long chains of conjugated carbon-carbon double bonds which are the light-absorbing chromophores of most pigments. The weakly-held orbital electrons of these chains can be removed by highly electronegative elements, such as oxygen, thereby breaking conjugation and preventing pigments from absorbing light in the visible wavelengths. In addition, all pigments are unstable toward light, heat, acids, and peroxides. As well, some carotenoids are labile in weak alkaline conditions, including fucoxanthin, peridinin, carotenol esters, α-ketols and some allenic carotenoids. Esters of carotenoic acids are only unstable in strongly alkaline conditions (Liaaen-Jensen & Jensen, 1971).

General precautions: Exposure to light causes isomeric conversions (*trans* into *cis*) of dissolved pigments which may result in the formation of chromatographic and spectrophotometric artefacts. In the presence of oxygen, light can also promote the addition of oxygen across chromophores and rapidly decolourize pigments. Because pigments are especially sensitive to photo-oxidation while adsorbed onto surfaces such as glass, all samples should be wrapped in aluminum foil and or covered when not in use (Davies, 1976). If pigment solutions are to be evaporated to dryness, we recommend use of rotary evaporation at reduced pressures (azeotropic distillation) and that the vacuum be released with a stream of oxygen-free nitrogen or other inert gas. Similarly, investigators should be cautious with acid-washed glassware or solvents with trace contaminant (e.g., chloroform with HCl, ethers with peroxides) to avoid oxidative decompositions, acid-catalysed *cis-trans* isomerization or interconversion of carotenoid epoxides (Davies, 1976). Although pesticide or HPLC-grade solvents are of adequate purity for some pigment manipulations, we recommend that investigators also employ additional solvent purification steps such as detailed in Vogel (1978) and Davies (1976).

Extraction: Due to the wide variety of sediments which contain pigments, no one method of extraction is universally applicable. Instead of reviewing the many options available to investigators (see Millie et al., 1993; Schiedt & Liaaen-Jensen, 1995; Wright et al., 1997), we present a simple protocol which we have found thorough, efficient and relatively free of artefacts for sediments from tropical, temperate, alpine, polar and saline environments. In most cases, sediments were stored frozen <6 months at $-20\,°C$, in the dark and under an atmosphere of >99.999% N_2 prior to pigment extraction. Freezing not only aids preservation, but helps disrupt the organic matter structure of sediments and improves extraction efficiency. Other methods of disruption such as sonication, grinding with abrasives and continuous freeze-thaw cycles do not appear to offer significant advantages when extracting lake sediments.

Efficient and reproducible pigment extraction requires that sediments contain a constant, low amount of water. Because sediment porosity and water content vary with burial depth, particle size and post-coring handling, this requirement is most easily achieved by first freeze-drying (lyophilizing) well-mixed sediment samples under a hard vacuum (<0.1 Pa). Air-drying oxidizes pigments and should be avoided, as should preliminary dehydration using water-miscible solvents (methanol, ethanol, acetone) which may also remove some polar pigments and cause hardening of remaining sediments. We recommend drying all samples simultaneously to assure comparable degrees of dehydration, and that vacuums be broken using N_2. Samples should be covered with a porous tissue held by elastics and dried in sturdy containers which do not implode in a high vacuum. The length of time required to freeze-dry samples increases directly with sediment water content, vial fullness, total sample load and duration of run, and inversely with sample area:volume ratio and freezer-dried capacity. Samples are completely dry when repeated determinations of individual mass produce no significant change through time.

Dried samples are extremely susceptible to pigment transformation or destruction and must be extracted immediately upon lyophilization. Because all sediments are now of low water content, and because pigments differ in their polarity and solubility in organic solvents, we extract all material with mixtures of acetone:methanol:water (80:15:5, by volume). Surveys of methods for extraction of pigments from algae usually find that 10% water in acetone represents the optimal extraction mixture, combining high overall efficiency, ease and safety (Schiedt & Liaaen-Jensen, 1995; Wright et al., 1997). However, our experience suggests the addition of methanol improves extraction efficiency for more complex samples typical of lake sediments. Investigators are cautioned to avoid excessive water or methanol content in extract solutions because of problems with derivative formation arising from algal enzymes (e.g., chlorophyllase in water), chemical transformations (methylation) or prolonged exposures during excessive subsequent drying. Similarly, we recommend the presence of low (5–10%) water content to improve extraction of highly polar chlorophyllides, scytonemin and some pheophorbide derivatives, and to soften sedimentary materials for multiple extractions. Finally, because oxygen is more readily dissolved in many organic solvents than it is in water (Vogel, 1978), investigators should thoroughly degas solvents prior to extraction of sediments. Oxygen can be removed by rotary evaporation or bubbling ('scrubbing') with pure N_2, and should be conducted using pure solvents rather than mixtures which rapidly change composition under vacuum due to different vapour pressures of component solvents.

Prior to extraction, samples can be re-lyophilized (2 h) to remove any residual water, frost or ice. Aliquots of dry sediments (25–100 mg dry weight) are weighed to 0.1 mg, added to 5 mL of oxygen-free extraction solvent mixture in 10 or 20 mL glass vials, and flushed with N_2 prior to storage in the dark at -10 to $-20\,°C$ for 12 h. Sample mass required varies inversely with organic C content of sediments, and may be as great as 1.5 g dry wt in highly inorganic materials (clay, sand, tephras). In general, sufficient mass has been used if extracts produce visible colour against a white background. Extracts are then filtered through an 0.2 μm pore chemically-resistant membrane filter, dried and stored under an inert atmosphere at least $-20\,°C$. Filters are washed with three 1 mL volumes of fresh acetone or acetone:methanol (10% by volume) and solutions combined to usually achieve $\sim$95% complete extraction and minimal derivative formation in comparison with more exhaustive protocols. Multiple extractions or those of longer duration may be required in cases of high organic matter content ($> 10\%$ organic C) or if an active algal mat is present. Extractions are always conducted under low, indirect lighting using degassed, purified HPLC-grade solvents.

After extraction, pigment oxidation may be reduced by the storage of extracts in the absence of oxygen and light, with antioxidants such as pyrogallol or quinol, or by freezing of solutions in liquid N (Mantoura et al., 1997a). In general, we try to avoid these latter steps by analyzing samples within 2–5 d of extraction. In this case, solutions are concentrated to dryness in the dark at room temperature by evaporation under a stream of pure N_2 gas or a vacuum. Dried extracts are dissolved into a precise volume (300–2000 μL) of injection solvent prior to pigment isolation and quantification by HPLC analysis. In general, injection solvent composition should be fixed and similar in composition to that of the mobile phase of the chromatographic system at the time of sample injection. Our system is comprised of 70% acetone: 25% ion-pairing reagent: 5% methanol (by volume). Ion pairing reagent (IPR) consists of 0.75 g tetrabutyl ammonium acetate and 7.7 g ammonium acetate in 100 mL HPLC- grade water (Mantoura & Llewellyn, 1983). In addition, this solution contains one or more internal standards which allows us to correct for errors in volume injected and which acts as a reference standard for comparing chromotographic behaviours among runs. This step is particularly important if the HPLC is not using an automated sampling device. Ideal reference standards should be stable compounds with pigment-like chemical characteristics and unique positions on chromatograms (Mantoura & Repeta, 1997), although in complex sedimentary mixtures, this latter condition can be difficult to achieve and synthetic dyes may be used. In particular, Sudan II (Sigma Chemical Corp., St. Louis, MO) at 3.2 mg L^{-1} of injection solution has carotenoid-like absorption characteristics (lambda max = 485, 442.5 nm in acetone) and runs at a unique position on our chromatograms.

Isolation of pigments

High Performance (Pressure) Liquid Chromatography: Since 1980, HPLC analyses have become the method of choice for rapid, quantitative determinations of carotenoid, Chl and derivative content in aquatic ecosystems and their sediments (reviewed in Millie et al., 1993; Pfander & Riesen, 1995; Mantoura et al., 1997b; Jeffrey et al., 1999). With the recent publication of excellent monographs dedicated to HPLC-based pigment analyses, the investigator is presented with a wealth of technical information concerning choice of analytical system, performance, analytical constraints (see Kost, 1988; Pfander & Riesen, 1995; Jeffrey et al., 1997a). Rather than repeat this information, we present a brief summary of a technique which we have used successfully and focus instead on practical considerations related to paleolimnological analyses. These comments are designed to supplement more formal treatments of HPLC technology by providing advice designed to facilitate ecological and limnological applications.

Carotenoid and Chl concentrations are most often quantified by reversed-phase high performance liquid chromatography (RP-HPLC), as reviewed by Pfander & Riesen (1995) or Jeffrey et al. (1999). The basic principle of chromatographic separation is simple; pigment mixtures introduced to a column of small particles are separated based on their differential attraction to the packing (stationary phase) and the solvent stream moving through the column (mobile phase). In RP-HPLC, packing material is usually a small (5 μm) particle coated with a non-polar (e.g., C8, C18) monomer or polymer, whereas mobile phases are variably polar and are composed of two or three solvent streams (e.g., A, B, C) mixed in succession to produce gradients of solvent polarity. Typical solvents used include methanol, acetone, water, and acetonitrile, often modified with ion-pairing reagents or other agents (ethyl acetate, pyridine). Separation of complex mixtures depends on relative attraction of individual pigments for the non-polar stationary phase (both coating and support material) and their solubility in the mobile phase. Typically, the sample is dissolved in injection solvent (see above) and is introduced to the column within the most polar of the mobile phases. This extract solution arrives at the column as a bolus of pigment which then interacts with the non-polar coating of the stationary phase. Because polar pigments are more strongly attracted to polar solvents than to the non-polar particles ('like attracts like'), they pass through the HPLC column rapidly and are the first compounds to be detected. In contrast, non-polar pigments attract strongly with non-polar packing and will not be re-dissolved into the mobile phase until its polarity is decreased, most often by altering the solvent composition (e.g., % A:B:C) in one or more linear gradients. Compounds are released from the stationary phase and flow to the detectors in sequence of decreasing polarity. Some small reversals of sequence can occur between very similar molecules as a result of the precise characteristics of the HPLC system (column characteristics, solvent brand and purity, gradient steepness). In addition, the chromatographic behaviour of some pigments may be slightly altered because of sediment-specific matrix effects, usually co-eluting substances. In all cases, investigators are cautioned to use rigorous independent methods to confirm the identity and chromatographic behaviour of all pigments for each lake and chromatographic system, and not to rely exclusively on chromatographic position and simple spectral characteristics in pigment identification.

Criteria and guidelines for selection of the HPLC hardware, software and gradient system for pigment analyses are summarized by Pfander & Riesen (1995), Wright & Mantoura (1997) and Jeffrey et al. (1999) among others. In most cases, efficient analysis of large sample numbers minimally requires an autosampler, high-pressure pumps (>20 kPA), zero dead-volume column switchers, high-resolution column (vertical orientation recommended), and in-line photo-diode array spectrophotometer (>300–800 nm range). Additional components may include in-line detectors of pigment fluorescence, radioactivity or mass-selective spectrometric detectors (see below), depending on application and availability of funds. It is critical to note that there is often an important, though unstated, interaction among hardware components (and possibly software) which influences system performance, and that each technique will require some optimization before effective, reproducible analyses can be achieved (see Wright & Mantoura, 1997a; Mantoura & Repeta, 1997). In addition, it should be recognized clearly that analytical separations of pigments from a complex sedimentary mixture of native compounds, derivatives and organic matrix will rarely be as precise as those from pure algal samples and that this feature will depend strongly on the specific lake to be studied. As a result, investigator experience with a specific HPLC system may be as important as the precise method being used to isolate and quantify pigment content.

Solvent composition and gradient systems are often optimized for type of material from which pigments are extracted and the analytical goal of the study (cf. Fig. 2 in Jeffrey et al., 1999). Some HPLC systems focus on separating a range of compounds with similar polarities, whereas ultra-high performance systems are designed for use with 'clean' environmental samples such as those arising from mixtures of unialgal cultures and from the mixed layer in the marine systems. Similarly, selection of analytical system for paleoecological samples must consider that sediments often contain two to five-fold more peaks than samples from overlying water, and that some degree of co-elution may occur regardless of investigator precautions. In this case, researchers may consider a trade-off between the ability to finely resolve many (unknown) compounds and the ability to better isolate key marker carotenoids and Chls while losing other less important markers. Hence, the analytical system detailed by Mantoura & Llewellyn (1983) and Mantoura et al. (1997b) and modified by Leavitt & Findlay (1994) has proven robust in isolating key pigments from sediments of 500 lakes (see below), despite some limitations in resolving power (Jeffrey et al., 1999). Preliminary results with the methods of Wright et al. (1991) and Wright & Jeffrey (1997) also indicate that this technique is also robust for a wide variety of lakes (Lami et al., 1994; Hodgson et al., 1997, 1998).

Highly reproducible separations of many key indicator pigments (Table I) have been achieved using the simple binary (two-stage) solvent system modified from Mantoura & Llewellyn (1983; Fig. 2). Presently, the hardware includes a Hewlett-Packard (HP) model 1050 or 1100 system, a Rainin Model 200 Microsorb C-18 column (5 μm particle size; 10 cm length), an HP model 1050/1100 scanning photodiode array spectrophotometer (435 nm detection wavelength), and an HP fluorescence detector (435 nm excitation wavelength, 667 nm detection wavelength). Analytical separation is achieved by isocratic

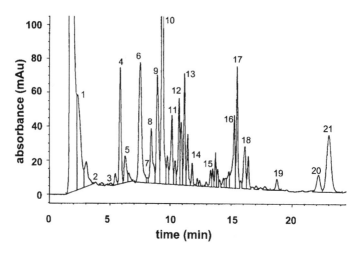

Figure 2. Chromatogram of pigments isolated by high performance liquid chromatography using the methods of Mantoura & Llewellyn (1983). Pigments include scytonemin (1), Chl *c* (2), fucoxanthin (3), UV-absorbing compound *b* (4) and *a* (5), Sudan II (6), myxoxanthophyll (7), alloxanthin (8), diatoxanthin (9), lutein-zeaxanthin (10), carotenoid isomers (11–13), canthaxanthin (14), Chl *b* (15), okenone (16), Chl *a* (17), echinenone (18), pheophytin *b* (19), pheophytin *a* (20) and β-carotene (21). See Table I for taxonomic affinity of individual compounds. Note: chromatograph has been loaded with excessive pigment for the purposes of presentation of minor components of the extract. Not all compounds appear in every lake sediment sample.

delivery (i.e., no gradient) of mobile phase A (10% ion-pairing reagent in methanol) for 1.5 min at 1.5 ml min^{-1} and 21,000 Pa pressure, followed by a linear ramp to 100% solvent mixture B (27% acetone in methanol) over 7 min and isocratic hold for an additional 12.5 min. The column is re-equilibrated by a continued isocratic delivery for 3 min, a linear return to 100% solution A over 3 min, and a further isocratic hold for 12.5 min. Total run time (injection-to-injection) is just over 30 min, and most major carotenoids are fully isolated from many types of lake sediments (Fig. 2).

Accurate quantification of pigment abundance requires fulfilment of three criteria: adequate separation of marker compounds from contaminants (see above), identification of a pigment's true identify (see below), and accurate calibration of the HPLC system with an authentic standard of known purity. Mantoura & Repeta (1997) detail the factors to be considered in choosing the calibration method. In our standard applications, systems are calibrated using a dilution series of each chromatographically-pure pigment, where the amount of pigment in each standard was determined spectrophotometrically within 30 min following its isolation (see below). Calibration curves are constructed by injecting 10 different volumes (e.g., 20–200 μL, 20 μL increments) of a stock solution of known pigment concentration (nmoles or μg of pigment μL^{-1}) and quantifying the area under the resulting peak on the chromatogram. Linear regressions of mass pigment injected (as volume × concentration) and peak area should explain >99.99%, although small negative intercepts often occur. Concentrations of pigments in sediment samples can then be estimated from peak areas from chromatograms of whole extracts as follows:

$$\text{Mass pigment (J)} \times (\text{mass sediment})^{-1} = \left(A_i \times S_j - I_j\right) \times \left(V_t \times V_i^{-1}\right) \times M_s^{-1} \times M_j^{-1}, \quad (1)$$

where A_i = peak area on sample chromatogram; S_j = slope of regression of calibration curve relating area of peak vs mass of pure pigment; I_j = intercept of calibration curve; V_t = total volume of final dilution containing entire original sample; V_i = volume of final dilution injected; M_s = mass of sediment extracted; and, M_j = molecular weight of pigment J.

Although some authentic samples can be obtained commercially, many key biomarkers can only be isolated from algal or bacterial stocks of known pigment composition (Repeta & Bjørnland, 1997; Jeffrey, 1997). In this case, unialgal cultures are grown, harvested and pigment extracted using standard procedures. Extracts can be highly concentrated to produce strong chromatographic peaks, and each peak collected following passage through a detector ('drop-catching'). Pigments can be further separated from contaminants (sedimentary material; ion-pairing reagents) by diluting isolates 50% with water and collecting on C-18 separation cartridges (cf. Hodgson et al., 1998). Care must be taken to keep isolates cold ($\leq 0\,^\circ C$), dark and under N_2 to avoid transformation or loss during collection or purification. Because isolates will be in a complex solvent for which extinction coefficients are usually unknown, pigments will have to be evaporated to dryness using N_2 or vacuum, then dissolved in a pure solvent or known mixture before quantification by either gravimetric (if a pure crystal) or spectrophotometric means (Davies, 1976; Mantoura & Repeta, 1997). This latter procedure is relatively simple for pigments with known specific extinction coefficients ($E_{1cm}^{1\%}$). Here, mass of pigment in a pure solution within a spectrophotometric cuvette is:

$$X = \frac{E \times Y}{(E_{1cm}^{1\%}) \times 100}, \tag{2}$$

where X= grams of pigment in Y ml of solution; E= absorbance at a given wavelength (absorbance maximum, λ_{max}); $E_{1cm}^{1\%}$ = extinction coefficient of a 1% solution (by weight) of the carotenoid with a path length of 1 cm for λ_{max} (see Davies, 1976; Jeffrey, 1997). Specific extinction coefficients (light absorbance per molecule or mass of pigment) for various pigments in solvents can be obtained from Jeffrey et al. (1997b), Davies (1976) and Foppen (1971), although some errors occur in earlier works.

Fossil pigment abundance has been presented using a variety of metrics including organic matter-specific concentration (moles or mass pigment g^{-1} organic matter), dry matter-specific concentration (moles or mass g^{-1} dry mass), accumulation rate (moles or mass area^{-1} y^{-1}) and relative (%) abundance. Each unit has advantages in different circumstances. For example, pigment accumulation rate is best suited to analyses when allochthonous organic matter inputs are highly variable (i.e., Younger Dryas, tephras, mass flows) or where investigations seek to compare algal abundance among lakes. Unfortunately, recent empirical (Leavitt & Findlay, 1994) and theoretical (Cuddington & Leavitt, 1999) analyses demonstrate that artefacts may arise in surface deposits and over post-glacial time due to rapid pigment losses and changes in lake morphometry, respectively. Similarly, while mass-specific analyses provide accurate estimates of algal abundance during high-resolution ($\sim$annual) reconstructions over several centuries (see Leavitt, 1993), these measures can be biased by sudden changes in either inorganic or organic matter influx independent of pigment deposition (i.e., allochthonous inputs, inorganic precipitates, etc.). Further, because bulk-sediment accumulation rates may vary widely among lakes of similar productivity, fossil concentrations may be strongly influenced by dilution effects which obscure relationships between algal and fossil abundance. Despite these caveats, we find that fossil concentrations can be accurately interpreted in cases where the primary objective is to describe historical variations in algal abundance relative to the magnitude of past events (Leavitt, 1993). In contrast, we note that it is usually difficult to interpret biologically changes in the relative abundance (%) of fossil pigments, mainly due to the high variability in the degree of pigment preservation (see Table I) and because the relative magnitude of taphonomic processes (see above) may vary greatly among lakes (see Cuddington & Leavitt, 1999).

Other methods: HPLC represents the latest in a long series of techniques developed to isolate component pigments from whole algal, plant or sedimentary extracts. Isler (1971) provides a thorough documentation of many of these historical developments. Although many of these early techniques have become somewhat outmoded, they still represent valuable analytical tools for isolating large quantities of pigment, or for preliminary separations of major sedimentary pigment components (e.g., polar and non-polar; Chls and carotenoids). Readers are referred to Liaaen-Jensen (1971) for an authoritative description of most analytical protocols. Here we review the most common approaches and illustrate their value.

When a crude pigment extract contains pigments of differing polarities, *phase partitioning* may be used to separate components (e.g., carotenes vs. xanthophylls) between two immiscible solvents (e.g., petroleum ether and 90% aqueous methanol). Phase separation was used extensively in early carotenoid isolations, prior to the introduction of chromatographic techniques (Karrer & Jucker, 1950). Fox et al. (1944) and subsequently Fogg & Belcher (1961) and Gorham & Sanger (1975) partitioned pigments before estimating total carotenes and xanthophylls by spectrophotometry. Phase partitioning has found some use in the structural identification of closely-related pigments (Hertzberg et al., 1971), although its most routine application is in separation of Chls from alkali-labile carotenoids. Foppen (1971) reviews the partition behaviour of many carotenoids.

Saponification is a useful procedure for separating alkali-stable carotenoids from unwanted Chls, lipids and components with carboxyl-like functional groups. It is unsuitable for isolating highly oxygenated carotenoids such as fucoxanthin or peridinin. The mechanism of saponification (literally 'soap-making') involves the hydrolysis of esters or carboxylate groups with strong bases to form the parent alcohol of the base and the salt of the parent acid (ester) which is water soluble and therefore separable from stable pigments. While not extensively used in paleoecology (Griffiths et al., 1969; Sanger & Crowl, 1979), this method is useful for removing background contaminants (e.g., sterols) from whole sediment extracts prior to isolation of components for structural analyses such as mass spectrometry.

Column, paper and thin-layer chromatography provides a versatile suite of procedures suitable for both preparatory separation of individual pigments and high-resolution quantitative analyses of fossil deposits. In all cases, the basis for pigment isolation remains similar to that of HPLC, with individual pigments exhibiting differential attraction for a mobile solvent phase and its stationary support (paper, inorganic or organic powders). As with HPLC, 'straight phase' chromatography uses a relatively polar stationary phase (adsorbants such as $Ca(OH)_2$, silica, alumina, even sucrose) and non-polar mobile solvents to retain polar pigments and elute non-polar compounds. In contrast, 'reversed-phase' chromatography (nonpolar stationary, polar mobile) has found favour to rapidly elute polar pigments from more strongly-retained non-polar compounds. In all cases, chromatograms may be developed with a single solvent or with a series of sequential mixtures (e.g., Züllig, 1981, 1982). Lists of available adsorbants and solvents are presented by Davies (1976), whereas Liaaen-Jensen (1971) and Bernhard (1995) consider the practical aspects of column chromatography. Although largely replaced by HPLC in modern investigations, these types of chromatography still retain utility as preparatory techniques (e.g., C-18 separation cartridges) or low-cost alternative to HPLC.

Early paleoecological applications of chromatography have varied considerably and often reflect the analytical limitations of individual techniques. While a valuable preparatory method to isolate greater than mg quantities of pigment in a single development, column chromatography has not seen wide use in fossil investigations (Vallentyne, 1957; Brown, 1968) due to high detection limits ($\sim$50 μg), long development times, and the superior fractionating capabilities of other approaches. Similarly, paper chromatography has proven valuable in early paleoecological studies involving fossil Chls and their derivatives (Daley et al., 1977; Brown et al.,1977), although the approach has not been widely adopted. While paper chromatography is comparatively simple to use (reviewed in Davies, 1976; Šesták, 1980), the technique can suffer from long development times ($>$2 hr), sensitivity to atmospheric moisture, only moderate sensitivity ($>$10 μg) and high potential for oxidation during final pigment elution. In comparison, thin-layer chromatography (TLC; Schiedt, 1995) offers the rapid development (15–60 min) and high sensitivity (10 ng) of many HPLC procedures without the associated high cost of hardware. Further, TLC also has the potential for two-dimensional analyses which are useful in eliminating the background coloured contamination characteristic of highly organic sediments (e.g., Sanger & Gorham, 1970; Gorham & Sanger, 1975). Not surprisingly, TLC rapidly found favour with both aquatic scientists and paleoecologists in diverse applications including evaluation of plankton composition (Jeffrey, 1967), Holocene lake production (Daley et al., 1977; Brown et al., 1984), human impacts on lakes (Griffiths, 1978; Züllig, 1981), and the relative contribution of terrestrial and aquatic organic matter to lakes (Gorham & Sanger, 1972). Unfortunately, while offering many of the advantages of HPLC methodologies, TLC-based research requires substantially greater efforts to assure high reproducibility and, as remarked by several experienced practitioners, 'is equal parts art and science'. Interested readers are referred to Liaaen-Jensen (1971), Davies (1976) and Schiedt (1995) for further information.

Pigment identification

Accurate identification of compound identity is one of the most serious constraints preventing widespread application of fossil pigment analyses in paleoecology. In principle, only detailed elucidation of chemical structure by mass spectroscopy (MS) and nuclear magnetic resonance (NMR) studies provide definitive proof of pigment identify. However, these definitions are impractical for paleoecological applications because of difficulties in isolating chemically-pure compounds from heterogenous mixtures characteristic of lake sediments and of applying these tests to all levels in every core. While we endorse the use of advanced techniques to confirm the identity of major, but problematic, pigments in sediments (e.g., Hodgson et al., 1998), we also feel that the more accessible procedures outlined below provide adequate rigour for many identifications and will allow the application of fossil pigment analyses to pressing environmental and social problems. Below we review means of identifying pigments, particularly carotenoids, on the basis of partition characteristics, adsorptive characteristics, partial synthesis (chemical means) and electronic absorption spectra. We conclude with a more detailed consideration of the role of mass spectroscopy in pigment identification.

Phase partitioning: Although phase partitioning is a useful preparatory method of separating carotenoids based on their functional groups, it is of limited usefulness in determining the actual structure of pigments. Partition ratios of selected carotenoids in light petroleum ether-95% aqueous methanol or 85% aqueous methanol systems has been determined under standard conditions by Krinsky (1963) and Foppen (1971). The actual partition ratio obtained is primarily a function of the presence of polar hydroxyl groups, with rather more subtle effects being exerted by carbonyl, ester, ether and epoxy functions. Liaaen-Jensen (1971) notes that since the partition ratio also depends on the chromophore properties and the type of hydroxyl group (primary, secondary), comparison of ratios is only significant for carotenoids of closely-related structure. Identification of carotenoids based on their partition behaviour has also found some use in reversed-phase thin layer chromatography, which operates on the principle of pigment partitioning between mobile phases and oil-impregnated stationary supports. Again, hydroxyl groups are the primary factor dictating pigment polarity although chromophore characteristics, molecular size and the type of cyclic end group (e.g., β, ε, etc.) present also affect the rate of carotenoid elution.

Chromatographic behaviour: Lack of separation of an isolate from an authentic pigment on a series of different chromatographic systems constitutes reasonable proof of pigment identity. In addition to partition characteristics, the chromatographic mobility of pigments is influenced by adsorption characteristics, or affinity for the stationary phase in a chromatographic system (Davies, 1976). Once again, functional groups, cyclic end-groups and chromophore features have important effects on pigment adsorption. For example, addition of functional groups to a carotenoid always increases adsorption affinity. The contributions of oxygen functions to increased adsorption affinity are summarized as follows: $-COOH > -OH > 2[-C=O] > -C=O > -OR$ where R may be a methyl, acetyl, or glycosyl group (Davies, 1976). This basic scheme is modified by the number and position of the functions. Similarly, comparisons of structurally-related carotenoids suggest that pigment affinity increases as the length of the chromophore increases, that geometric (*cis-trans*) isomerization may either increase or decrease adsorption affinity (poly-*cis* less strongly adsorbed; peripheral *cis* more adsorbed than all-*trans*), and that formation of a cyclic end group from a linear end group will decrease adsorption affinity (Davies, 1976).

Electronic absorption spectra: Preliminary pigment identification is most often made using electronic absorption spectra (Britton, 1995; Jeffrey et al., 1997b; Bjørnland, 1997). In particular, the ultraviolet or visible light absorption properties of a carotenoid provide an excellent indication of the chromophoric system present, although not necessarily the types or positions of functional groups (Moss & Weedon, 1976). The position of the long wave bands (usually three) of the carotenoid spectrum is a function of several features including the number of conjugated double bonds, the type of solvent used and, to a lesser extent, the presence of functional groups, especially conjugated ketones (Davies, 1976). For example, addition of a conjugated double bond (up to a maximum of 11) causes a shift in absorption maxima to higher wavelengths (bathochromic shift) and an increase in specific extinction coefficient relative to the unmodified parent pigment. In contrast, end-group type affects the shape of the spectrum causing a decrease in fine structure in the case of cyclic end groups, as well as a (hypsochromatic) shift to shorter wavelengths (8–17 nm) for each end cyclized (Moss & Weedan, 1976). The presence of a conjugated carbonyl function causes a bathochromatic shift (12–25 nm) in the maxima of absorption as well as a complete loss in fine structure such that a symmetrical absorption peak replaces the triple banded spectrum. Although the introduction of hydroxyl or alkoxyl functions to a carotenoid has little effect on the absorption spectrum, formation of a 5,6-epoxide causes a hypsochromatic shift (approximately 3 nm per epoxide) and a decrease in fine structure. The effect of isomerization on the absorption spectrum is four-fold: a) to cause a small hypsochromatic shift in the wavelength of maximum absorption; b) to decrease the extinction coefficient to a value less than that of the all-*trans* compound; c) to decrease the fine structure of the spectrum, and; d) to cause the appearance of a distinct *cis* peak in the region of 320 to 380 nm (Davies, 1976). Foppen (1971), Davies (1976), Kost (1988), Wright & Shearer (1984) and Jeffrey et al. (1997b) all present comprehensive lists of pigment absorption maxima in various solvents and mixtures.

Microscale chemical derivitizations: Once a preliminary molecular identity has been established (e.g., polar carotenoid, non-polar Chl derivative, etc.), microscale chemical derivitizations can be used to establish the presence of functional groups, particularly with carotenoids (Liaaen-Jensen, 1971; Davies, 1976; Eugster, 1995). Critically, most of the important reactions may be carried out with less than mg quantities of pigment isolated by HPLC protocols. Detailed protocols for most procedures are presented in Liaaen-Jensen (1971) and Eugster (1995). Here we review the applications of these procedures to carotenoid analyses.

The presence of *hydroxyl groups* can be determined qualitatively by comparisons of pigment partition and chromatographic behaviour (see above), as well as through use of infra-red (IR) spectrophotometry (Bernhard & Grosjean, 1995). More quantitative identification of primary and secondary hydroxyl groups can also be made using acetylation (esterification) by acetic anhydride in dry pyridine at room temperature. In contrast, silylation procedures provide evidence of tertiary hydroxyl groups (Liaaen-Jensen & Jensen, 1971). In both cases, the number of chromatographically-distinct products indicates the number of hydroxyl group present (Eugster, 1995). Similarly, secondary allylic hydroxyl groups can be selectively methylated and isolated from parent pigments by HPLC (Liaaen-Jensen & Hertzberg,1966) while all allylic hydroxyl groups can be transformed to ketones with attendant changes in electronic absorbance characteristics (see above; Liaaen-Jensen, 1965). Finally, hydroxyl functions can be completely eliminated from carotenoids by dehydration with chloroform and HCl-saturated $CHCl_3$ in intense light (Davies, 1976). When dehydration is 100% complete, there is a 16 nm increase in absorbance maximum for each allylic hydroxyl group eliminated.

Carotenol esters can be easily identified by reduction of the ester to the free carotenol using lithium aluminium hydride ($LiAlH_4$). Alternatively saponification of the ester in methanolic KOH solution may be used to produce a chromatographically-distinct derivative. Similarly, reductions of *carbonyl* functions (including those present as ketones, aldehydes, carboxylic acids, carboxylic acid methylesters and lactones) can be conducted using either $NaBH_4$ (cabonyl only) or $LiAlH_4$ (all; Eugster, 1995), while *alkoxy (methyl ether)* functions can be converted first to the corresponding carotenol using LiAl$_4$, then quantified using the acetylation or silylation procedures mentioned above. Finally, reductions are also useful for identification of the presence of secondary or tertiary *glycosides*.

Although carotenoids with *epoxy* functions are often not well preserved in lake sediments, identification of molecules with this functional group is comparatively easy. In particular, addition of dilute acid (HCl, citric, etc.) will rearrange epoxides formed between adjacent carbons (usually position 5, 6) to those separated by three carbons (5, 8 position). Such a *furanoid rearrangement* will cause a dramatic hypsochromatic shift of 17–22 nm per epoxy group and is easily visible in a spectrophotometric cuvette (Davies, 1976). Although epoxides can also be converted to alcohols by reduction, this procedure can also introduce artefacts and can be difficult to interpret (Liaaen-Jensen, 1971).

An essential part of carotenoid identification is the spectroscopic characterization of the predominant *cis* isomer. Although the all-*trans* isomer predominates in most living systems (Davies, 1976), carotenoids recovered from lake sediments may be predominantly *cis* isomeric forms (e.g., Brown & Colman, 1963). Consequently, *iodine-catalysed isomerization* may be useful for interconversion among the isomers. In principle the number of possible isomers depends on whether a pigment is structurally symmetrical, as well as the number of sterically-effective double bonds; however, because of steric hindrance between lateral methyl groups and olefinic protons, only 2–3 different *cis* isomers are usually observed in equilibrium with the all-*trans* form. Davies (1976) details a modification of Zechmeister's (1962) original procedure in which a concentrated carotenoid solution is illuminated in hexane or benzene in the presence of iodine. The resulting *cis* isomers will show decreased definition in the electronic absorption spectra, decreased stability (increased lability), increased solubility and different adsorptive characteristics (see above).

Identification of pigments using mass spectrometry

In HPLC studies of pure pigments, the shape of absorption spectra, and retention time can reveal several pieces of important structural information. The main absorption peak (λ_{max}) is determined by the linear chromophore (carotenoids) or tetrapyrole ring and side groups (chlorophylls). Subsidiary peaks, and the ratio of their peak heights (Soret:red for chlorophylls; % band III:II for carotenoids), add additional diagnostic information which has been extensively listed (e.g., Jeffrey et al., 1997b; Britton, 1995). Despite these diagnostic characteristics, the absorption spectrum of a pigment mostly gives information about the carotenoid chromophore or chlorophyll macrocyclic structure but little or no information on substituent groups belonging to the carotenoids and specific changes in the functional side groups of chlorophylls. Thus, where pigment degradation has occurred there is often a shift in retention time but little change in the absorbance spectrum. Further information about the substituent and functional side groups is required therefore to identify the degraded pigments and facilitate their use as valuable biomarkers. In the case of degraded pigments, once identified they can be aggregated to give information on the relative abundance of the parent pigment, or retained in the analyses as a paleolimnological indicator of the transformation processes themselves, and hence, conditions at the time of deposition (e.g., Eckardt et al., 1991a).

Mass spectrometry (MS) is ideally suited to identify the nature of molecular modifications because it confirms the molecular mass of the pigments, their characteristic fragmentation patterns, and the presence of key functional groups or co-eluting, but un-coloured contaminants. In cases where mass spectroscopic data (mass ion, fragments) are unknown, MS can be combined with NMR technologies to indicate clearly the presence of possible a novel pigment. General descriptions of the use of NMR for pigment analyses are given in Englert (1995).

Sample preparation for mass spectrometry: Investigators using MS must take precau-tions to minimise pigment degradation during extraction and preparation of samples. As detailed above, limited exposure to heat, light and oxygen are all essential to avoid pigment derivitization and introduction of artefacts to MS analyses.

The mass of dry sediment required for successful MS varies considerably between studies. In sediments consisting of highly pigmented microbial mats 500 mg wet mass of sediment can be sufficient, while in sediments consisting of a high proportion of in-organic matter >5 g may be required. Regardless, gross extracts should display strong visible colour, normally green-brown or yellow-red. Finally, we recommend that glass containers be used for all stages of pigment analysis as polythene plastics can introduce plasticisers during extraction which interfere with MS determination of pigment mass and fragmentation patterns.

Most MS techniques require pure isolates for quantifiable analysis of pigment character-istics. Most commonly, target pigments are collected from the HPLC eluent immediately after the peak has passed through the detector (drop-catch method) either manually or using a fraction collector. Here the detector is monitored in real-time to identify when the target pigment has eluted from the HPLC column. Typically, several HPLC runs will be required to collect sufficient quantities of isolated pigment for MS analysis. If necessary, isolated extracts can be further purified by a combination of RP-HPLC, or normal (straight) phase HPLC and post-chromatographic concentration using solid-phase extraction. In solid-phase extraction, pigment solutions are diluted with water (intermediate and high polarity molecules) or methanol (native Chls, carotenes or single ketone functions) and passed through a C_{18} cartridge previously conditioned with acetone to activate the C_{18} bed (stationary phase). The cartridge bed is then blown dry with nitrogen gas and the pigment eluted using methanol. In cases where the hydrophobic pigments are incompletely eluted by methanol, diethyl ether can be used. Prior to MS analysis, isolated extracts should be concentrated using N_2 gas or centrifugal evaporation using normal precautions against pigment loss. Such preliminary purifications simplify the detection of individual pigment molecular ions under MS and ensure that the sample is continually refreshed at the surface of the liquid matrix, resulting in a greater signal peak half-width.

Mass spectrometric methods: *Electron Ionisation* (EI) is the 'traditional' method of MS involving bombardment of a sample in gas phase with a beam of electrons. EI has been applied to the analysis and identification of pigments for several decades. It is well suited to the analysis of carotenoids because it produces strong molecular ions ($[M]^{+}$) that can be used to assign molecular weights to isolates. Carotenoid samples of *ca.* 100–600 ng are typically introduced into the electron beam on the glass tip of a direct insertion probe at room temperature, and then heated to a maximum of 180–200 °C. This causes the carotenoid to be vapourized at the very low pressures that are maintained in the system (Enzell & Back, 1995). Relative intensities of the ions in a spectrum can be manipulated by changing the temperature and ionising voltage, with lower voltages favouring the production of a prominent molecular ion with reduced fragmentation (Enzell & Back, 1995). Accurate mass measurements (to $+/-4$ ppm accuracy) can be undertaken at high resolution by peak matching using perfluorokerosene (PFK) as an internal reference. EI spectra can also be acquired using a desorption chemical ionisation probe to minimise thermal degradation of the sample. The conditions for this are as above, except for a source temperature increasing from $50\,°C$ to $500\,°C$ $10\,°C\,s^{-1}$.

Carotenoid mass spectra from EI will usually show clear molecular ion clusters from which molecular weights can be estimated. Carotenoids also produce many characteristic fragment ions that can be used for structural confirmation of the carbon skeleton of the molecule and the presence of particular functional groups. For example, β, β-carotene produces a strong molecular ion at m/z 536 under EI conditions (Fig. 3), and a characteristic fragment ion at m/z 444 (Hodgson et al., 1997). Similarly, alloxanthin produces a molecular ion at m/z 564, a fragmentation product at m/z 441 ($[M-123]^+$) arising from the cleavage of the C-6,7 bond, and a derivative at m/z 426, indicating a 3-hydroxy-ε-ring. Characteristic fragment ions of the major algal carotenoids and Chls are presented in Young & Britton (1993) and Enzell & Back (1995).

Liquid Secondary Ion MS (LSIMS) method is essentially identical to *Fast Atom Bombardment MS* (FAB MS), differing only in the use of a beam of heavy ions (e.g., Cs) rather than a beam of heavy atoms (typically Xe) as in FAB MS. Both of these soft ionization methods have significantly improved the capability of MS for the analysis of thermally labile chlorophylls (e.g., Mukaida & Nishikawa, 1990; van Breemen et al., 1991) and carotenoids (e.g., van Breemen et al., 1993; van Breemen, 1996). These techniques use identical probes and an atom beam to ionize compounds gently from the surface of a liquid matrix, making it possible to obtain spectra of large, non-volatile organic molecules (Enzell & Back, 1995).

Hodgson et al. (1997) acquired LSIMS spectra using a static LSIMS probe with a stainless steel sample well and samples dissolved in m-nitrobenzyl alcohol (mnba) as the liquid matrix (Fig. 4). With low levels of pigment available for analysis, the matrix was loaded as a very thin film at the target point of the LSIMS probe to maximise the sample to matrix ratio. Successive $1\,\mu l$ aliquots (up to a total of $10\,\mu l$) were then loaded, with some solvent allowed to evaporate between addition of each aliquot. The primary beam was $10\,keV$ cesium ions, and the ion source accelerating voltage was $5.3\,kV$. Low resolution spectra were acquired from m/z 1200 to 100 at 1sec/decade. Accurate mass measurements (±6 ppm accuracy) were taken by peak matching at 8000 resolution, using polyethylene glycol (PEG 600) and mnba as internal references. An unheated probe was used to minimize thermal degradation of pigments.

Where low concentrations of pigment are analysed, both Chls and bacteriochlorophylls (BChl) produce pigment molecular ions immediately after the LSIMS probe is inserted. In these cases, sample peak half-widths (width at 50% maximum signal intensity) are as short as 3–10 seconds and contrast sharply with signals from matrix and background ions that can last up to 2 min. Using this procedure, the molecular ion for Chl a has been observed at m/z 892 (Fig. 4), together with a significant fragment ion at m/z 614 corresponding to the loss of the phytol group (Hodgson et al., 1997). Pheophorbide a was also identified (Chl a $-Mg^{2+}-$phytol) and a strong molecular ion at m/z 812 has been observed for pyropheophytin a (Chl a $-Mg^{2+}$ $-$ COOCH$_3$; Fig. 2). The presence in sediments of several

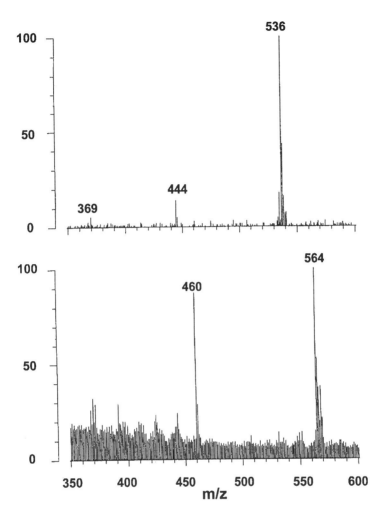

Figure 3. EI mass spectrum of $\beta\beta$-carotene (top) and LSIMS mass spectrum of alloxanthin (bottom). Signal intensity is plotted as a function of mass/charge ratios (*m/z*).

bacteriopheophytin *c* isomers and derivatives BChl *c* have also been identified from the pigments' mass spectra.

The mass spectra of both carotenoids and chlorophylls under LSIMS (and FAB MS) conditions are unusual in that they are represented as 'odd electron' molecular ions ($[M]^+$.) rather than as protonated molecular ions ($[M + H]^+$). The latter are typical of most organic compounds, but are either absent from carotenoid analyses (Vetter & Meister, 1985; Caccamese & Garozzo, 1990; van Breemen et al., 1993) or only weakly present when mnba is used as the liquid matrix (van Breemen et al., 1991). M-nitrobenzyl alcohol generally provides a better signal to noise than glycerol for the relatively hydrophobic carotenoids and chlorophylls in static analytic mode (Hodgson et al., 1997). In contrast,

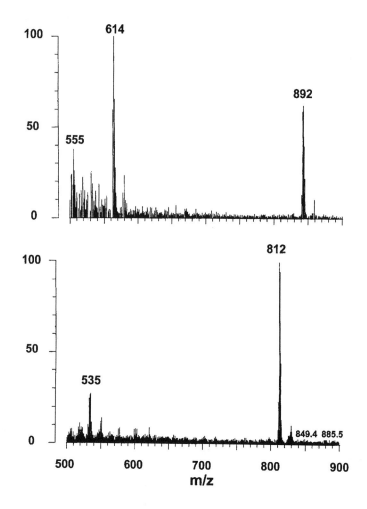

Figure 4. LSIMS mass spectra of chlorophyll *a* (top) and pyropheophytin *a* (bottom). Signal intensity is plotted as a function of mass/charge ratios (*m/z*).

other studies have shown that good $[M^+H]^+$ signals can be achieved in continuous flow mode with glycerol as the matrix (van Breemen et al., 1991). While the $[M + H]^+$ ion for Chls predominates over the $[M]^{\overset{+}{\cdot}}$ ion using mnba as the liquid matrix, when corrections are made for the contributions from heavy isotopes, the $[M]^{\overset{+}{\cdot}}$ ion is more abundant. Similarly, small amounts of $[M - H]^+$ are observed together with complex clusters of ions in the region of the Chl molecular weight due to the normal presence of heavy stable isotopes of C and N. Further matrix combinations which promote changes in the proportion of the radical cation $[M]^{\overset{+}{\cdot}}$ against $[M + H]^+$ are discussed in Keely & Maxwell (1990).

Tandem MS (MS-MS) has also been used in conjunction with HPLC for the identification of carotenoids and their derivatives (Van Breemen et al., 1993, 1995; van Breemen, 1996). In tandem MS, a precursor ion is mass-selected by an initial MS unit, typically fragmented by collision-induced dissociation (also called collisionally-activated dissociation), and analyzed for product ions by a second MS in series. Similar analyses can be conducted with a single mass analyser used in sequential mode. Tandem MS provides key structural information by establishing the relationships between precursor ions and their fragmentation products (Borman, 1998). In particular, the combination of the newer soft ionization methods and collision-induced dissociation gives tandem MS unprecedented analytical power. For example, in cases where FAB spectra show extensive fragmentation of Chls, MS-MS has been used to quantify the fragmentation behaviour and chemical relation among product ions. This approach has been especially useful for identification of chlorins in recent lake sediments (Keely & Maxwell, 1990).

Continuous flow HPLC-MS is one of the most powerful methods for examining complex mixtures of pigments in lake sediments and is achieved through the direct coupling of HPLC to MS or tandem MS using an HPLC column installed between the injection loop and a continuous flow probe. With this on-line approach, pigments that are only partially separated by the interfaced HPLC can still produce clean electron impact (EI) spectra. When fast atom bombardment MS (FAB MS) or liquid secondary ion MS (LSIMS) is employed, a continuous flow probe is used to inject the HPLC effluent into a stream of solvent containing some matrix, and then introduce the sample into the downstream MS (Caprioli et al., 1986; van Breemen, 1996). Continuous flow methods have the advantage of improving the signal to noise ratios for poorly-resolved samples, elucidating small differences in their mass spectrograms and eliminating inconsistencies involved in the manual collection of HPLC (coelution, poor resolution). The development of continuous flow methods is discussed in van Breemen (1996).

On-line coupling of HPLC with MS has been refined through the use of interfaces that use the mobile phase to effect ionization. For example, negative ion thermospray (TS) interfaces (Eckardt et al., 1991b) which link a quaternary HPLC system and quadrupole MS via a TS interface have been used to identify pigments in lacustrine sediments from Priest Pot (Cumbria, UK). The improved resolution afforded by this approach permitted the identification of biomarkers indicative of the photosynthetic bacteria, and an unexpected suite of chlorins which revealed variations in the physical structure of the ecosystem. An atmospheric Pressure Chemical Ionisation (APCI) interface has also been used successfully to quantify the presence of algal-derived chlorophyll transformation products in invertebrate and marine samples (Harris et al., 1995).

More recent developments for the study of large biomolecules include electrospray ionization MS (ESI MS) and matrix-assisted laser desorption/ionization MS (MALDI MS). In ESI MS, highly charged droplets dispersed from a capillary in an electric field and are evaporated before the resulting ions are drawn into an MS inlet. In MALDI MS, sample molecules are laser-desorbed from a solid or liquid matrix containing a highly UV-absorbing substance (Borman, 1998). Finally, detailed studies of pigment structures and transformation pathways have recently been carried out using surface enhanced resonance Raman spectroscopy coupled with NMR (Wooley et al., 1998). Although these methods have not yet been applied to studying pigments in lake sediments, we anticipate that they will allow valuable investigations of post-depositional degradation processes and diagenetic patterns.

Applications: HPLC-MS has been used in several limnological studies to identify pigments in the water column (Eckardt et al., 1991a), their degradation products (Eckardt et al., 1991b) and pigment transformations after incorporation into the sediments (Naylor & Keely, 1998). For example, Villanueva et al. (1993) used HPLC-MS to quantify pigment degradation and to determine how accurately sediments recorded historical variations in Chl and BChl content of suspended particulate matter in of Lake Ciso (Catalonia, Spain) during periods of variable lake stratification. Here bacterial pigments were better preserved than were their algal counterparts reflecting the occurrence of prokaryotic populations adjacent to the anoxic hypolimnion, while algae were restricted to the aerobic epilimnion. HPLC-MS has also been used to demonstrate that carotenoid preservation was good despite extensive biodegradation of the mat filaments from which they originate (Villanueva et al., 1994). Similarly, HPLC-MS was used recently to unravel the gradual evolution of meromixis in a riverine lake in Tasmania, Australia. Here signatures of transformed pigments belonging to the green anaerobic phototrophic sulphur bacteria (BChl *c* derivatives, isorenieratene) were used to identify past presence of chemocline micro-habitats where the overlap of light and anoxic water allowed bacteria to proliferate (Hodgson et al., 1997). The identification of pigments throughout an 8000 year old sediment core permitted reconstruction of past stratification and anoxia, an assessment of the sensitivity of microbiological communities to changes in the physical and chemical environment, and recommendations for a management strategy to prevent the further decay of meromixis by modifying hydro-electric operations in the upstream catchment (Hodgson & Tyler, 1996).

Conclusions

In principle, fossil pigments can be used in any paleoecological application in which historical changes in lake production or primary producer composition are a key response (e.g., eutrophication, acidification, climate change, food-web interactions, human impacts). Recent advances in analytical techniques (HPLC, MS, diode-array detection, low-cost automation) and a more sophisticated understanding of the pigment biogeochemistry have lead to significant advances in the potential for fossil pigments to be broadly used in paleolimnology. However, this potential is as yet unfulfilled. We hope that this chapter provides the basic information needed for investigators to evaluate the usefulness of biochemical fossils and that it stimulates incorporation of this approach into a broader scientific agenda.

Summary

Fossil pigments often preserve in lake sediments long after the morphological remains of most algae and bacteria are lost. In principle, analyses of sedimentary carotenoids, chlorophylls, their derivatives and other lipid-soluble pigments can be used to reconstruct historical changes in primary-producer community abundance and composition, so long as biomarkers are accurately isolated, identified and quantified. This chapter summarizes a series of practical techniques in order to familarize investigators with the potential and pitfalls inherent in fossil pigment analyses. First we describe the common uses of sedimentary pigments in paleolimnology and summarize knowledge of pigment biogeochemistry and taphonomy, especially as concerns water-column processes. Second we review a series of practical procedures to collect, isolate and quantify pigments, particularly by high performance liquid chromatography. We conclude with a summary of recent advances in pigment identification using various mass spectrometric techniques.

Acknowledgements

We thank Dr. Brendan Keely for advice on MS methods, and Dr. Pierro Guilizzoni and the editors for scientific reviews. Manuscript preparation supported by NSERC Canada and the British Antarctic Survey.

References

Belcher, J. H. & G. E. Fogg, 1964. Chlorophyll derivatives in the sediments of two English lakes. In Miyaka, Y. & T. Koyama (eds.) Recent Researches in the Field of Hydrosphere, Atmosphere and Nuclear Geochemistry. Maruzen co. Tokyo: 39–48.

Bernhard, K., 1995. Column chromatography. In Briton, G., S. Liaaen-Jensen & H. Pfander (eds.) Carotenoids. Vol. 1A. Isolation and Analyis. Birkhäuser, Boston: 117–130.

Bernhard, K. & M. Grosjean, 1995. Infared spectroscopy. In Britton, G., S. Liaaen-Jensen & H. Pfander (eds.) Caroteniods: Volume 1B: Spectroscopy. Birkhäuser Verlag, Boston: 117–134.

Bianchi, T. S., S. Findlay & R. Dawson, 1993. Organic-matter sources in the water column and sediments of the Hudson River estuary—the use of plant pigments as tracers. Est. Coast. Shelf Sci. 36: 359–376.

Bjørnland, T., 1997. UV/Visible spectroscopy of carotenoids. In Jeffrey, S. W., R. F. C. Mantoura & S. W. Wright (eds.) Phytoplankton Pigments in Oceanography. UNESCO Publishing, Paris: 578–594.

Borman, S., 1998. Chemistry crystallizes into modern science. Chemical & Engineering News, 12: 39–75.

Britton, G., 1995. In Britton, G., S. Liaaen-Jensen & H. Pfander (eds.) Caroteniods: Volume 1B: Spectroscopy. Birkhäuser Verlag, Boston: 13–62.

Britton, G., S. Liaaen-Jensen & H. Pfander, 1995a. Carotenoids: Volume 1A: Isolation and Analysis. Birkhäuser Verlag, Boston, 328 pp.

Britton, G., S. Liaaen-Jensen & H. Pfander, 1995b. Carotenoids: Volume 1B: Spectroscopy. Birkhäuser Verlag, Boston, 360 pp.

Brown, S. R., 1968. Bacterial carotenoids from freshwater sediments. Limnol. Oceanogr. 13: 233–241.

Brown, S. R., 1969. Paleolimnological evidence from fossil pigments. Mitt. Internat. Verein. Limnol. 17: 95–103.

Brown, S. R. & B. Colman, 1963. Oscillaxanthin in lake sediments. Limnol. Oceanogr. 8: 352–353.

Brown, S. R., R. J. Daley & R. N. McNeely, 1977. Composition and stratigraphy of fossil phorbin derivatives of Little Round Lake. Ontario. Limnol. Oceanogr. 22: 336–348.

Brown, S. R., H. J. McIntosh & J. P. Smol, 1984. Recent paleolimnology of a meromictic lake: Fossil pigments of photosynthetic bacteria. Int. Ver. Theor. Angew. Limnol. Verh. 22: 1357–1360.

Caccamese, S. & D. Garozzo, 1990. Odd-electron molecular ion and loss of toluene in fast atom bombardment mass spectra of some carotenoids. Org. Mass Spectrom. 25(3): 137–140.

Caprioli, R. M., T. Fan & J. S. Cottrel, 1986. Continuous flow sample probe for fast atom bombardment mass spectrometry. Anal. Chem. 58: 2949–2954.

Carpenter, S. R. & P. R. Leavitt, 1991. Temporal variation in a paleolimnological record arising from a trophic cascade. Ecology 72: 277–285.

Carpenter, S., M. Elser & J. Elser, 1986. Chlorophyll production, degradation and sedimentation: Implications for palaeolimnology. Limnol. Oceanogr. 31: 112–124.

Carpenter, S. R., P. R. Leavitt, J. J. Elser & M. M. Elser, 1988. Chlorophyll budgets: Response to food web manipulation. Biogeochemistry 6: 79–90.

Cohen, A. S., 2002. Paleolimnology: History and Evolution of Lake Systems. Oxford University Press, Oxford, 350 pp.

Cuddington, K. & P. R. Leavitt, 1999. An individual-based model of pigment flux in lakes: Implications for organic biogeochemistry and paleoecology. Can. J. Fish. Aquat. Sci. 56: 1964–1977.

Daley, R. J., S. R. Brown & R. N. McNeely, 1977. Chromatographic and SCDP measurements of fossil phorbins and the postglacial history of Little Round Lake. Ontario. Limnol. Oceanogr. 22: 349–360.

Damste, J. S. S. & M. P. Koopmans, 1997. The fate of carotenoids in sediments: a Review. Pure Appl. Chem. 69: 2067–2074.

Davies, B. H., 1976. Carotenoids. In Goodwin, T. W. (ed.) Chemistry and Biochemistry of Plant Pigments. Volume I. Academic Press, N.Y., 870 pp.

Descy, J. P., T. M. Frost & J. P. Hurley, 1999. Assessment of grazing by the freshwater copepod *Diatpomus minutus* using carotenoid pigments: a Caution. J. Plankton Res. 21: 127–145.

Eckardt, C. B., B. J. Keely & J. R. Maxwell, 1991a. Identification of chlorophyll transformation products in a lake sediment by combined liquid chromatography-mass spectrometry. J. Chromatog. 557: 271–278.

Eckardt, C. B., G. E. S. Pearce, B. J. Keely, G. Kowalewska, R. Jaffé & J. R. Maxwell, 1991b. A widespread chlorophyll transformation pathway in the aquatic environment. Adv. Org. Geochem. 19: 217–227.

Englert, G., 1995. NMR spectroscopy. In Britton, G., S. Liaaen-Jensen & H. Pfander (eds.) Caroteniods: Volume 1b, Spectroscopy. Birkhäuser Verlag, Boston: 147–260.

Enzell, C. R. & S. Back, 1995. Mass Spectrometry. In Britton, G., S. Liaaen-Jensen & H. Pfander (eds.) Caroteniods: Volume 1b, Spectroscopy. Birkhäuser Verlag, Boston: 261–320.

Eugster, C. H., 1995. Chemical derivatization: Microscale tests for the presence of common functional groups in carotenoids. In Briton, G., S. Liaaen-Jensen & H. Pfander (eds.) Carotenoids. Vol. 1A. Isolation and Analyis. Birkhäuser, Boston: 71–80.

Fogg, G. E. & J. H. Belcher, 1961. Pigments from the bottom deposits of an English lake. New Phytol. 60: 129–138.

Foppen, F. H., 1971. Tables for the identification of carotenoid pigments. Chromatog. Rev. 14: 133–298.

Fox, D. L., 1944. Biochemical fossils. Science 100: 111–113.

Fox, D. L., D. M. Updegraff & G. D. Novelli, 1944. Carotenoid pigments in the ocean floor. Arch. Biochem. 5: 1–23.

Goodwin, T. W., 1980a. The Biochemistry of the Carotenoids. Vol. 1. Plants. Chapman and Hall, N.Y., 377 pp.

Goodwin, T. W., 1980b. The Biochemistry of the Carotenoids. Vol. 2. Animals. Chapman and Hall, N.Y., 224 pp.

Gorham, E. & J. E. Sanger, 1975. Fossil pigments in Minnesota lake sediments and their bearing upon the balance between terrestrial and aquatic inputs to sedimentary organic matter. Vehr. Inernat. Verein. Limnol. 19: 2267–2273.

Griffiths, M., 1978. Specific blue-green algal carotenoids in the sediments of Esthwaite Water. Limnol. Oceanogr. 23: 777–784.

Griffiths, M., P. S. Perrot & W. T. Edmonson, 1969. Oscillaxanthin in the sediments of Lake Washington. Limnol. Oceanogr. 14: 317–326.

Guilizzoni, P. & A. Lami, 1992. Historical records of changes in the chemistry and biology of Italian lakes. Mem. Ist. ital. Idrobiol. 50: 61–77.

Guilizzoni, P., A. Lami & A. Marchetto, 1992. Plant pigment ratios from lake-sediments as indicators of recent acidification in alpine lakes. Limnol. Oceanogr. 37: 1565–1569.

Hall, R. I., P. R. Leavitt, R. Quinlan, A. S. Dixit & J. P. Smol, 1999. Effects of agriculture, urbanization and climate on water quality in the northern Great Plains. Limnol. Oceanogr. 43: 739–756.

Harris, P. G., J. F. Carter, M. Head, R. P. Harris, G. Eglinton & J. R. Maxwell, 1995. Identification of chlorophyll transformation products in zooplankton fecal pellets and marine sediment extracts by liquid-chromatography mass-spectrometry atmospheric-pressure chemical-ionization. Rap. Comm. Mass Spec. 9: 1177–1183.

Hendry, G. F., J. D. Houghton & S. R. Brown. 1987. The degradation of chlorophyll—a biological enigma. New Phytol. 107: 255–302.

Hertzberg, S., S. Liaaen-Jensen & H. W. Siegelman, 1971. The carotenoids of blue-green algae. Phytochemistry 10: 3121–3127.

Hodgson, D. A. & P. A. Tyler, 1996. The impact of a hydro-electric dam on the stability of meromictic lakes in south west Tasmania. Australia. Arch. Hydrobiol. 137: 310–323.

Hodgson, D. A., S. W. Wright & N. Davies, 1997. Mass spectrometry and reverse phase HPLC methods for the identification of degraded fossil pigments in lake sediments and their application in palaeolimnology. J. Paleolimnol. 18: 335–350.

Hodgson, D. A., S. W Wright, P. A Tyler & N. Davies, 1998. Analysis of fossil pigments from algae and bacteria in meromictic Lake Fidler, Tasmania, and its application to lake management. J. Paleolimnol. 19: 1–22.

Hurley, J. P. & D. E. Armstrong, 1990. Fluxes and transformations of aquatic pigments in Lake Mendota, Wisconsin. Limnol. Oceanogr. 35: 384–398.

Hurley, J. P. & P. J. Garrison, 1993. Composition and sedimentation of aquatic pigments associated with deep plankton in lakes. Can. J. Fish. Aquat. Sci. 50: 2713–2722.

Hurley, J. P. & C. J. Watras, 1991. Identification of bacteriochlorophylls in lakes via reverse-phase HPLC. Limnol. Oceanogr. 36: 307–315.

Isler, O. (ed.), 1971. Carotenoids. Birkhauser-Basel, 932 pp.

Jeffrey, S. W., 1967. Quantitative thin layer chromatography of chlorophylls and carotenoids from marine algae. Biochim. Biophys. Acta 162: 271–285.

Jeffrey, S. W., 1997. Chlorophyll and carotenoid extinction coefficients. In Jeffrey, S. W., R. F. C. Mantoura & S. W. Wright (eds.) Phytoplankton Pigments in Oceanography. UNESCO Publishing, Paris: 595–596.

Jeffrey, S. W., R. F. C. Mantoura & S. W. Wright, 1997a. Phytoplankton pigments in oceanography: guidelines to modern methods. UNESCO Publishing, Paris, 661 pp.

Jeffrey, S. W., R. F. C. Mantoura & T. Bjørnland, 1997b. Data for the identification of 47 key phytoplankton pigments. In Jeffrey, S. W., R. F. C. Mantoura & S. W. Wright (eds.) Phytoplankton Pigments in Oceanography. UNESCO Publishing, Paris: 447–559.

Jeffrey, S. W., S. W. Wright & M. Zapata, 1999. Recent advances in HPLC pigment analysis of phytoplankton. Mar. Freshwater Res. 50: 879–896.

Jensen, A. & S. Liaaen-Jensen, 1968. Quantitative paper chromatography of carotenoids. Acta. Chem. Scanda. 13: 1863–1868.

Karrer, P. & E. Jucker, 1950. Carotenoids. Elsevier Publishing, NY, 384 pp.

Keely, B. J. & J. R. Maxwell, 1990. Fast atom bombardment and tandem mass spectrometric studies of some functionalised tetrapyrroles derived from chlorophylls a and b. Energy and Fuels 4: 737–741.

Koopmans, M. P., J. Koster, H. M. E. vanKannPeters, F. Kenig, S. Schouten, W. A. Hartgers, J. W. deLeeuw & J. S. S. Damste, 1996. Diagenetic and catagenetic products of isorenieratene: Molecular indicators for photic zone anoxia. Geochim. Cosmochim. Acta 60: 4467–4496.

Kost, H.-P., 1988. Handbook of Chromatography. Plant pigments. Vol. I Fat-soluble pigments. CRC Press, 328 pp.

Krinsky, N. I., 1963. A relationship between partition coefficients of carotenoids and their functional groups. Anal. Biochem. 6: 293–302.

Lami, A., F. Niessen, P. Guilizzoni, J. Masaferro & C. A. Belis, 1994. Paleolimnological studies of the eutrophication of volcanic Lake Albano (central Italy). J. Paleolimnol. 10: 181–197.

Leavitt, P. R., 1993. A review of factors that regulate carotenoid and chlorophyll deposition and fossil pigment abundance. J. Paleolimnol. 9: 109–127.

Leavitt, P. R. & S. R. Carpenter, 1989. Effects of sediment mixing and benthic algal production on fossil pigment stratigraphies. J. Paleolimnol. 2: 147–158.

Leavitt, P. R. & S. R. Carpenter, 1990a. Aphotic pigment degradation in the hypolimnion: Implications for sedimentation studies and paleolimnology. Limnol. Oceanogr. 35: 520–534.

Leavitt, P. R. & S. R. Carpenter, 1990b. Regulation of pigment sedimentation by herbivory and photo-oxidation. Can. J. Fish. Aquat. Sci. 47: 1166–1176.

Leavitt, P. R. & D. L. Findlay, 1994. Comparison of fossil pigments with 20 years of phytoplankton data from eutrophic Lake 227, Experimental Lakes Area, Ontario. Can. J. Fish. Aquat. Sci. 51: 2286–2299.

Leavitt, P. R., S. R. Carpenter & J. F. Kitchell, 1989. Whole-lake experiments: The annual record of fossil pigments and zooplankton. Limnol. Oceanogr. 34: 700–717.

Leavitt, P. R., P. R. Sanford, S. R. Carpenter & J. F. Kitchell, 1994a. An annual fossil record of production, planktivory and piscivory during whole-lake experiments. J. Paleolimnol. 11: 133–149.

Leavitt, P. R., D. E. Schindler, A. J. Paul, A. K. Hardie & D. W. Schindler, 1994b. Fossil pigment records of phytoplankton in trout-stocked alpine lakes. Can. J. Fish. Aquat. Sci. 51: 2411–2423.

Leavitt, P. R., B. J. Hann, J. P. Smol, B. A. Zeeb, C. C. Christie, B. Wolfe & H. J. Kling, 1994c. Paleolimnological analysis of whole-lake experiments: An overview of results from Experimental Lakes Area Lake 227. Can. J. Fish. Aquat. Sci. 51: 2322–2332.

Leavitt, P. R., R. D. Vinebrooke, D. B. Donald, J. P. Smol & D. W. Schindler, 1997. Past ultraviolet radiation environments in lakes derived from fossil pigments. Nature 388: 457–459.

Leavitt, P. R., D. L. Findlay, R. I. Hall & J. P. Smol, 1999. Algal responses to dissolved organic carbon loss and pH decline during whole-lake acidification: Evidence from paleolimnology. Limnol. Oceanogr. 44: 757–773.

Liaaen-Jensen, S., 1965. Studies on allylic oxidation of carotenoids. Acta. Chem. Scanda. 19: 1166–1174.

Liaaen-Jensen, S., 1971. Isolation, reactions. In Isler, O. (ed.) Caroteniods. Birkhauser-Verlag Bassel: 61–179.

Liaaen-Jensen, S., 1979. Carotenoids: a Chemosystematic approach. Pure Appl. Chem. 51: 661–675.

Liaaen-Jensen, S. & S. Hertzberg, 1966. Selective preparations of the lutein monomethyl ethers. Acta. Chem. Scanda. 20: 1703–1709.

Liaaen-Jensen, S. & A. Jensen, 1971. Quantitative determination of carotenoids in photosynthetic tissues. In San Petro, A. (ed.) Methods in Enzymology. Vol. XXIII. Photosynthesis. Academic Press, N.Y.: 586–602.

Louda, J. W., J. Li, L. Liu, M. N. Winfree & E. W. Baker, 1998. Chlorophyll a degradation during cellular senescence and death. Org. Geochem. 29: 1233–1251.

Mantoura, R. F. C. & C. A. Llewellyn, 1983. The rapid determination of algal chlorophyll and carotenoid pigments and their breakdown products in natural waters by reversed-phase high-performance liquid chromatography. Anal. Chim. Acta 151: 297–314.

Mantoura, R. F. C. & D. J. Repeta, 1997. Calibration methods for HPLC. In Jeffrey, S. W., R. F. C. Mantoura & S. W. Wright (eds.) Phytoplankton Pigments in Oceanography. UNESCO Publishing, Paris: 383–406.

Mantoura, R. F. C., S. W. Wright, S. W. Jeffrey, R. G. Barlow & D. E. Cummings, 1997a. Filtration and storage of pigments from microalgae. In Jeffrey, S. W., R. F. C. Mantoura & S. W. Wright (eds.) Phytoplankton Pigments in Oceanography. UNESCO Publishing, Paris: 283–305.

Mantoura, R. F. C., S. W. Jeffrey, C. A. Llewelln, H. Claustre & C. E. Morales, 1997b. Comparison between spectrophotometric, fluorometric and HPLC methods for chlorophyll analysis. In Jeffrey, S. W., R. F. C. Mantoura & S. W. Wright (eds.) Phytoplankton Pigments in Oceanography. UNESCO Publishing, Paris: 361–380.

McElroy-Etheridge, S. L. & G. B. McManus, 1999. Food type and concentration affect chlorophyll and carotenoid destruction during copepod feeding. Limnol. Oceanogr. 44: 2005–2011.

Millie, D. F., H. W. Pearl & J. P. Hurley, 1993. Microalgal pigment assessments using high-performance liquid chromatography: a Synopsis of organismal and ecological applications. Can. J. Fish. Aquat. Sci. 50: 2513–2527.

Moss, G. P. & B. C. L. Weedon, 1976. Chemistry of carotenoids. In Goodwin, T. W. (ed.) Chemistry and Biochemistry of Plant Pigments. Academic Press, N.Y.: 149–224.

Mukaida, N. & Y. Nishikawa, 1990. Chromatographic separation of protochlorophylls and their structural analysis by fast atom bombardment mass spectrometry. Nippon Kagaku Kaishi 11: 1244–1249.

Murphy, M. T. J., A. McCormick & G. Eglington, 1967. Perhydro-β-carotene in Green River Shale. Science 157: 1040–1042.

Naylor, C. C. & B. J. Keely, 1998. Sedimentary purpurins: Oxidative transformation products of chlorophylls. Org. Geochem. 28: 417–422.

Ostrovsky, I. & Y. Z. Yacobi, 1999. Organic matter and pigments in surface sediments: Possible mechanisms of their horizontal distributions in a stratified lake. Can. J. Fish. Aquat. Sci. 56: 1001–1010.

Overmann, J., G. Sandmann, K. J. Hall & T. G. Northcote, 1993. Fossil carotenoids and paleolimnology of meromictic Mahoney Lake, British Columbia, Canada. Aquat. Sci. 55: 31–39.

Pfander, H. & R. Riesen, 1995. High-performance liquid chromatography. In Briton, G., S. Liaaen-Jensen & H. Pfander (eds.) Carotenoids. Vol. 1A. Isolation and Analyis. Birkhäuser, Boston: 145–190.

Poister, D., D. E. Armstrong & J. P. Hurley, 1999. Influences of grazing on temporal patterns of algal pigments in suspended and sedimenting algae in a north temperate lake. Can. J. Fish. Aquat. Sci. 56: 60–69.

Repeta, D. J. & T. Bjørnland, 1997. Preparation of carotenoid standards. In Jeffrey, S. W., R. F. C. Mantoura & S. W. Wright (eds.) Phytoplankton Pigments in Oceanography. UNESCO Publishing, Paris: 239–260.

Sanger, J. E., 1988. Fossil pigments in paleoecology and paleolimnology. Palaeogeog. Palaeoclim. Palaeoecol. 62: 343–359.

Sanger, J. E. & G. H. Crowl, 1979. Fossil pigments as a guide to the postglacial history of Kircchner Marsh, Minnesota. Limnol. Oceanogr. 17: 840–854.

Sanger, J. E. & E. Gorham, 1970. The diversity of pigments in lake sediments and it's ecological significance. Limnol. Oceanog. 15: 59–69.

Schiedt, K., 1995. Thin layer chromatography. In Briton, G., S. Liaaen-Jensen & H. Pfander (eds.) Carotenoids. Vol. 1A. Isolation and Analyis. Birkhäuser, Boston: 131–144.

Schiedt, K. & S. Liaaen-Jensen, 1995. Isolation and analysis. In Briton, G., S. Liaaen-Jensen & H. Pfander (eds.) Carotenoids. Vol. 1A. Isolation and Analyis. Birkhäuser, Boston: 81–108.

Scheer, H., 1991. Chlorophylls. CRC Press, Boston, 950 pp.

Šesták, Z., 1980. Paper chromatography of chloroplast pigments (chlorophylls and carotenoids)—Part 3. Photosynthetica 14: 239–270.

Steenbergen, C. L. M., H. J. Korthals & E. G. Dobrynin, 1994. Algal and bacterial pigments in non-laminated lacustrine sediment: Studies of their sedimentation, degradation and stratigraphy. FEMS Microbiol. Ecol. 13: 335–352.

Steinman, A. D., K. E. Havens, J. W. Louda, N. M. Winfree & E. W. Baker, 1998. Characterization of the photoautotrophic algal and bacterial communities in a large, shallow, subtropical lake using HPLC-PDA based pigment analysis. Can. J. Fish. Aquat. Sci. 55: 206–219.

Swain, E. B., 1985. Measurement and interpretation of sedimentary pigments. Freshwat. Biol. 15: 53–75.

Tett, P., 1982. The Loch Eil project: Planktonic pigments in sediments from Loch Eil and the Firth of Lorne. J. Exp. Mar. Biol. Ecol. 56: 111–114.

Vallentyne, J. R., 1954. Biochemical limnology. Science 119: 605–606.

Vallentyne, J. R., 1956. Epiphasic carotenoids in postglacial lake sediments. Limnol. Oceanogr. 1: 252–262.

Vallentyne, J. R., 1957. Carotenoids in 20,000-year old sediment from Searles Lake. California. Arch. Biochem. Biophys. 70: 29–34.

van Breemen, R. B., F. L. Conjura & S. J. Schwartz, 1991. High performance liquid chromatography-continuous-flow fast atom bombardment mass spectrometry of chlorophyll derivatives. J. Chromatogr. 542: 373–383.

van Breemen, R. B., H. H. Schmitz & S. J. Schwartz, 1993. Continuous flow fast atom bombardment liquid chromatography/mass spectrometry of carotenoids. Anal. Chem. 65(8): 965–969.

van Breemen, R. B., H. H. Schmitz & S. J. Schwartz, 1995. Fast atom bombardment mass spectrometry of carotenoids. J. Agric. Food Chem. 42(2): 384–389.

van Breemen, R. B., 1996. Innovations in carotenoid analysis. Anal. Chem. 68: 299A–304A.

Vetter, M. & W. Meister, 1985. Fast atom bombardment mass spectrum of β-carotene. Org. Mass Spectrom. 20: 266–267.

Villanueva, J., J. O. Grimalt, R. Dewit, B. J. Keely & J. R. Maxwell, 1993. Sources and transformations of chlorophylls and carotenoids in a monomictic sulphate-rich karstic lake environment. Adv. Org. Geochem. 22: 739–757.

Villanueva, J., J. O. Grimalt, R. Dewit, B. J. Keely & J. R. Maxwell, 1994. Chlorophyll and carotenoid-pigments in solar saltern microbial mats. Geochim. Cosmochim. Acta 58: 4703–4715.

Vogel, A., 1978. Textbook of Practical Organic Chemistry. 4th Edition. Longman, N.Y., 1368 pp.

Watts, D. C. & J. R. Maxwell, 1977. Carotenoid diagenesis in a marine sediment. Geochim. Cosmochim. Acta 41: 493–497.

Winfree, N. M., J. W. Louda, E. W. Baker, A. D. Steinman & K. E. Havens, 1997. Application of chlorophyll and carotenoid pigments for the chemotaxonomic assessment of seston, periphyton, and cyanobacterial mats of Lake Okeechobee. Florida. Molec. Mark. Environ. Geochem. 671: 77–91.

Wright, S. W., S. W. Jeffrey, R. F. C. Mantoura, C. A. Llewellyn, T. Bjørnland, D. Repeta & N. A. Welschmeyer, 1991. Improved HPLC method for analysis of chlorophylls and carotenoids from marine phytoplankton. Mar. Prog. Ecol. Ser. 77: 183–196.

Wright, S. W. & S. W. Jeffrey, 1997. High resolution HPLC system for chlorophylls and carotenoids of marine phytoplankton. In Jeffrey, S. W., R. F. C. Mantoura & S. W. Wright (eds.) Phytoplankton Pigments in Oceanography. UNESCO Publishing, Paris: 327–341.

Wright, S. W. & R. F. C. Mantoura, 1997. Guidelines for setting up an HPLC system and laboratory. In Jeffrey, S. W., R. F. C. Mantoura & S. W. Wright (eds.) Phytoplankton Pigments in Oceanography. UNESCO Publishing, Paris: 383–406.

Wright, S. W. & J. D. Shearer, 1984. Rapid extraction and high-performance liquid chromatography of chlorophylls and carotenoids from marine phytoplankton. J. Chromatogr. 294: 281–295.

Wright, S. W., S. W. Jeffrey & R. F. C. Mantoura, 1997. Evaluation of methods and solvents for pigment extraction. In Jeffrey, S. W., R. F. C. Mantoura & S. W. Wright (eds.) Phytoplankton Pigments in Oceanography. UNESCO Publishing, Paris: 261–282.

Yacobi, Y. Z., W. Eckert, H. G. Trüper & T. Berman, 1990. High Performance Liquid Chromatography detection of phototrophic bacterial pigments in aquatic environments. Microb. Ecol., 19: 127–136.

Yacobi, Y. Z., R. F. C. Mantoura & C. A. Llewellyn, 1991. The distribution of chlorophylls and carotenoids and their breakdown products in Lake Kineret (Israel) sediments. Freshwat. Biol. 26: 1–10.

Young, A. & G. Britton, 1993. Carotenoids in Photosynthesis. Chapman and Hall, London, 498 pp.

Zechmeister, L., 1962. *Cis-trans* isomeric carotenoids, vitamin A and aryl polyenes. Academic Press, N.Y.

Züllig, H., 1981. On the use of carotenoid stratigraphy in lake sediments for detecting past developments of phytoplankton. Limnol. Oceanogr. 26: 970–976.

Züllig, H., 1982. Untersuchungen über die Stratigraphie von Carotinoiden im geschichteten Sediment von 10 Schweizer Seen zur Erkundung früherer Phytoplankton-Entfaltungen. Schweiz. Z. Hydrol. 44: 1–98.

Glossary, acronyms and abbreviations

^{210}Pb dating: Method of absolute age determination based on the decay of radioactive lead-210, which is formed in the atmosphere. It rapidly attaches itself to aerosol particles, and settles to the Earth's surface. Lead-210 activity in lake sediments provides an indication of their age for the last $\sim$150 years.

absorption maximum: Wavelength at which light is absorbed most intensely; used for pigment identification.

acarid mite: Mite in the suborder Acariformes, subgroup Oribatei, within the order Acari of the Arachnida. The chitinous bodies of free-living terrestrial and aquatic species are found in lake and bog sediments.

acetylation: Chemical process to convert a primary or secondary hydroxyl group to an ester by addition of acetate.

acid neutralizing capacity: The capability of natural waters to resist the influence of added acids; specifically measured as the amount of a strong acid that is necessary to reduce the pH of a water sample to a specific point.

acrostom: A term to describe the apertural morphology of testate amoebae. In an acrostom test, the aperture is terminal.

adnate: Used to describe a diatom whose valve face is attached to the host substrate.

agglutinate: Stick together.

agglutinated shells: Shells of testate amoebae made from mineral particles or diatoms, which are glued together by an organic cement.

akinetes: Asexual spores (resting cells) which form through thickening of the walls of a normal vegetative cell.

algae: Plant-like organisms of any of several phyla, divisions, or classes of chiefly aquatic, usually chlorophyll-containing, nonvascular organisms of polyphyletic origin. They usually include the green, yellow-green, brown, and red algae in the eukaryotes and the blue-green algae in the prokaryotes.

Allerød: Warm interstadial period that occurred at the close of the Weichselian (or latest) Glacial Stage in Europe, approximately 11,800 to 11,000 ^{14}C years before present (about 12,700–13,900 calibrated yr B.P.).

Allerød-Younger Dryas transition: The period in geological time (ca. 12,900–12,700 calibrated years BP) between the mild Allerød period and the cold Younger Dryas period.

allochthonous: From elsewhere, as opposed to autochthonous (produced *in situ*). Used, for example, for organic matter that may be imported from elsewhere, or for organisms.

allomer: Organic compounds sharing the same empirical formula.

amoeba: Unicellular naked protozoan, typically of indefinite shape, moving and feeding with pseudopodia. The several groups of amoebae are not closely related to each other.

amorphous silica: A non-crystalline isotropic mineral ($SiO_2 \cdot nH_2O$). It is often found in the siliceous skeletons of various aquatic organisms, including diatoms. It can also be found in phytoliths and inorganically precipitated in highly alkaline lakes.

AMS: Accelerator mass spectrometry.

AMS ^{14}C dating: Method of absolute age determination based on the radioactive decay of the carbon-14 isotope in organic material; method can use accelerator mass spectrometry, which allows the dating of 1–2 mg of carbon. Method can cover the last ~50,000 years.

analyte: A compound being analyzed.

anemophilous pollen dispersal: Pollen dispersal by wind. Plants adapted for this generally produce large amounts of pollen, and these pollen types are dominant in pollen assemblages in lake sediments. Trees such as *Betula, Pinus, Alnus, Corylus,* and *Quercus* are anemophilous, and the most common herb types are grasses (Gramineae or Poaceae), sedges (Cyperaceae), Chenopodiaceae, *Artemisia,* and *Plantago.*

anemophilous: Pollinated by wind-dispersed pollen.

angiosperm: A subdivision of seed plants that includes flowering plants. Features used to separate angiosperms from other seed plants include the enclosed nature of the ovary (the carpel), the presence of flowers, specialised conducting cells in the xylem and phloem, ovules that are surrounded by a double-layered seed coat, and the structure of the pollen grain.

annual laminations: In undisturbed conditions at the sediment surface (e.g., in a meromictic lake) sediment components that are deposited at different times of the year may lead to the formation of discrete layers. If the components are visibly different, one year's deposition often consists of a couplet of a light and dark layer.

annulus of cyst: Shelf of silica between the inner and outer pore margin of a chrysophyte cyst.

anterior lip: Front lip of the aperture of a test of testate amoebae.

anther: The terminal part of a flower's stamen-bearing pollen sacs that contain pollen.

apertural: Near an opening; for example, in testate amoeba, apertural plates are those near the opening of the test.

aphotic: Without light, or sufficient light for a given process (e.g., photosynthesis).

apical: Belonging to the apex; situated at the tip.

aplanospores: Spores formed directly from the contents of separate vegetative cells when the protoplast contracts and a new cell wall is formed inside the enclosing vegetative cell wall.

aquatic succession: (= hydrosere). Plant succession from colonisation of open-water habitats, through various aquatic plant assemblages, leading to fen communities as the water becomes shallower, and then to either bog or woodland development above the water table. Evidence of the succession is preserved in the sediments below it. Many factors affect aquatic successions, and the directions of development can be diverse, and even retrogressive.

arcellaceans: Testate amoebae belonging to the order Arcellinida.

arctic tree-line: The northward limit of tree occurrence in the northern hemisphere; the northern limit of the tree growth-form. Beyond is usually shrub-tundra (low-arctic vegetation).

areola: Simple or chambered pore-like perforation through the diatom wall.

areolate surface: Shell surface of *Arcella* species (testate amoebae) with a honeycomb structure generated by the regular arrangement of small organic building units.

Ascomycetes: Class of higher fungi with septate hyphae and spores formed in an ascus.

ascospores: Spores produced in the ascus of an Ascomycete.

ascus: The membranous oval or tubular spore case of an Ascomycete.

athalassic waters: Inland or continental saline water bodies (cf. thalassic).

atmospheric pressure chemical ionization: Process of production of mass ions using chemical means at atmospheric pressure in an inert gas, rather than in a vacuum, or by using high-energy molecular collisions.

autecology: That division of ecology which treats the relations of a single species to its environment.

autochthonous: From within, as opposed to allochthonous, which is material from elsewhere. Used, for example, to describe material, such as algal cells, which are produced within a lake basin.

autotrophy: Self-nourishing; ability to feed on simple substances.

azeotropic distillation: Method to remove small amounts of water. Small volumes of benzene are added to methanolic solutions containing water and subject to a mild vacuum. Water is evaporated in conjunction with the benzene by rotating the solution, often with mild external warming, under vacuum.

background charcoal component: Slowly varying component of a charcoal record, which may reflect changes in fuel accumulation and its influence on charcoal production over long time periods.

baculate spines: Cylindrical spines with a blunt apex on chrysophyte cysts.

bathochromatic: Towards longer wavelengths (e.g., red, infrared).

benthic: Living attached to or associated with a substrate.

Beringia: The unglaciated lands of eastern Siberia, Alaska, and western Yukon and the intervening continental shelves comprise Beringia. During the glacial periods of lowered sea-level, the Bering and Chukchi seas were dry land, known as the Bering Land Bridge, that allowed the intercontinental migration of plants, animals, and people.

bilobate: A type of two-lobed phytolith found only in grasses. They are also commonly referred to in the literature as dumbbells.

binary solvent system: Chromatographic system in which two solvents or mixtures of differing polarity are used sequentially to elute compounds from a chromatographic substrate.

biogenic silica (BSi): An amorphous form of Si biogenically precipitated by a variety of organisms including chrysophytes, diatoms, radiolarians, silicoflagellates, and sponges. In most sedimentary environments diatoms are the major contributor to the biogenic silica measured in the sediments.

biogeographic indicator taxa: Taxa with a restricted geographical distribution (e.g., Gondwanaland species).

biome: A geographic zone characterized by similar animal and plant life.

biotope: An area, usually of uniform environmental conditions, where a characteristic assemblage of organisms lives.

bioturbation: Mixing of sediments by animal activity.

blue-green algae: See Cyanobacteria.

bog: Wetland environment often with large peat deposits. Raised bogs have an elevated central part and are usually dominated by *Sphagnum*. Blanket bogs cover flat or gently sloping areas in upland areas with a highly oceanic climate. Both types are usually fed by rainwater exclusively.

bootstrapping: A computer intensive procedure to estimate parameters or derive confidence intervals by resampling the original sample with replacement.

Boreal: An interval in Holocene time, from 10,500 to 7800 calibrated years BP, with a relative warm and dry climate.

BP: Before Present (AD 1950).

brackish water: Water with a salinity between that of freshwater and seawater (approximately between 3 and 30 g l^{-1}).

bright field microscopy: Standard compound light microscopy using transmitted light.

BSi: Biogenic silica.

bulbil: An axillary bud modified for dispersal and asexual reproduction. *Polygonum viviparum* bulbils have been found preserved in lake sediments.

C3: A photosynthetic pathway that is part of a suite of biochemical and anatomical properties of plants generally adapted to low light intensities and/or cool growing-season temperatures. C3 plants include all woody and most non-woody genera.

C4: A photosynthetic pathway of plants generally adapted to high light intensities and/or warm growing season temperatures. With the exception of bamboos, most tropical grasses are C4 plants.

CA: Correspondence analysis.

calibration training set: A dataset of micro- or macrofossil counts from modern surface sediments together with their associated water chemistry or other environmental data that are used to generate transfer functions.

canal raphe: A diatom raphe with a tubular inner passage separated from the cell interior by bridge-like fibulae, also called a fibulate raphe.

canonical correspondence analysis (CCA): A constrained ordination technique based on correspondence analysis, but where the ordination axes are constrained to be linear combinations of the supplied environmental (predictor) variables.

carbonyl: An oxygen double-bonded to a carbon atom.

carotenoid: A group of accessory pigments involved in photosynthesis and photoprotection of cells. Chemically, a class of hydrocarbons and their oxygenated derivatives consisting of eight isoprenoid units in a linear arrangement.

caudal scales: Chrysophyte scales near the caudal (tail) end of the organism.

CCA: Canonical correspondence analysis.

cellulose: A long-chain polysaccharide made up of glucose units. A fundamental constituent of plant cell walls.

Cenozoic: The geologic era, which began 65 million years ago, and continues to the present.

central nodule: A zone of thickened silica lying between the raphe ends in the centre of some pennate diatoms.

centric diatoms: Diatoms with radial symmetry of the valve pattern (cf. pennate diatoms).

CHAR: Abbreviation for charcoal accumulation rate (number of charcoal particles cm^{-2} yr^{-1}).

charcoal: Any black-coloured plant-derived material that has had its chemical composition and structure altered as a result of heating in a fire, and retains recognizable anatomic structure of the plant material, even if only in a fragmentary form.

charophyte: (= stonewort). A member of the green alga (Chlorophycota) family Charophyceae, characterised by a large thallus differentiated into rhizoids and stems bearing whorls of branches. The thallus can be coated with calcareous deposits and can be a component of lake marl. *Chara* spp. generally grow in base-rich or even brackish water, whereas *Nitella* spp. generally grow in base-poor, oligotrophic water. Charophytes can be an important component of the macrophyte biomass and often occur abundantly in open, pioneer situations. Their oospores can be very abundant in lake sediments.

chemical ionization: Process to create a molecular ion during mass spectrometry by using a chemical reaction.

chi-squared distance: A mathematical coefficient that measures the dissimilarity between two fossil assemblages.

chlamydospores: Thick-walled (usually resting) fungal spores.

chlorarachniophyte algae: Group of amoeboid algae.

chlorins: Derivatives of chlorophyllous molecules.

chlorophyll: The main pigment(s) involved in light collection and electron transfer during photosynthesis of algae, plants, and bacteria. Usually, composed of four ring structures (pyrols), a central Mg^{2+} atom, and a hydrocarbon (phytol) chain.

chlorophyllase: A class of enzymes or molecules which selectively remove the phytol chain from a chlorophyll molecule.

chlorophyllide: Chlorophyll molecule which has lost only its phytol chain.

chloroplast: The membrane-bound organelle within eukaryote cells that contains chlorophyll.

chromophore: Light-absorbing region of a pigment.

chrysophytes: Common term for algae in the Chrysophyceae and Synurophyceae.

chrysophycean cysts: Siliceous resting stages of chrysophyte algae; see stomatocysts.

ciliates: A phylum of protozoa that swim with numerous cilia. A common representative is *Paramecium*.

cingulum: One or more girdle bands attached to a diatom valve forming the connection between the two valves of a diatom cell.

circulus: Ridge that forms a closed ring (formerly known as 'flange') on a chrysophyte cyst.

collar: Siliceous thickening surrounding a chrysophyte cyst pore.

community structure: The composition of a community of organisms (e.g., the relative abundances of (indicator) species, size classes, feeding types, etc.).

complex collar: More than one separate collar surrounding the chrysophyte cyst pore.

confounding effects: In a statistical analysis with many environmental (i.e., explanatory) variables, where the effects of one variable cannot be separated from the effects of another.

conidia: Asexual spores produced on conidiophores.

conidiophores: Fungal hyphae bearing conidia.

conjugation: In biological terms, fusion of usually similar gametes with ultimate union of their nuclei that, among lower thallophytes, replaces the typical fertilisation of higher forms. In chemical terms, a series of carbon-carbon bonds in which outer orbital electrons circulate among several C atoms.

continuous flow: In mass spectrometry, process by which an HPLC or other system of molecular isolation is interfaced directly to an MS unit, allowing a continuous input of analyte to the spectrometer.

conula: Siliceous projection on a chrysophyte cyst with circular basal diameter of $\leq 0.2\ \mu$m.

convective column: Plume of upwelling air and smoke created by ground heating during a fire.

coprophilous fungi: Fungi growing or living on dung.

correlation coefficient (r): A statistical index that quantifies the linear relationship between two variables.

correspondence analysis (CA): An ordination method that simultaneously ordinates samples and variables (species), and maximizes the correlation between sample and variable scores. It is widely used in ecology because it assumes a unimodal response of species variables to underlying gradients.

cosmopolitan: Widely distributed.

cross-correlation: Numerical approach to assess the dependence of one temporal stratigraphical sequence on the past or future values of another, providing samples are spaced at even intervals of time.

cryptogams: Lower plants lacking flowers or seeds.

cryptostom: A term to describe the apertural morphology of testate amoebae. Cryptostom tests have a largely reduced slit-like aperture at the ventral side. This structure is characteristic of soil testate amoebae.

cuticle: A thin, waxy, protective layer covering aerial parts of plants (leaves, stems, flowers, etc.).

cuvette: A translucent glass or quartz container used to hold a sample solution, often for determination of light absorbance properties in a spectrophotometer.

Cyanobacteria: (formerly Cyanophyceae = blue-green algae). A large group of prokaryotes that possess chlorophyll *a* and carry out photosynthesis with the concomitant production of oxygen. Many species fix atmospheric nitrogen. Colonies of *Gloeotrichia* and *Nostoc* have been found in lake sediments.

cysts: Resting stage.

cytoplasm: All the plasma of a cell except that in the nucleus. Usually one distinguishes a hyaline ectoplasm and a more granular endoplasma.

DAR: Diatom accumulation rate.

Dematiaceae: Hyphomycetes (Fungi Imperfecti: no sexual state known) with dark conidia or conidiophores.

dendrochronology: A method of dating using annual tree rings; the use of annual tree rings to determine the age of a tree. Matching annual rings between trees, including dead and fossil wood, can be used to make a chronology. Several long series have extended into the early Holocene, and even into the Younger Dryas ($>11,500$ calibrated years BP). Radiocarbon dating of series of wood samples of known age is used to calibrate radiocarbon ages to calendar years, so-called calibrated years.

depression: Circular indentation $>0.2\,\mu$m in diameter on a chrysophyte cyst.

desorption: To remove a compound adhering (adsorbed) to a surface.

Devensian: See Weichselian.

diaspore: A spore, seed, fruit, bulbil, or other structure that functions in plant dispersal: a propagule.

DIATCODE: Diatom Codes; a coded list of diatom names.

diatom accumulation rate (DAR): The number of diatom valves or cells accumulating each year in sediments.

diatom concentration: The number of diatom valves or cells found in a known weight or volume of sediment.

diatom dissolution index (F): The ratio of perfectly preserved diatom valves to partly dissolved valves.

DIC: Differential interference contrast; also used as an abbreviation for dissolved inorganic carbon.

dicotyledon: Plants that have seeds with two cotyledons, the first leaves of the embryo plant (= dicot).

diene: A C − C double bond system.

differential interference contrast (DIC): A light microscopy system that uses a combination of polarizer, analyzer, and prisms to convert differences in specimen thickness and refractive index into differences in light intensity in the image, producing a three-dimensional effect.

dinoflagellates: Planktonic algae in which the motile stage has two flagella, one ribbon-like with multiple waves, which beats to the cell's left, the other beating towards the posterior, with only one or a few waves. As fossils, they are found in rocks as early as the Permian.

dome: Distal area above the shield of a *Mallomonas* (chrysophyte) scale.

drop-catch: To isolate compounds by collecting specific portions of the elutant from the outflow of an HPLC system.

DRSi: Dissolved reactive silica (SiO_2).

duripan: Amorphous Si that occurs as cemented horizons in soils in sub-humid Mediterranean and arid climates. These soils have moisture regimes where Si is solubilized and translocated to a lower position in the soil profiles.

dystrophic lake: A lake with brownish, often acidic water due to the high concentration of humic substances.

echinate spine: Spine (chrysophyte) with a pointed apex.

Eemian: Penultimate interglacial period.

electron ionization: To create molecular ions during mass spectrometry by impacting compounds with a beam of high energy electrons in an inert gas atmosphere.

electronegative: The atomic property of having a high degree of attraction for electrons.

electrospray ionization: Process of mass ion production in which a solution containing the analyte is introduced into the mass spectrometer as a very fine mist within an electric field.

Ellenberg's ecological index numbers: Ordinal values (1–9 or 1–12) assigned to a species depending on its occurrence in a range of ecological and environmental gradients such as soil pH, light, soil humidity, soil nutrient-content, growing season temperature, oceanicity, and salinity.

elution: Process by which analytes adsorbed to a chromatographic substrate are removed by a solvent.

EM: Electron microscopy or electron microscope.

endogenous rods, nails: Biogenic siliceous components of the shells of certain species of testate amoebae (e.g., *Netzelia, Lesquereusia*).

endogenously: Growing or originating from within.

entomophilous: Pollinated by insect-dispersed pollen.

ephippia: Parthenogenetic eggs encased in a resistant coat. *Daphnia* (Cladocera) ephippia typically contain two 'eggs', whereas other genera have only one 'egg'. Ephippia resist adverse conditions and hatch with the return of favourable conditions. The fossil record of *Daphnia*, in particular, relies on small body parts that are difficult to find and identify, whereas the ephippia may be very abundant in lake sediments.

epidermal tissue: A thin layer of cells that forms the outer covering of leaves.

epilithon: The assemblage growing attached to stones or rocks.

epimer: A type of isomerism (see isomer) shown by substances which contain several asymmetric centres but differ in the configuration of only one of these.

epipelon: The assemblage living on and in muds and sands.

epiphyton: The assemblage growing attached to plants.

epipsammon: The assemblage growing attached to sand grains.

epoxy: An oxygen-bearing functional group in which the oxygen atom is bonded to two carbon atoms (also epoxide).

esterfication: Chemical process to produce an ester of a molecule.

euhaline: Degree of salinity in brackish to marine waters; water having $>1500\,\text{mg}\,\text{l}^{-1}$ salinity (cf. mesohaline, oligohaline).

eukaryote: An organism composed of one or more cells containing a nucleus or nuclei and membrane-bound organelles.

Euparal®: Mounting medium made from natural resins. Refractive index: 1.53. Supplier: CHROMA, Münster, Germany, http//www.chroma.de.

euplankton: Plankton whose life cycle takes place entirely in the water column.

euplanktonic: Living exclusively within the water column; not attached to a substrate.

eutrophic lake: A nutrient rich, highly productive lake. In temperate regions usually anoxic profundal water during summer stagnation.

eutrophication: The biological effects of increased nutrients (usually nitrogen and phosphorus) on aquatic ecosystems, usually resulting in a change in species composition, an increase in biomass, oxygen depletion due to decay and microbial decomposition, and increased sediment accumulation rate. Increased turbidity and reduced light may eventually reduce photosynthesis and biomass. Eutrophication can be caused by natural hydroseral infilling or by human addition of nutrients from sewage or agricultural fertilization.

excystment: Process of emerging from a cyst.

extralocal fires: Fires that occur in adjacent watersheds to a study site but not in the immediate watershed.

F: Diatom dissolution index.

false complex collar: Single collar on a chrysophyte cyst with an apical groove and hence the appearance of two separate collars.

fast atom bombardment mass spectrometry: Method of mass spectrometry in which the mass ion of analyte is produced by bombardment using a high-energy beam of heavy atoms (e.g., Xe).

fen: A soligenous or topogenous mire or wetland dominated by herbaceous plants where nutrients are supplied by lateral ground-water flow. The species present vary with water base-status and the rate of water flow. Fens may originate by succession from wetter reed-swamp vegetation.

fibulae: Siliceous struts that separate the canal raphe from the cell interior in some pennate diatoms, previously termed keel punctae or carinal pores.

fibulate raphe: See canal raphe.

filopodia: Thin, finely pointed, not anastomosing, ectoplasmatic pseudopodia which, during retraction, fold in a jack-knife manner.

filose amoebae: Amoebae with filopodia.

fire event: A single fire or series of fires within an area at a particular time.

flagella: Plural of flagellum.

flagellum: A whip-like structure providing an organism the means of locomotion or facilitating feeding.

flagellated cells: Motile cells with flagella.

florin structure: The structure formed by the subsidiary cells of a stomata.

fluorescence lighting: Microscope illumination at narrow wavelengths that produce fluorescence in the object being studied.

foot: Proximal end of *Mallomonas* (chrysophyte) bristle.

fossa: Elongated indentations that may be straight or curved.

freeze-dry: Removal of water as vapour from a frozen object by creating a vacuum (syn. sublimation).

frustule: The siliceous component of the diatom cell wall.

338

full-glacial: Refers to environmental conditions or vegetation type at the height of glaciation. It is usually applied to deposits and fossil sequences beyond the glacial limits.

fultoportula: A tube-like opening on the valve face or mantle of some centric diatoms, surrounded on the inner side by two to five satellite pores (also called strutted processes).

functional group: Chemical group responsible for characteristic chemical properties (e.g., polarity) of a compound.

fungal spores: Spores of fungi.

fungi: Saprophytic and parasitic spore-producing organisms usually classified as plants that lack chlorophyll and include moulds, rusts, mildews, smuts, mushrooms, and yeasts.

Gaussian plume models: Theoretical models used to describe the airborne transport of charcoal particles away from a fire. The models make simplifying assumptions about the release of particles at a particular height from a buoyant convection plume, their lateral transport by winds of prescribed velocity, and settling rates and distances based on particle size.

gemmoscleres: Sponge spicules associated with gemmules that are essential in the identification of species.

gemmules: Asexually produced resting stages of sponges that are frequently resistant to adverse environmental conditions, such as cold, dry or hot conditions.

generation time: In unicellular organisms the time needed to double the number of individuals.

geometric isomers: Isomers in which the position of functional groups across a point of reference (e.g., a C = C bond) differs. *Cis* isomers have functional groups on the same side of a bond, *trans* isomers have the functional groups in transverse position.

germ slit: Long narrow cut or opening playing a role when spores start to grow.

girdle: Part of the frustule between the two valves comprising siliceous bands connecting the valves.

glycerol: An alcohol with the formula $C_3H_8O_3$; it binds fatty acids to form lipids. Used to store concentrates of specimens after preparation, and as a medium to examine specimens with the microscope.

Golgi apparatus: Organelle present in the cytoplasm of cells, believed to have a secretory function.

Gondwana (= Gondwanaland): One of the two ancient supercontinents produced by the first split of the even larger supercontinent Pangaea about 200 million years ago, comprising chiefly what are now Africa, South America, Australia, Antarctica, and the Indian subcontinent.

GRIP: Greenland Ice Core Project.

gymnosperm: A sub-division of seed plants. Features used to separate gynmosperms from other seed plants include unprotected ovules carried on a seed-leaf.

half-life: Length of time for a process to run 50% to completion.

hard-water error: When ancient geological carbon derived from the catchment of a lake (e.g., limestone) is incorporated into lake water and produces anomalously old ^{14}C dates for the lake sediments.

hard-water effect: The utilisation by aquatic organisms of carbon from a source older than the life of ^{14}C (e.g., HCO_3^{-1} from carbonate rock or coal), or the utilisation of non-contemporaneous carbon dilutes the contemporary amount of ^{14}C and results in a radiocarbon age older than the true age of the sample.

heliozoan: A group of predominantly freshwater protozoa, some of which are covered with siliceous scales.

heterotrophy: Deriving nourishment from outside sources.

high performance liquid chromatography: (also high pressure liquid chromatography, HPLC). Chromatographic method in which compounds are separated on a narrow-bore column packed with particles which differentially attract analytes moving past them, dissolved in a stream of solvent. Isolation of molecules is based on the chemical properties of the compound which cause differential and characteristic attraction of the analyte for both the particles (stationary phase) and solvents (mobile phase).

holoplankton: See euplankton.

HPLC: See high performance liquid chromatography.

humus microcosm: Small experimental container with humus. Used to study biological processes in soil.

hyaline: Describes areas of diatom silica not perforated by areolae.

hydrocarbon: An organic compound of hydrogen and carbon.

hydrosere: See aquatic succession.

hydroxyl: OH functional group.

hyper-oligotrophic lake: Extremely nutrient-poor lake.

hyphae: Threads that make up the mycelium of a fungus.

hyphal attachment: Short mycelium appendage of a fungal spore, broken off, but originally forming the connection between fungal spore and mycelium.

hyphopodia: Short lateral appendages of fungal hyphae which adhere firmly to the surface of a host plant.

Hypsithermal: Interval in the Holocene with mean temperatures higher than the present ones. It includes the Boreal, Atlantic, and Subboreal climatic intervals, roughly from 10,500 BP to 2500 calibrated years BP. It varies in its age in different parts of the world.

hypsochromatic: Towards shorter wavelengths (e.g., blue, ultraviolet).

ice core: Core taken from the ice of a glacier or (more usually) an ice-sheet. Ice cores from Greenland and Antarctica cover several hundred thousand years and their isotopic and particulate composition are increasingly influential for our understanding of late-Quaternary environmental change.

idiosomes: Biogenic calcareous or siliceous platelets or scales produced by testate amoebae. They are often group or species specific.

image analysis: Method that uses a computer to capture a video image of a sample and then analyzes the optical density of the image's pixels. Charcoal particles are typically the densest objects in an image.

index α: The ratio diatoms found predominantly in acidic waters to those found predominantly in alkaline waters in a sample (developed by G. Nygaard in 1956).

indicator species: A species whose presence is a marker for a specific habitat condition or environmental factor value.

infra-red analysis: A technique used to identify minerals by exposing a sample to infra-red light and studying the resulting vibrational patterns of the chemical bonds.

infra-red spectrophotometry: (also IR spectrophotometry) Process by which patterns of absorbance of infra-red irradiance are used to identify the functional groups present on a molecule of interest.

interference-contrast: Microscopical contrasting method.

interglacial: A long (ca. 10–20,000 yr) period of relatively warm climate between glacial periods. Forest usually develops on formerly glaciated land.

internal standard: A compound dissolved within a sample and used to either correct for analytical errors or to calibrate an analytical process.

International Statospore Working Group (ISWG): An international body of experts formed in 1986 to study and attempt to standardize chrysophyte statospore (stomatocyst) nomenclature and taxonomy.

inverted microscope: A microscope in which the positions of the objectives and condenser are reversed in relation to the stage, thereby allowing specimens (e.g., phytoplankton) to be observed through the glass bottom of settling chambers. These microscopes were originally developed for plankton enumeration (Utermöhl technique) but are now widely used in cell biology.

ion-pairing reagent: A solution of ionic compounds used in HPLC solvents to improve the chromatographic resolution and isolation of individual analytes from a mixture.

Ipswichian interglacial: The last interglacial in Britain, equivalent to the European Eemian and the North American Sangamon interglacials, and marine oxygen isotope stage 5e (ca. 130–115 ka BP).

IR: Infra-red.

isochrone: Contours on a map defining areas at which events occurred at similar times. Commonly used to display times of spread of trees (as indicated by pollen analyses) across large geographical areas.

isocratic: At constant conditions (e.g., pressure, chemistry, rate).

Isoetid plants: Aquatic macrophytes with a short, tufted growth-form of narrow, linear leaves, typified by *Isoetes*. They usually form a characteristic community in shallow, oligotrophic water at stony and gravelly lake shores (e.g., *Subularia aquatica, Limosella aquatica, Isoetes lacustris, Lobelia dortmanna, Littorella uniflora*).

isomer: Compounds possessing the same elemental composition and molecular weight, but differing in their chemical structure.

isopoll: Contours on a map defining areas of similar abundance of a pollen type of interest. Commonly used in a series (one map for each of a series of ages) to display rates of spread and rates of change of abundance for trees across large geographical areas.

ISWG: International Statospore Working Group.

jackknifing: A computer intensive procedure to estimate parameters or derive confidence intervals by calculating the estimate with one observation omitted in turn, then averaging these "trimmed" estimates.

keel punctae: See fibula.

krummholz conifers: Coniferous trees with dwarfed, decumbent growth forms. Generally found at arctic and alpine treelines.

labiate process: See rimoportula.

lacuna: The space between ridge walls in a reticulum of a chrysophyte cyst.

lamellae: The intercellular surface of stomata guard cells.

late-glacial: The period at the end of a glacial period transitional to the following interglacial. The last late-glacial period occurred ca 15,000–11,500 calibrated yr BP. It was a time of rapid climate fluctuations, including the relatively warm late-glacial interstadial (Bølling/Allerød; GI-1) and the subsequent cold stadial (Younger Dryas; GS-1).

least-squares regression: A statistical technique to model the quantitative relationship between two variables, Y and X. Once this relationship is established, one can predict the value of the Y variable from the value of the X variables.

lentic habitats: Still, inland aquatic habitats in which there is no systematic, directional water movement (e.g., lakes and ponds). Opposite of lotic habitats.

lignification: The process by which lignin, a cellulose-like substance, is impregnated within cell walls of plants.

lignified: Combination of lignin and cellulose to form wood tissue.

limnic stage: Refers to a stage with an open body of freshwater.

lipophilic: Able to be dissolved in lipids; non-polar.

liquid secondary ion mass spectrometry: Method of mass spectrometry in which the mass ion of the analyte is produced by bombardment using a high-energy beam of heavy ions (e.g., Cs).

Little Ice Age: A cool period recognised in several parts of the world and occurring from about A.D. 1550 to 1850, the timing varying from area to area.

littoral: Shallow water zone of a lake that typically supports rooted aquatic macrophyte growth.

liverwort: (= hepatic). The common name for a bryophyte belonging to the Class Hepaticae. They may be thallose (undifferentiated into stems and leaves) or leafy, with nerveless, often lobed, 1-cell thick leaves arranged in 2–3 ranks. The lidless capsule splits into 4 valves. Liverworts usually frequent damp habitats.

LM: Light microscopy or light microscope.

lobopodia: Fingerlike pseudopodia.

lobose amoebae: Amoebae with lobopodia.

local fires: Fires that occur in the immediate watershed of a study site.

lotic habitats: Flowing freshwater habitats in which there is directional water movement downhill (e.g., rivers and streams). Opposite of lentic habitats.

Lugol's iodine: A preservative solution typically used for algae, prepared by dissolving 10 g pure iodine and 20 g potassium iodide in 200 ml distilled water and 20 ml concentrated glacial acetic acid. The solution should be stored in a ground-glass stoppered, darkened bottle.

lyophilization: See freeze-dry.

MacDonald-Pfitzer rule: The supposition that vegetatively reproducing diatoms become smaller as daughter valves are laid down within parent cells.

macroalgae: Algae with large multicellular thalli with an organised structure. They include marine Phaeophyceae and Rhodophyceae (sea-weeds) and fresh-water Charophyceae (stoneworts).

macrofossil: A fossil that is large enough to see with the naked eye and to be manipulated by hand under a stereomicroscope. The size range is about 0.5 mm to ca. 100 mm or more. Plant macrofossils are typically seeds, fruits, oospores, sporangia, cones, and other vegetative parts such as twigs, buds, budscales, wood fragments, moss, etc. Animal macrofossils found in lake sediments include fish bones and scales, Mollusca, Ostracoda, Cladocera ephippia, Acarid mites, insect remains (e.g., Chironomidae, Trichoptera, Coleoptera), Bryozoa statoblasts, etc.

macroscopic charcoal: Refers to particles of charcoal that are >100 microns in minimum diameter; often counted as the residue after washing sediment through analytic sieves.

magnetic susceptibility: The property of a material that determines the size of an applied magnetic field required to generate a certain level of magnetism in the material.

mantle: See valve.

marine reservoir effect: Radiocarbon dates of marine organisms and sediments are older than their true age because they have incorporated older, non-contemporary carbon with reduced ^{14}C derived from CO_2 or carbonates that have resided in the deep ocean for several hundred years during ocean circulation. The size of the marine reservoir effect may be established by dating modern organisms (e.g., mollusc shells), by dating marine and terrestrial material from the same sediment sample, and by dating volcanic tephras of known terrestrial age in marine sediments. The modern and Holocene marine reservoir effect is in the order of 400 years.

marker grains: Internal standard for microscopical enumeration procedures. Often *Lycopodium* spores are used. These are available in tablet form (T. Persson, University of Lund, Sweden).

marsh: Wetland, swamp.

mass spectrometry (MS): Measurements of the mass/charge ratio of molecules made on a mass spectrometer. A mass spectrometer is a device in which molecules are ionised and the accelerated ions are separated according to their mass/charge ratio. Used in identification to determine molecular weight and investigate characteristics of molecular breakdown.

MAT: Modern Analogue Technique.

matrix-assisted laser desorption/ionization: Process by which the mass ion of the analyte is produced by directing a beam of high-energy photons onto the surface of a probe on which an analyte has been adsorbed.

mean fire interval (MFI): Arithmetic average of all fire intervals in a given area over a given time period.

megafossil: A very large fossil compared to a macrofossil. The term is usually applied to large plant remains such as tree stumps and logs, but it can also be applied to smaller parts, such as conifer cones.

megascleres: Large sponge spicules that serve as primary structural elements.

meroplankton: Planktonic algae whose life cycle includes a benthic stage.

mesohaline: Degree of salinity in brackish to marine waters; water having 500–1500 mg l^{-1} salinity (cf. euhaline, oligohaline).

mesotrophic lake: A moderately nutrient-rich lake.

MFI: Mean fire interval.

microbial mat: A complex planar community of algae, bacteria and other organisms, usually located on the sediments of aquatic ecosystems.

microscale chemical derivitization: Conversion of a molecule to a product using only very small quantities (e.g., g) of an analyte.

microscleres: Small sponge spicules that provide secondary structure.

microscopic charcoal: Refers to particles of charcoal that are < 100 microns in diameter; often counted on microscopic slides as part of routine pollen analysis.

Miocene: An epoch in Earth's history from about 24 to 5 million years ago.

mitochondria: Membrane-bound organelles responsible for respiration containing enzymes used in converting food to energy.

mitosis: Nuclear division in which the number of chromosomes in the daughter cells is the same as that of the parent cell.

modern analogue technique (MAT): A method of environmental reconstruction based on matching fossil samples to modern surface sediment assemblages of similar taxonomic composition.

monomer: The smallest repeating unit of a molecule within a polymer.

monophyletic: A group of taxa that contains an ancestor and all of its descendants.

Monte Carlo permutation test: A type of randomisation test where the actual data values are maintained, but they are randomly permuted in order to obtain the distribution of a test statistic.

moss: The common name for a bryophyte belonging to the Class Musci. The gametophyte bears simple, sessile leaves, usually one cell thick and often with a midrib. The sporophyte bears a capsule with a lid. Mosses can be tufted (acrocarpous) or creeping (pleurocarpous) and occur in a wide range of habitats, from under-water to dry rocks and arid soil.

MS: Mass spectrometry.

mucilaginous hairs: Hairs secreting a gelatinous substance.

muco-polysaccharide: Organic material secreted by diatoms and used in some species to form attachment pads or to make links between adjacent cells in the formation of colonies.

mutualism: A type of symbiosis in which both members depend on each other for their nutrients or other services.

mycology: The study of fungi.

mycorrhizal fungus: Fungus forming with its mycelium a symbiotic association with the roots of a seed plant.

Naphrax®: Mounting medium with high refractive indices used for diatoms. Supplier: Northern Biological Supplies, U.K.

naviculoid raphe: A pair of longitudinal slits separated by a central nodule and running along or near the apical axis of some pennate diatom valves.

Neolithic: (= New Stone Age). A prehistoric archaeological period in Europe and the Near East from ca. 7000–3000 calibrated years BC. It is characterised by the use of stone tools and the development of agriculture and a settled life-style.

neritic: The aquatic environment overlying the sublittoral zone, (i.e., the water over the continental shelves). Paleoecologists have also applied the word to the environment of the bottom itself.

neutral formaldehyde: Formaldehyde solution that is free from formic acid by the addition of calcium carbonate or phosphate buffer. This is used to prevent erosion of the structures of tests or valves (diatoms) of the preserved organisms.

NMR: Nuclear magnetic resonance.

NOAA: National Oceanic and Atmospheric Administration, part of the US Department of Commerce. Among other functions, hosts important data archives, accessible via the Internet at http://www.ngdc.noaa.gov.

no-analogue: An assemblage of fossil taxa that is different from any assemblage of species found at the present day. No-analogue assemblages have been used to infer no-analogue past climates that were different from any known present-day climate. However, the composition of fossil assemblages can be affected by several factors that may not affect modern assemblages, involving preservation and taphonomy, particularly transport from other locations and incorporation of non-contemporaneous fossils, so the interpretation as a no-analogue climate must be made with caution.

non-pollen palynomorphs (NPP): Various microfossils (algae, fungi, etc.) occurring in microscope slides made for pollen analyses. Records of such fossils can result in additional ecological information.

NPP: Non-pollen palynomorphs.

nuclear magnetic resonance: (NMR) A technique for studying nuclear magnetic moments to elucidate the chemical structure and functional groups of an analyte.

nucleoli: Term used for RNA-containing bodies in the nucleus that are associated with particular chromosomes.

nucleus: A spheroidal body surrounded by a membrane in the cells of animals and plants. It contains the chromosomes that carry the genetic information.

nunatak hypothesis: Also called theory of per-glacial survival, proposes that certain mountain plants survived glacial periods in ice-free refugia at the coast or on nunataks on coastal and inland mountains. Fossil evidence has shown that some of these species survived beyond the ice-sheet margins and readily migrated and colonised recently deglaciated terrain as the glaciers retreated. The nunatak hypothesis is probably unnecessary. Isolated mountain populations are probably the result of post-glacial habitat restriction.

Nymphaeaceae: Water-lily family.

obligate aquatic plants: Plants or groups of plant taxa that are all unable to exist away from an aquatic environment.

'old' carbon: Organic material may contain a lower proportion of ^{14}C than the contemporary proportion because the carbon source has been sequestered away from contemporaneous exchange. The 'old' carbon may be older than (e.g., coal, carbonate rocks) or within (e.g., marine CO_2, CO_2 released from lake sediments) the radioactive life-span of ^{14}C. In either case, a radiocarbon date from the organic material will be older than its true age.

oligohaline: Degree of salinity in brackish to marine waters; water having $50-1500 \, \text{mg} \, \text{l}^{-1}$ salinity (cf. euhaline, mesohaline).

oligotrophic lake: A nutrient-poor lake.

oligotrophication: Process characterised by a decline in available nutrients.

oligotrophy: Condition in which few nutrients are present.

oospore: Resting algal diaspore produced by sexual fertilisation of an oosphere in an oogonium. Oospores of Characeae are ca. 1 mm long and can be well preserved and abundant in lake sediments. They are identifiable to genera and sometimes to species.

opaline silica: Amorphous, hydrated silica; this forms the main constituent of the diatom cell wall.

organelle: A structurally discrete compartment within a cell (e.g., mitochondria).

outwash plain: A more or less flat area of sands and gravels deposited by glacial meltwater and traversed by channels that frequently migrate across the plain.

ovular nucleus: A nucleus with several to many nucleoli.

Paleozoic: An era of geologic time lasting from 544 to 248 million years ago.

paludification: The process of waterlogging of mineral soil and the development of fen or bog ecosystems that form peat.

palynology: The study of living and fossil pollen and spores; often used to reconstruct past vegetation dynamics in order to infer environmental change.

papillae: See scabrae.

parasitism: A type of symbiosis in which one member depends on another for its nutrients or other services.

partial canonical correspondence analysis: A type of canonical correspondence analysis in which the effects of specific environmental variables (covariables) are "factored out".

pathogen: Parasite that causes disease.

PCA: Principal components analysis.

peak half-width: Wavelength in photometry, or mass/charge ratio in mass spectrometry, at which one-half of maximum signal strength is attained.

peaks component: The high frequency component of a charcoal record, which is usually interpreted as the fire events.

peat bog: A bog in which peat has been formed by accumulation and degradation of plants such as *Sphagnum* mosses.

pennate diatoms: Diatoms with a valve pattern that exhibits bilateral rather than radial symmetry (cf. centric diatoms).

perhydrocarotene: A degraded carotenoid whose carbon skeleton is intact, but which contains no double bonds.

periphytic: Living on a substrate.

Permount®: A resin-based mounting medium for microscope slides.

phagotrophy: An organism that is capable of engulfing (eating) food particles.

348

phase contrast: A light microscopy system that enhances specimen contrast through converting differences in specimen thickness and refractive index into differences in light intensity in the image through interference between the light that passes through the specimen and a reference beam retarded by half a wavelength.

phase partitioning: The distribution of a dissolved substance between two immiscible (unmixed) liquids. The precise ratio of distribution of a molecule among solvents depends on its functional groups and chemical properties.

phenol: Popularly called carbolic acid. A chemical substance (C_6H_5OH) that is used for preservation or as a disinfectant.

phenolic substances: Organic molecules that include a 6-member ring of carbon atoms that has at least one hydroxide group.

pheophorbide: Chlorophyll molecule which has lost both its central Mg atom and phytol chain.

pheophytin: Chlorophyll molecule which has lost only its central Mg atom.

photic zone: The illuminated upper zone of the water column in lakes.

photoprotectant: A compound which protects organisms against cellular damage arising from exposure to energetic photons.

phototroph: An organism which uses energy from light for cellular metabolism.

phycobiliprotein: Water-soluble, protein-containing accessory pigments.

phycology: The study of algae.

phylla: Thin sheet of silica that drapes across chrysophyte cyst surface.

phylogeny: The evolutionary history of a group of organisms.

phytolith: Literally, any element, such as calcium or silica, deposited within various structures and tissues of plants. Opaline phytoliths are deposits of silicon dioxide ($SiO_2 \cdot nH_2O$) formed within plants (= opal phytolith).

phytolith accumulation rate: The estimated rate at which phytoliths accumulate.

pigment: A compound which absorbs light, usually in the visible range (400–700 nm), and hence imparts colour. Organic pigments often contain double-bonded carbon atoms (i.e., $C = C$).

PIRLA: Paleoecological Investigation of Recent Lake Acidification.

piston corer: A device for taking cores of lake sediment. A tight-fitting piston at the bottom of the core tube can be manipulated to allow a core to be taken at a chosen sediment depth. The vacuum created by the piston as the core is taken prevents the sediment from falling out.

plagiostom: A term to describe the apertural morphology of testate amoebae. In plagiostom tests the aperture lies eccentrically on the plane ventral side.

plankton: Organism suspended freely in water.

planktonic: Term to describe organisms that are floating or drifting in water. Planktonic organisms cannot swim against the current.

podsolisation: The leaching of an acidic brown-earth soil to form a podsol (peaty, mor humus overlying bleached sand and a humus and/or iron pan). It occurs in areas of humid climate and high precipitation, often under coniferous forest or acid heath vegetation.

point count method: Method of charcoal analysis that measures the area of microscopic charcoal particles in a grid by selecting random points on the pollen slide and determining the percentage of points that overlie charcoal.

pollen accumulation rate: The rate at which pollen grains accumulate in a sedimentary environment. Usually expressed as the number of grains falling on a unit area of sediment surface per year, with units of grains cm^{-2} yr^{-1}. Also referred to (incorrectly) as pollen influx.

pollen influx: See pollen accumulation rate.

pollen source area: The geographic area that is responsible for a particular pollen assemblage in a lake-sediment record.

polyhalobous: Referring to seawater with a salinity of 30 g l^{-1} and more, according to the salinity-system of F. Hustedt.

polymer: A chemically-bonded series of repeating subunit molecules.

polyphyletic: A group of organisms that has more than one line of descent.

polysaccharide: A large organic molecule that consists of many molecules of a monosaccharides such as glucose. Examples include cellulose and starch.

population dynamics: Fluctuations of a population as a result of growth and mortality.

post-glacial: (= Holocene, Flandrian). The period since ca. 11,500 calibrated yr BP of warm, interglacial-type climate and much reduced extent of glaciers.

primary charcoal: Sedimentary charcoal that was deposited in a lake during or shortly after a fire.

primary collar: Collar nearest the pore in a chrysophyte cyst.

principal components analysis (PCA): A numerical technique for the analysis of a multidimensional data set based on the identification of orthogonal linear combinations of variables that are selected to capture as much of the total variance in the data as possible.

profundal: The deep parts of a lake.

proteinaceous: Made from proteins. Term used, for example, with testate amoebae with organic shells.

Protista: Kingdom that includes unicellular, colonial, and simple multicellular eukaryotes.

protozoa: Microscopic, single-celled organisms with a nucleus.

psammobiontic: Living in the interstitial pore system of sandy beaches.

pseudoannulus: Thin shelf of silica projecting from the inner pore margin of a chrysophyte cyst.

pseudoplankton: See tychoplankton.

pseudopodial strands: Thin pseudopodial fibers connecting thicker elements of the reticulolobose pseudopodia of certain amoebae.

pseudopodium: "False foot"; a generally temporary, retractile cytoplasmatic protrusion that is used for moving and food uptake by amoeboid organisms.

psila: Circular indentation in chrysophyte cyst with a diameter of $\leq 0.2\ \mu$m.

pyrolysis: Combustion and degradation of molecules using thermal energy.

quadrapole mass spectrometry: Method by which mass ions are isolated after their production through the use of two apposing pairs of dielectric poles $(+/-)$. Ions are accelerated into the electric field and differential charges across the two pairs of dipoles are used to attract and extinguish ions which do not possess the mass/charge ratios of interest.

quaternary HPLC: An HPLC system using four solvents or a mixture to separate and elute analytes.

r: Correlation coefficient.

r^2 value: The square of the correlation coefficient, a measure of dependence between two variables. It gives the proportion of the variation in one variable that is accounted for by the other. Also known as the coefficient of determination.

radical cation: A chemically-reactive molecule which contains one electron less than its normal complement.

radiocarbon dating: When organisms cease to exchange carbon with the atmosphere (i.e., they die), the proportion of the radioactive isotope ^{14}C decreases according to its decay rate. Measurement of the residual ^{14}C radioactivity in an organic sample therefore allows its age to be estimated. The limit of radiocarbon dating is ca. 50,000 years.

radiocarbon plateau: The amount of ^{14}C (radioactive isotope of carbon) in atmospheric CO_2 depends upon its rate of formation by solar activity and the rate of dilution by inert carbon or carbon with much reduced ^{14}C activity derived from deep-ocean water, glacial ice, terrestrial recycling, and recently, the burning of fossil fuels. If the rate of dilution increases, younger samples may show the same ^{14}C activity as older samples, so that serial samples through a stratigraphic sequence have indistinguishable radiocarbon ages (i.e., a plateau of the same age). Long (ca. 400 yr) plateaux occur during the late-glacial and early Holocene, probably mainly caused by glacial melting and changes in ocean circulation strength, whereas smaller Holocene plateaux and irregularities are probably caused mainly by changes in atmospheric ^{14}C production related to changes in solar activity.

Radiolaria: Exclusively holoplanktonic marine protists with siliceous skeletons most of which exist as fossil forms. They are members of the microzooplankton, feeding on other small protists, copepods, etc.

raised bog: Rainwater-fed mire, developed mainly in lowlands where peat-forming plants, principally *Sphagnum* mosses, produce a domed surface above the level of the surrounding ground.

raised plateau: Low siliceous projection on a chrysophyte cyst with a smooth, flat top.

raphe: An elongated fissure or pair of fissures through the valve wall of a diatom.

refractive index: A material property equal to the reciprocal of the relative speed of light in a medium. By definition air has a refractive index of 1. As the speed of light is slower in more dense media, their refractive index always exceeds one. It is also related to the change or direction of a ray of light in passage from one medium to another of different density. To make the fine structures of, for example, a testate amoebae or a diatom better visible under the microscope, it is important to create a difference between the refraction index of the organism and the mounting medium.

regional fires: Fires that are within a broad biogeographic area, but not in the immediate or adjacent watershed of a study site.

reservoir effect: The addition of carbon depleted in ^{14}C to a sample will result in its radiocarbon age being older than its true age by a certain amount known as the reservoir effect. The main old carbon reservoirs are in calcareous rocks and coal (hard-water effect), in deep ocean water (marine reservoir effect), and in fossil fuels (Suess effect).

residence time: The amount of a given element or compound in a given body (for example, a lake or the ocean) divided by the average removal rate, assuming the system is in a steady state.

resource-ratio theory: Relates to the ability of organisms to compete for essential nutrients at different relative concentrations. For diatoms the Si : P ratio is especially important.

resuspension: Disturbance of surface sediments and their suspension in the water column.

reticule: A finely divided measuring scale inserted or built into one microscope eyepiece. The reticule is calibrated for each objective using a micrometer slide placed directly on the stage.

reticulolobopodia: Ectoplasmatic lobopodia that can merge and have tips. They are intermediate between lobopodia and filopodia and characteristic for the suborder Phryganellina (testate amoebae).

reticulum: Network of ridges, as on some chrysophyte cysts and pollen grains.

reversed phase HPLC: (see also high performance liquid chromatography). HPLC analysis in which non-polar compounds are strongly attracted to the non-polar packing of a chromatographic column. Polar molecules pass easily through the column, dissolved in a polar solvent or mixture of solvents.

rhizome: A horizontally creeping underground stem bearing roots and leaves, usually persisting for several seasons and sometimes modified for food storage.

rhizopod: Group of protists that move with pseudopodia. They are not a monophyletic taxon.

ridge: Siliceous projection on a chrysophyte cyst with an elongated base.

rimoportula: A tube running through the valve wall with the internal opening appearing as a longitudinal slit surrounded by two lips (also called a labiate process).

riparian vegetation: Vegetation growing along the banks of streams, rivers, and lakes.

RMSEP: Root mean squared error of prediction.

root mean squared error of prediction (RMSEP): In environmental reconstruction, a measure of the prediction error of the inferred environmental values, usually obtained by cross-validation using jackknifing or bootstrapping.

Rotifera: Minute, usually microscopic, but many-celled, chiefly freshwater aquatic invertebrates having the anterior end modified into a retractile disk bearing circles of strong cilia that often give the appearance of rapidly revolving wheels. Commonly referred to as rotifers.

rotules: The rounded ends of dumbbell-shaped sponge spicules.

saddle: A type of phytolith found only in grasses. They are, for the most part, restricted to bamboos and members of the Chloridoid (short grass) sub-family of the Poaceae (= Gramineae).

saponification: A chemical process by molecules with numerous oxygen-bearing functional groups (hydroxyl, ester, carbonyl) which are selectively precipitated from an organic solution using a strong base.

scabrae: Circular siliceous projection on a chrysophyte cyst with a basal diameter of ≤0.2 μm.

scanning electron microscope (SEM): An electron microscope where the electron beam is reflected off the surface of an object rather than being transmitted through it.

SDV: Silica deposition vesicle.

secondary charcoal: Sedimentary charcoal that is deposited in a lake as a result of surface or fluvial processes during non-fire years.

secondary collar: Collar second nearest the pore on a chrysophyte cyst.

sediment accumulation rate: The rate of accumulation of sediment (cm yr^{-1}) in a lake or mire. It is measured by age-depth modelling from a series of radiocarbon dates, or, if appropriate, by counting annual laminations. It is the inverse of sediment deposition time (yr cm^{-1}).

seed coat: (= testa, integument). The outer layer of an ovule develops into the seed coat or testa after fertilisation. Characteristics of the seed coat are used in the identification of seeds, and the structure can be closely examined with the SEM.

SEM: Scanning electron microscopy or scanning electron microscope.

senescence: Processes by which living organisms age or degrade.

separation valve: A specialised valve produced within a diatom filament (e.g., *Aulacoseira*) allowing the filament to divide into two shorter filaments. Separation valves thus become terminal valves in divided filaments.

septate: Divided by or having a septum (septate mycelium; septate fungal spores).

shaft: Gently curved middle section of a *Mallomonas* (chrysophyte) bristle.

shelled amoebae: See testate amoebae.

signal-to-noise ratio: The ratio of the "signal" (i.e., systematic) component of a statistical measure to the "noise" component (i.e., that due to random effects).

silica deposition vesicle (SDV): Vesicles where silica are transported in a cell; used in, for example, chrysophyte cyst and scale development.

siliceous tests: A test of shelled amoebae that is made endogenously (biogenically) by siliceous plates or scales.

silicification: The process by which soluble silica occurring in groundwater and taken up by plants is converted to a solid form of amorphous silica inside plant cells.

silicoflagellates: Tiny marine single-celled flagellated algae plankton with silica skeletons made of hollow tubular rods in a net-like pattern. They are both photosynthetic and heterotrophic and are widely distributed throughout the world's oceans.

silylation: Chemical process to convert a tertiary hydroxyl group to a trimethylsilyl ether.

size-class method: Method of charcoal analysis that measures the area of each microscopic charcoal particles by use of a gridded eyepiece in the microscope.

skip distance: Theoretical distance between the point of origination and the point of deposition for charcoal particles that are carried aloft during a fire.

slash and burn: A system of agricultural production typical in tropical regions of the world, whereby the vegetation is cut and burned before crops are planted. This process allows nutrients in the living vegetation to be transferred to the soils of the agricultural plots, thus increasing agricultural productivity in areas where soils are nutrient poor (= swidden farming).

sodium polytungstate: A commercially available compound that can be made up with distilled water to produce solutions of differing specific gravity and used in diatom preparations to remove unwanted debris from diatom and other samples by flotation.

soft ionization: Process used during mass spectrometry in which ions are created by means other than high-energy molecular collisions.

soil testate amoebae: Testate amoebae living in soils. They have often developed a characteristic shell morphology. See: cryptostom and plagiostom.

soligenous: Mire whose main water supply is emergent drainage water from springs, rills, or flushes. Mainly, but not exclusively, in upland areas.

sonication: The treatment of samples in an ultrasonic bath used to disaggregate material and, for example, split diatom colonies and cells into individual valves. However, this treatment also causes valves to fracture.

spicules: Needle-shaped structures that in invertebrate animals (e.g., sponges) occur in a wide variety of shapes and sizes.

sporopollenin: A complex polymer that includes saturated and unsaturated hydrocarbons and phenolic substances. Extremely resistant to chemical and physical attack, and responsible for the resilience of the walls of pollen grains to these forms of attack.

ssu RNA gene: "Small subunit of the Gene for the Ribonucleic Acid". This is the 18S subunit of the ribosomal RNA. The genes for the 18S (small sub unit), the 5.8S and the 28S (large sub unit) RNA occur in large numbers in the genome. The sequences of these genes are relatively conserved and are increasingly used to reconstruct organism phylogenies.

stand-age analysis: The use of tree ages to reconstruct when a fire initiated the establishment of an individual forest stand, followed by statistical analysis of the distribution of the ages of different forest stands to estimate how frequently fire occurred in that region.

stand-replacement fire: Fires of high-intensity that kill almost all standing trees.

statoblast: The asexual means of reproduction in Bryozoa, formed by internal budding and the encasement of the bud in a capsule consisting of two tight-fitting sclerotised and sculptured valves. Statoblasts may float and be dispersed by water currents, or they may be attached to the substrate. Statoblasts are highly resistant structures capable of surviving adverse conditions and germinating to form new colonies at the onset of favourable conditions. They can be very abundant in lake sediments. The large hooked statoblasts of *Cristatella mucedo* are characteristic. The other common fossil types are smaller with no hooks and often reddish brown, and belong most commonly to *Plumatella* spp.

statospore: Siliceous resting stage of chrysophyte algae (also cyst or stomatocyst).

stenohaline: Refers to species that can only tolerate narrow changes in environmental salinity, and thus are unable to tolerate a wide variation in osmotic pressure of the environment.

stoma: See stomata.

stomata: Minute intracellular fissures and specialized guard cells that are found on the epidermis of leaves, stems, ovules and some flowers.

stomatal complex: The opening, guard cells, and subsidiary cells associated with a stomata.

stomatal density: The number of stomata per unit area of leaf surface.

stomatocyst: Siliceous resting stage of chrysophyte algae (also statospore or cyst).

stria: A row of areolae.

striae density: An important taxonomic character in diatom research relating to the spacing of striae and measured as the number in 10 m.

strutted process: See fultoportula.

Styrax®: Mounting medium with high refractive indices used for diatoms. Supplier: CHROMA (Münster, Germany. http//www.chroma.de).

Subatlantic: The present interval in the Holocene time, from about 2600 calibrated years BP, during which the climate became somewhat cooler and wetter.

Subboreal: An interval in the Holocene time, from about 5700 to 2600 calibrated years BP, preceding the Subatlantic. The climate was subject to fluctuations, but generally somewhat warmer and drier than in present times.

suberized (cell walls): Infiltrated with suberin so that corky tissue is formed.

subfossil: A specimen that is found under fossil conditions (e.g., sediments), but is relatively young on the geological time scale (i.e., post-Pleistocene). Usually belongs to an extant taxon.

submarginal ridge: Two ribs continuous with, or originating near, the distal ends of the V-rib arms of a *Mallomonas* (chrysophyte) scale.

subsidiary cells: Specially shaped epidermal cells that surround some stomata.

succession: Temporal changes in the presence, abundance or dominance of different species in a community. It is initiated by one disturbance and often terminated by the next.

supernatant: Floating on the surface of a liquid. The part of a mixture that can be poured off (and kept or discarded as appropriate) after centrifugation.

SWAP: Surface Water Acidification Programme.

synecology: Part of ecology that deals with the relation of whole assemblages or communities to their environment.

taphonomy: The study of what happens to a fossil, from the time of its initial creation (e.g., the death of an organism or the imprint left by the movement of an organism) to the time that the fossil is discovered.

Tardigrada: Phylum of microscopic arthropods with four pairs of stout legs that usually live in water or damp moss—also called water bears.

tectum: A flexible, secreted covering over the body surface of certain amoebae.

telescoping effect: The decreasing accuracy with time for fire-history reconstructions that are based on the ages of living trees.

tephra: (= volcanic ash). Aerially dispersed volcanic glass particles produced by a volcanic eruption. Layers of tephra in sediments are time markers, and can often be used for correlation or for determining the magnitude of reservoir effects on ^{14}C dating.

terminal valve: See separation valve.

terrestrial testate amoebae: See soil testate amoebae.

tertiary collar: Collar third nearest the pore on a chrysophyte cyst.

testaceans: See testate amoebae.

testate amoebae: Amoebae (rhizopods) that are protected by an external test or shell composed of proteinaceous, calcareous, agglutinate, or siliceous material. Also referred to as testaceans. It is now clear that testate amoebae are polyphyletic.

tetrapyrole: Containing four pyrol ring structures.

thalassic: Water bodies of marine origin, dominated by NaCl and with a more or less stable salinity of 30–35 g l^{-1}.

theca: Comprises a diatom valve and its associated cingulum.

Thecamoebae: See testate amoebae.

thermophilous taxa: Taxa that are adapted to a warm or hot environment.

thermophily: To prefer warm or hot environments.

thin layer chromatography (TLC): Process by which compounds are isolated using a thin layer of finely-divided particles spread over an inert support and a solvent or mixture which mores through the substrate via capillary action.

tip: Distal end of a *Mallomonas* (chrysophyte) bristle.

TLC: See thin layer chromatography.

topogenous: Mire formed where local topography results in a permanently high water-table.

transfer function: A mathematical function that describes the relationship between biological taxa and environmental variables used to infer past values of an environmental variable from fossil assemblages (also called an inference model).

tree-limit: The latitudinal or altitudinal extent of a tree species. At a tree-line it may be stunted (krummholz).

tree-line: The latitudinal or altitudinal extent of trees beyond the forest zone; the limit of the tree growth-form. Beyond is shrub tundra and/or alpine vegetation. Tree-lines may be artificially altered by logging, grazing, burning, etc.

trichosclereids: Star-shaped bodies consisting of cells with thick walls as occurring in leaf tissue of Nymphaeaceae.

turgor: Rigidity in plant cells caused by increased internal pressure.

tychoplankton: Benthic taxa resuspended in the water column and sampled accidentally as plankton.

ubiquistic: See ubiquitous.

ubiquitous: Existing or seemingly existing everywhere. In ecology an organism is called ubiquistic if it shows no restriction to a specific habitat.

ultrastructure: A structure that can only be seen with the aid of an electron microscope.

unicellular: Made of one cell.

upright microscope: A standard microscope with illumination below and objective lens and ocular above the specimen.

UV radiation: Light with wavelengths 280–400 nm, often distinguished as UV-A (320–400 nm) and UV-B (280–320 nm).

valves: The dish or boat-like end parts of the diatom frustule that usually carry the intricate ornamentation used for identification purposes, often separated into a valve face (the flat or domed top of the valve) and mantles (the down-turned sides that link with the cingulum).

vascular plants: Plants that contain water-conducting cells with differentially thickened cell walls (provided by lignin). This reinforces mechanically the walls from implosion as a consequence of negative internal pressure created when fluid moves rapidly through a tube and thus provides support. Angiosperms, gymnosperms, and ferns are vascular plant groups.

velum: A thin, delicate, perforated layer of silica that occludes the opening of an areola.

verrucae: Circular siliceous projections on a chrysophyte cyst with a basal diameter of >0.2 μm.

vesicular nucleus: A nucleus with one central nucleolus.

visor: A projecting part of the dorsal lip of the shell of testate amoebae that protects the aperture below.

v-rib: Prominent rib pointing towards the posterior rim of a *Mallomonas* (chrysophyte) scale.

WA: Weighted averaging.

WA-PLS: Weighted averaging partial least squares.

Weichselian: Last glacial stage (Devensian, Wisconsin).

weighted averaging (WA): In environmental reconstruction, a numerical technique to derive transfer functions by estimating species optima from a calibration training set (WA regression), and then using these optima to estimate the past value of an environmental variable from a fossil assemblage (WA calibration). The species optima and tolerances are estimated from the weighted averages and weighted standard deviations, respectively. Used in developing organism-environment transfer functions.

weighted averaging partial least squares (WA-PLS): An extension of two-way weighted averaging and calibration where two or more components, which utilize the residual structure in the modern biological data, are used to improve the predictive power of the transfer function.

wiggle matching: A process by which a series of radiocarbon dates is statistically matched against a reference series of radiocarbon dates that has been calibrated against a calendar chronology (tree-ring series, series of annually laminated sediments). The new series is thus converted to a calendar/calibrated chronology and the ages should be denoted as 'cal yr BP'.

window: Area between arms of the V-rib on a *Mallomonas* (chrysophyte) scale when devoid of ornamentation.

Wisconsian: See Weichselian.

xenosomes: External test building material collected by testate amoebae. See also idiosomes.

x-ray diffraction: A technique used in mineralogy to identify mineral structures by exposing a sample to X-rays and studying the resulting diffraction patterns.

Younger Dryas: A cold reversal in the late-glacial warming trend during the glacial/Holocene transition. It occurred between ca. 12,700 and 11,500 calibrated yr BP (ca. 11,000–10,000 ^{14}C yr BP). The climatic changes occurred rapidly in decades or less. They are registered most strongly around the North Atlantic region, in ice cores and in the biological and sedimentary records in marine and terrestrial sediments. The term 'Younger Dryas' is now more precisely defined as Greenland Stadial 1 (GS-1) in the GRIP ice core. Named after the plant *Dryas octopetala*, which has now an arctic-alpine distribution.

zygnemataceous spores: Spores of Zygnemataceae (family of filamentous green algae).

zygospore: Thick-walled spore of some algae and fungi that is formed by a union of two similar sexual cells, usually serving as a resting spore.

INDEX